T0211369

LONDON MATHEMATICAL SOCIETY LECTURE NOTE SERIES

Managing Editor: Professor M. Reid, Mathematics Institute,
University of Warwick, Coventry CV4 7AL, United Kingdom

The titles below are available from booksellers, or from Cambridge University Press at
http://www.cambridge.org/mathematics

London Mathematical Society Lecture Note Series: 423

Inequalities for Graph Eigenvalues

ZORAN STANIĆ

University of Belgrade, Serbia

CAMBRIDGE
UNIVERSITY PRESS

CAMBRIDGE
UNIVERSITY PRESS

University Printing House, Cambridge CB2 8BS, United Kingdom

Cambridge University Press is part of the University of Cambridge.

It furthers the University's mission by disseminating knowledge in the pursuit of education, learning and research at the highest international levels of excellence.

www.cambridge.org
Information on this title: www.cambridge.org/9781107545977

First published 2015

A catalogue record for this publication is available from the British Library

Library of Congress Cataloguing in Publication data
Stanić, Zoran, 1975–
Inequalities for graph eigenvalues / Zoran Stanic, Univerzitetu Beogradu, Serbia.
pages   cm. – (London Mathematical Society lecture note series ; 423)
Includes bibliographical references and index.
ISBN 978-1-107-54597-7 (Paper back : alk. paper)
1. Graph theory.   2. Eigenvalues   I. Title.
QA166.S73 2015
512.9´436–dc23   2015011588

ISBN 978-1-107-54597-7 Paperback

Contents

Preface

This book has been written to be of use to mathematicians working in algebraic (or more precisely, spectral) graph theory. It also contains material that may be of interest to graduate students dealing with the same subject area. It is primarily a theoretical book with an indication of possible applications, and so it can be used by computer scientists, chemists, physicists, biologists, electrical engineers, and other scientists who are using the theory of graph spectra in their work.

The rapid development of the theory of graph spectra has caused the appearance of various inequalities involving spectral invariants of a graph. The main purpose of this book is to expose those results along with their proofs, discussions, comparisons, examples, and exercises. We also indicate some conjectures and open problems that might provide initiatives for further research.

The book is written to be as self-contained as possible, but we assume familiarity with linear algebra, graph theory, and particularly with the basic concepts of the theory of graph spectra. For those who need some additional material, we recommend the books [58, 98, 102, 170].

The graphs considered here are finite, simple (so without loops or multiple edges), and undirected, and the spectra considered in the largest part of the book are those of the adjacency matrix, Laplacian matrix, and signless Laplacian matrix of a graph. Although the results may be exposed in different ways, say from simple to more complicated, or in parts by following their historical appearance, here we follow the concept of *from general to specific*, that is, whenever possible, we give a general result, idea or method, and then its consequences or particular cases. This concept is applied in many places, see for example Theorem 2.2 and its consequences, the whole of Subsection 2.1.2 or Theorem 2.19 and its consequences.

We briefly outline the content of the book. In Chapter 1 we fix the terminology and notation, introduce the matrices associated with a graph, give the necessary results, select possible applications, and give more details about the content. In this respect, the last section of this chapter can be considered as an extension of this Preface. In Chapters 2–4 we consider inequalities that include the largest, the least, and the second largest eigenvalue of the adjacency matrix of a graph, respectively. The last section of Chapter 4 contains the lists of graphs obtained, together with some additional data. The remaining, less investigated, eigenvalues of the adjacency matrix are considered in Chapter 5. Chapters 6 and 7 deal with the inequalities for single eigenvalues of the Laplacian and signless Laplacian matrix. The inequalities that include multiple eigenvalues of any of three spectra considered before are singled out in Chapter 8. In Chapter 9 we consider the normalized Laplacian matrix, the Seidel matrix, and the distance matrix of a graph.

Each of Chapters 2–9 contains theoretical results, comments (including additional explanations, similar results or possible applications), comparisons of inequalities obtained, and numerical or other examples. Each of these chapters ends with exercises and notes. The exercises contain selected problems or a small number of the previous results whose proofs were omitted. The notes contain brief surveys of unmentioned results and directions to the corresponding literature.

Spectral inequalities occupy a central place in this book. Mostly, they are lower or upper bounds for selected eigenvalues. Apart from these, we consider some results written rather in the form of an inequality that bounds some structural invariant in terms of graph eigenvalues (and possibly some other quantities) or, as we have already said, inequalities that include more than one eigenvalue. All inequalities exposed are listed at the end of the book.

In an informal sense, *extremal graph theory* deals with the problem of determining extremal graphs for a given graph invariant in a set of graphs with prescribed properties. In the context of the theory of graph spectra, the invariant in question is a fixed eigenvalue of a matrix associated with a graph or a spectral invariant based on a number of graph eigenvalues (like the graph energy). Extremal graphs for a given spectral invariant in various sets of graphs are widely considered.

The terminology and notation are mainly taken from [98, 102], and they can also be found in similar literature. However, since there is some overlap in the wider notation used, we have made some small adjustments for this book only.

The author is grateful to Dragoš Cvetković and Vladimir Nikiforov, who read the manuscript and gave valuable suggestions. In addition, these colleagues – together with Kinkar Chandra Das, Martin Hasler, and Slobodan K. Simić – gave permission to use some of their proofs with no significant change. Finally, Sarah Lewis helped with correcting language and technical errors, which is much appreciated.

1

Introduction

In order to make the reading of this book easier, in Section 1.1 we give a survey of the main graph-theoretic terminology and notation. Section 1.2 deals with matrix theory and graph spectra. In Section 1.3 we emphasize some more specific results of the theory of graph spectra that will frequently be used. Once we have fixed the notation and given all the necessary results, in Section 1.4 we say more about the applications of the theory of graph spectra and give some details related to the content of the book.

1.1 Graph-theoretic notions

Let G be a finite undirected graph without loops or multiple edges on n vertices labelled $1, 2, \ldots, n$. We denote the set of vertices of G by V (or $V(G)$). We say that two vertices i and j are *adjacent* (or *neighbours*) if they are joined by an edge and we write $i \sim j$. We denote the set of edges of G by E (or $E(G)$), where an edge ij belongs to E if and only if $i \sim j$. In this case we say that the edge ij is incident with vertices i and j. A graph consisting of a single vertex is called the *trivial graph*. Two edges are said to be *adjacent* if they are incident with a common vertex. Non-adjacent edges are said to be mutually *independent*. The number of vertices n and edges m in a graph are called the *order* and *size*, respectively.

Two graphs G and H are said to be *isomorphic* if there is a bijection between $V(G)$ and $V(H)$ which preserves the adjacency of their vertices. The fact that G and H are isomorphic we denote by $G \cong H$, but we also use the simple notation $G = H$. A graph is *asymmetric* if the only permutation of its vertices which preserves their adjacency is the identity mapping.

We say that G is the unique graph satisfying given properties if and only if any other graph with the same properties is isomorphic to G.

1

A graph H obtained from a given graph G by deleting some vertices (together with their edges incident) is called an *induced subgraph* of G. In this case, we also say that H is induced in G, and that G is a *supergraph* of H. We say that a graph G is *H-free* if it does not contain H as an induced subgraph. A *subgraph* of G is any graph H satisfying $V(H) \subseteq V(G)$ and $E(H) \subseteq E(G)$. If $V(H) = V(G)$, H is called a *spanning subgraph* of G.

If $U \subset V(G)$, then we write $G[U]$ to denote the induced subgraph of G with vertex set U and two vertices being adjacent if and only if they are adjacent in G. Similarly, an induced subgraph of G obtained by deleting a set of vertices $V' \subseteq V(G)$ is denoted by $G - V'$ (rather than $G[V(G) \backslash V']$). If V' consists of a single vertex v, we simply write $G - v$ (instead of $G - \{v\}$). Similarly, $G - E'$ and $G - e$ designate the deletion of a subset of edges E' and a single edge e, respectively. By $G + e$ we denote a graph obtained from G by inserting a single edge. If V_1 and V_2 are disjoint subsets of $V(G)$, then $m(V_1)$ and $m(V_1, V_2)$ stand for the number of edges in $G[V_1]$ and the number of edges with one end in V_1 and the other in V_2, respectively.

The *degree* d_u of a vertex u (in a graph G) is the number of edges incident with it. In particular, the minimal and the maximal vertex degrees are denoted by δ and Δ, respectively. We say that a graph G is *regular* of degree r (or r_G) if all its vertices have degree r. If so, then we usually say that G is r-regular. The *complete graph* on n vertices, K_n, is a graph whose every pair of vertices is joined by an edge. A regular graph of degree 3 is called a *cubic graph*. The unique $(2n - 2)$-regular graph on $2n$ ($n \geq 1$) vertices is called a *cocktail party*, and is denoted by $CP(n)$. Obviously, it is an $(n - 1)$-regular graph. A *bidegreed graph* has exactly two distinct vertex degrees. The *edge degree* of an edge uv is defined as $d_u + d_v - 2$ (i.e., it is the number of edges that have a common vertex with uv).

The set of neighbours (or the open neighbourhood) of a vertex u is denoted by $N(u)$. The closed neighbourhood of u is denoted by $N[u]$ ($= \{u\} \cup N(u)$). The average degree of vertices in $N(u)$ is denoted by m_u, and it is also called the *average 2-degree* of u.

A graph is said to be properly coloured if each vertex is coloured so that adjacent vertices have different colours. G is *k-colourable* if it can be properly coloured by k colours. The *chromatic number* χ is k if G is k-colourable and not $(k - 1)$-colourable. G is called *bipartite* if its chromatic number is 1 or 2. The vertex set of a bipartite graph can be partitioned into two parts (or colour classes) X and Y such that every edge of G joins a vertex in X with a vertex in Y. A graph is called *complete bipartite* if every vertex in one part is adjacent to every vertex in the other part. If $|X| = n_1$ and $|Y| = n_2$, the complete bipartite graph is denoted by K_{n_1,n_2}. In particular, if $n_1 = 1$, it is called a *star*. More generally, a *k-partite graph* is a graph whose set of vertices is discomposed into k disjoint sets such that no two vertices within the same set are adjacent. If there are $n_1, n_2, \ldots, n_k$ vertices in the k sets, and if each two vertices which belong to different sets are adjacent, the graph is called *complete k-partite* (or simply *complete multipartite*) and denoted by $K_{n_1,n_2,\ldots,n_k}$.

A graph is called *semiregular bipartite* if it is bipartite and the vertices belonging to the same part have equal degree. If the corresponding vertex degrees are, say, r and s, the graph is referred to as (r,s)-*semiregular bipartite*.

A vertex of degree 1 (in a graph G) is called an *endvertex* or *pendant vertex*. The edge incident with such a vertex is a *pendant edge*.

A *k-walk* (or simply *walk*) in a graph G is an alternative sequence of vertices and edges $v_1, e_1, v_2, e_2, \ldots, e_{k-1}, v_k$ such that each edge e_i is incident with v_i and v_{i+1} $(1 \leq i \leq k-1)$. The walk is closed if v_1 coincides with v_k. The number of k-walks is denoted by w_k. Similarly, the number of k-walks starting with u (resp. starting with u and ending with v) is denoted by $w_k(u)$ (resp. $w_k(u,v)$).

If all vertices of a walk are distinct, it is called a *path*. A graph which is itself a path on n vertices is denoted by P_n. An endvertex of P_n is often called an *end* of P_n. By joining the ends of P_n by an edge we get a *cycle* C_n. In particular, C_3 is called a *triangle* and C_4 is called a *quadrangle*. The number of triangles in a graph G is denoted by $t(G)$. The *length* of a path P_n or a cycle C_n is equal to the number of edges contained in it. A graph is *Hamiltonian* if it contains a spanning subgraph which is a cycle, while any such cycle is referred to as a *Hamiltonian cycle*.

We say that a graph G is *connected* if every two distinct vertices are the ends of at least one path in G. Otherwise, G is *disconnected* and its maximal connected induced subgraphs are called the *components* of G. A component consisting of a single vertex is called an *isolated vertex* (or *trivial component*) of G. A graph is *totally disconnected* if it consists entirely of isolated vertices. If G has exactly one non-trivial component, this component is called the *dominant component*.

The *distance* $d(u,v)$ between the vertices u and v is the length of the shortest path between u and v, and the *girth* $gr(G)$ is the length of the shortest cycle induced in G. The *diameter* D of a graph G is the longest distance between two vertices of G. A shortest path between any pair of vertices u and v such that $d(u,v) = D$ is called a *diametral path*.

A connected graph G whose number of edges m equals $n-1$ is called a *tree*. Furthermore, if $m = n - 1 + k$ $(k \geq 1)$, G is said to be *k-cyclic*.

For $k = 1$, the corresponding graph is called *unicyclic*; for $k = 2$, it is called *bicyclic*. Clearly, any unicyclic graph contains a unique cycle as an induced subgraph. If this cycle has odd length then the graph is said to be *odd unicyclic*.

Any complete induced subgraph of a graph G is called a *clique*. The *clique number* ω is the number of vertices in the largest clique of G. Similarly, any totally disconnected induced subgraph is called a *co-clique*. The vertices of a co-clique make an *independent set* of vertices of G, and the number of vertices in the largest independent set is called the *independence number*, denoted by α.

A *matching* in G is a set of edges without common vertices. A matching is *perfect* if each vertex of G is incident with an edge from the matching. The *matching number* μ is the maximal size of a matching in G.

A *vertex* (resp. *edge*) cover of a graph G is a set of vertices (resp. edges) such that each edge (resp. vertex) of G is incident with at least one vertex (resp. edge) of the set. The *vertex* (resp. *edge*) *cover number* of G, denoted by β (resp. β'), is the minimum of the cardinalities of all vertex (resp. edge) covers.

A *dominating set* for a graph G is a subset D of $V(G)$ such that every vertex not in D is adjacent to at least one vertex in D. The *domination number* φ is the number of vertices in a smallest dominating set for G.

A *cut vertex* (resp. *cut edge*) of a connected graph is any vertex (resp. edge) whose removal yields a disconnected graph. The *vertex* (resp. *edge*) *connectivity*, denoted by c_v (resp. c_e), of a connected graph is the minimal number of vertices (resp. edges) whose removal gives a disconnected graph.

A *rooted graph* is a graph in which one vertex has been distinguished as the *root*. A pendant vertex of a rooted tree is often called a *terminal vertex*.

For two graphs G and H we define $G \cup H$ to be their disjoint union.[1] In addition, we use kG to denote the disjoint union of k copies of G. The *join* $G \nabla H$ is the graph obtained by joining every vertex of G with every vertex of H. In particular, $K_1 \nabla G$ is called the *cone* over G.

The *complement* of a graph G is a graph $\overline{G}$ with the same vertex set as G, in which any two distinct vertices are adjacent if and only if they are non-adjacent in G.

---

[1] With no confusion, we use the same symbol to denote the union of two sets. In addition, $\sqcup$ will stand for the union of disjoint sets.

1.1.1 Some graphs

The vertex with maximal degree in the star $K_{1,n}$ is called the *centre* of the star. The *double star* $DS(n_1, n_2)$ is a graph obtained from the stars K_{1,n_1-1} and K_{1,n_2-1} by inserting an edge between their centres.

A *starlike tree* $S_{i_1,i_2,...,i_k}$ is a tree with exactly one vertex of degree greater than two such that the removal of this vertex gives rise to paths $P_{i_1}, P_{i_2}, ..., P_{i_k}$. For $k = 3$, the corresponding starlike tree is often called a *T-shape tree*.

A *caterpillar* is a tree in which the removal of all pendant vertices gives a path. Let the vertices of a path P_k $(k \geq 3)$ be labelled $1, 2, ..., k$ (in natural order), then $T(m_2, m_3, ..., m_{k-1})$ denotes the caterpillar obtained by attaching m_i pendant vertices at the ith vertex of P_k $(2 \leq i \leq k - 1)$. If a caterpillar is obtained by attaching just a few pendant vertices at the same path, we use the shorter notation $T_n^{i_1,i_2,...,i_l}$, where n denotes the order and i_j $(1 \leq j \leq l)$ indicates attaching a pendant vertex at the vertex labelled i_j $(2 \leq i_j \leq k - 1)$. A *closed caterpillar* is a unicyclic graph in which the removal of all pendant vertices gives a cycle.

An *open quipu* is a tree with maximal vertex degree 3 such that all vertices of degree 3 lie on a path. A *closed quipu* is a unicyclic graph with maximal vertex degree 3 such that all vertices of degree 3 lie on a cycle.

A *cactus* is a connected graph G such that any two cycles induced in G have at most one common vertex.

The *comet* $C(k,l)$ is a tree obtained by attaching k pendant vertices at one end of the path P_l. The *double comet* $DC(k,l)$ is a tree obtained by attaching k pendant vertices at one end of the path P_l and another k pendant vertices at the other end of the same path.

The *kite* $K(k,l)$ is a graph obtained by identifying one end of the path P_{l+1} with a vertex of the complete graph K_k. The *double kite* $DK(k,l)$ is a graph obtained by identifying one end of the path P_{l+2} with a vertex of the complete graph K_k and the other end of the same path with a vertex of another complete graph K_k.

The *pineapple* $P(k,l)$ is a graph obtained by attaching l pendant vertices at a vertex of K_k.

$C(k,l)$, $K(k,l)$, and $P(k,l)$ have $k + l$ vertices, while $DC(k,l)$ and $DK(k,l)$ have $2k + l$ vertices.

Let $G^D = G^D(n_1, n_2, ..., n_{D+1})$ denote the graph defined as follows: $V(G^D) = \bigcup_{i=1}^{D+1} V_i$, where $G^D[V_i] \cong K_{n_i}$ $(1 \leq i \leq D + 1)$ and

$$G^D[V_i \cup V_j] \cong \begin{cases} K_{n_i+n_j}, & \text{if } |i-j| = 1, \\ K_{n_i} \cup K_{n_j}, & \text{otherwise.} \end{cases}$$

The graph G^D consists of a chain of $D+1$ cliques $K_{n_1}, K_{n_2}, \ldots, K_{n_{D+1}}$, where neighbouring cliques are fully interconnected (i.e., each vertex in one is adjacent to all vertices in the other). According to this, we name this graph the *clique chain graph*. Its order is $n = \sum_{i=1}^{D+1} n_i$.

Similarly, let $G_*^D = G_*^D(n_1, n_2, \ldots, n_{D+1})$ denote the graph of the same order defined as follows: $V(G_*^D) = \bigcup_{i=1}^{D+1} V_i$, where $G_*^D[V_i] \cong n_i K_1$ ($1 \le i \le D+1$) and

$$G_*^D[V_i \cup V_j] \cong \begin{cases} K_{n_i,n_j}, & \text{if } |i-j| = 1, \\ (n_i+n_j)K_1, & \text{otherwise.} \end{cases}$$

This graph consists of a chain of $D+1$ co-cliques $n_1 K_1, n_2 K_1, \ldots, n_{D+1} K_1$, where neighbouring co-cliques are fully interconnected. We name it the *co-clique chain graph*. Observe that G_*^D is bipartite.

Recall that a *multigraph* includes the possible existence of multiple edges between any two vertices or loops (i.e. edges with both endvertices identical). We say that a *petal* is added to a graph when we add a pendant vertex and then duplicate the edge incident with it.

The *line graph*[2] Line(G) of a multigraph G is the graph whose vertices are the edges of G, with two vertices adjacent whenever the corresponding edges have exactly one common vertex.

Let G be a graph with vertex set $V = \{v_1, v_2, \ldots, v_n\}$, and let $a_1, a_2, \ldots, a_n$ be non-negative integers. The *generalized line graph* Line($G; a_1, a_2, \ldots, a_n$) is the graph Line($\widehat{G}$), where $\widehat{G}$ is the multigraph $G(a_1, a_2, \ldots, a_n)$ obtained from G by adding a_i petals at vertex v_i ($1 \le i \le n$).

We introduce two classes of graphs called nested split graphs and double nested graphs. For these two classes of graphs we use the common name *nested graphs*.

A *nested split graph* (NSG for short) is a graph which does not contain any of the graphs P_4, C_4 or $2K_2$ as an induced subgraph. This name is derived from its structure; it is also called a *threshold graph* (for more details, see [409]). We describe the structure of connected NSGs. The vertex set of any such graph consists of a co-clique and a clique, where both the co-clique and the clique are partitioned into h cells $U_1, U_2, \ldots, U_h$ and $V_1, V_2, \ldots, V_h$, respectively. Then

---

[2] The line graph is often denoted by $L(G)$, but in this book $L(G)$ is reserved for the Laplacian matrix (see Section 1.2).

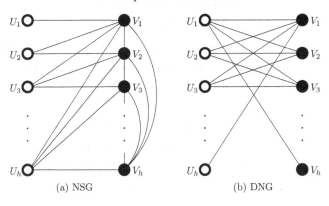

(a) NSG   (b) DNG

Figure 1.1 The structure of nested graphs.

all vertices in the co-clique U_i $(1 \leq i \leq h)$ are joined to all vertices in $V_1 \cup V_2 \cup \cdots \cup V_i$, so if $u \in U_i$ and $v \in U_{i+1}$ then $N(u) \subset N(v)$, which explains the nesting property.

A *double nested graph* (DNG for short) is a bipartite graph which does not contain any of the graphs P_4, C_4 or $2K_2$ as an induced subgraph (in [42] these graphs are called *chain graphs*). The nesting property of connected DNGs can be described in a similar way: the vertices of two parts are partitioned into the same h cells as above, and all vertices in U_i $(1 \leq i \leq h)$ are joined by cross edges to all vertices in $V_1 \cup V_2 \cup \cdots \cup V_{h+1-i}$.

The structure of nested graphs is illustrated in Fig. 1.1. In both cases let $m_i = |U_i|$ and $n_i = |V_i|$ $(1 \leq i \leq h)$, then any nested graph is determined by the following $2h$ parameters:

$$(m_1, m_2, \ldots, m_h; n_1, n_2, \ldots, n_h).$$

If we denote

$$M_s = \sum_{i=1}^{s} m_i, \ \ N_t = \sum_{i=1}^{t} n_i \ (1 \leq s, t \leq h),$$

we can easily see that the order of a nested graph is $n = M_h + N_h$ and the size of an NSG (resp. DNG) is $m = \sum_{i=1}^{h} m_i N_i + \binom{N_h}{2}$ (resp. $m = \sum_{i=1}^{h} m_i N_{h+1-i}$). For an NSG, the degree of a vertex $u \in U_i$ is equal to N_i and the degree of a vertex $v \in V_i$ is equal to $n - 1 - M_{i-1}$. For a DNG, these degrees are equal to N_{h+1-i} and M_{h+i-1}, respectively.

1.2 Spectra of graphs

In this section we introduce three matrices associated with graphs (the adjacency matrix, the Laplacian matrix, and the signless Laplacian matrix) and give some basic results related to their spectra. Another three matrices are considered later in Chapter 9.

We start with a small portion of the matrix theory. Let M be an $n \times n$ real matrix. If the eigenvalues of M are real (which, e.g., occurs if M is symmetric), they can be given in non-increasing order as

$$v_1 \; (= v_1(M)) \geq v_2 \; (= v_2(M)) \geq \cdots \geq v_n \; (= v_n(M)).$$

Here are some results that come from the Perron–Frobenius theory of non-negative matrices (for more details, see [317, Chapter 8]).

A symmetric matrix M is *reducible* if there exists a permutation matrix P such that $P^{-1}MP = \begin{pmatrix} B_1 & 0 \\ 0 & B_2 \end{pmatrix}$, where B_1 and B_2 are square matrices. Otherwise, M is called *irreducible*.

Theorem 1.1   ([98, Theorem 0.6] or [102, Theorem 1.3.6]) *Let M be a real irreducible symmetric matrix with non-negative entries. Then its largest eigenvalue v_1 is simple, while the coordinates of any eigenvector of v_1 are non-zero and of the same sign. In addition, $|v| \leq v_1$ for all eigenvalues v of M.*

Moreover, the largest eigenvalue of any principal submatrix of M is less than v_1.

Theorem 1.2   (cf. [317, p. 668]) *For any real symmetric matrix $M = (m_{ij})$ with non-negative entries,*

$$\min_{1 \leq i \leq n} \sum_{j=1}^{n} m_{ij} \leq v_1(M) \leq \max_{1 \leq i \leq n} \sum_{j=1}^{n} m_{ij}.$$

If M is irreducible then each inequality holds if and only if all row sums in M are equal.

Following [102, pp. 11–12], we give a technique which is frequently used for bounding graph eigenvalues. The *Rayleigh quotient* for a real symmetric matrix M is a scalar

$$\frac{\mathbf{y}^T M \mathbf{y}}{\mathbf{y}^T \mathbf{y}}, \tag{1.1}$$

where $\mathbf{y}$ is a non-zero vector in $\mathbb{R}^n$. The supremum of such scalars is the largest eigenvalue of M, and the infimum is the least eigenvalue of M. In other words,

$$v_1 = \sup\{\mathbf{y}^T A \mathbf{y} \ : \ \mathbf{y} \in \mathbb{R}^n, \|\mathbf{y}\| = 1\}, \tag{1.2}$$

$$v_n = \inf\{\mathbf{y}^T A \mathbf{y} \ : \ \mathbf{y} \in \mathbb{R}^n, \|\mathbf{y}\| = 1\}. \tag{1.3}$$

The *Rayleigh principle* can be stated as follows. If $\{\mathbf{x_1}, \mathbf{x_2}, \dots, \mathbf{x_n}\}$ is an orthonormal basis of the eigenvectors of M and if a non-zero vector $\mathbf{y}$ is spanned by $\{\mathbf{x_i}, \mathbf{x_{i+1}}, \dots, \mathbf{x_n}\}$, then

$$v_i \geq \frac{\mathbf{y}^T M \mathbf{y}}{\mathbf{y}^T \mathbf{y}},$$

with equality if and only if $M\mathbf{y} = v_i \mathbf{y}$; if a non-zero vector $\mathbf{y}$ is spanned by $\{\mathbf{x_1}, \mathbf{x_2}, \dots, \mathbf{x_i}\}$, then

$$v_i \leq \frac{\mathbf{y}^T M \mathbf{y}}{\mathbf{y}^T \mathbf{y}},$$

with equality if and only if $M\mathbf{y} = v_i \mathbf{y}$.

We establish some inequalities with wide application in the theory of graph spectra.

Theorem 1.3 (Courant–Weyl Inequalities, cf. [102, Theorem 1.3.15]) *Let M and N be $n \times n$ real symmetric matrices. Then*

$$v_i(M+N) \leq v_j(M) + v_{i-j+1}(N) \quad (n \geq i \geq j \geq 1),$$

$$v_i(M+N) \geq v_j(M) + v_{i-j+n}(N) \quad (1 \leq i \leq j \leq n).$$

We use I, J, and O to denote a unit, all-1, and all-0 matrix, respectively. Somewhere, the size of any of these matrices will be given in subscript. The $n \times 1$ all-1 vector is denoted $\mathbf{j}$.

If a graph G has n vertices and m edges, then the *vertex–edge incidence matrix* $R\ (= R(G))$ is an $n \times m$ matrix whose rows and columns are indexed by the vertices and edges of G, that is, the (i,e)-entry is

$$r_{i,e} = \begin{cases} 1, & \text{if } i \text{ is incident with } e, \\ 0, & \text{otherwise.} \end{cases}$$

We are now in a position to consider the above introduced matrices as they are associated with graphs.

1.2.1 Spectrum of a graph

The *adjacency matrix* A (or $A(G)$) of G is defined as $A = (a_{i,j})$, where

$$a_{i,j} = \begin{cases} 1, & \text{if } i \sim j, \\ 0, & \text{otherwise.} \end{cases}$$

The characteristic polynomial of A, $P_G = \det(xI - A)$, is also called the *characteristic polynomial* of G, while its roots are just the *eigenvalues* of G. The collection of eigenvalues of G (with repetition) is called the *spectrum* of G. We denote the eigenvalues of G by

$$\lambda_1 \; (= \lambda_1(G)), \lambda_2 \; (= \lambda_2(G)), \ldots, \lambda_n \; (= \lambda_n(G)),$$

and we assume that $\lambda_i \geq \lambda_j$ whenever $i < j$. According to this, the spectrum of G may be denoted $[\lambda_1'^{i_1}, \lambda_2'^{i_2}, \ldots, \lambda_k'^{i_k}]$, where $\lambda_1', \lambda_2', \ldots, \lambda_k'$ are the distinct eigenvalues while the exponents stand for their multiplicities.

The largest eigenvalue λ_1 is called the *spectral radius* or the *index* of G. For any non-negative integer k, the kth *spectral moment* of G is $M_k = \sum_{i=1}^{n} \lambda_i^k$. Note that M_k is equal to the trace $\operatorname{tr}(A^k)$.

The *spectral spread* σ of a graph G is the difference between its largest and least eigenvalue, that is, $\sigma = \lambda_1 - \lambda_n$. Similarly, the *spectral gap* η is the difference between the two largest eigenvalues, $\eta = \lambda_1 - \lambda_2$.

Any eigenvalue λ of G satisfies

$$A\mathbf{x} = \lambda\mathbf{x}, \tag{1.4}$$

for some non-zero vector $\mathbf{x} \in \mathbb{R}^n$. Each vector that satisfies (1.4) is called an *eigenvector* of G, and for $\mathbf{x} = (x_1, x_2, \ldots, x_n)^T$ it is usually assumed that the coordinate x_i (which is also called the *weight*) corresponds to the vertex i ($1 \leq i \leq n$). The *eigenvalue equation* of G follows from (1.4) and reads

$$\lambda x_i = \sum_{j \sim i} x_j \quad (1 \leq i \leq n).$$

If $\mathbf{x_1}, \mathbf{x_2}, \ldots, \mathbf{x_n}$ is a complete system of mutually orthogonal normalized eigenvectors of A belonging to the spectrum $[\lambda_1, \lambda_2, \ldots, \lambda_n]$, $X = (\mathbf{x_1}, \mathbf{x_2}, \ldots, \mathbf{x_n})$, and B is the diagonal matrix $\text{diag}(\lambda_1, \lambda_2, \ldots, \lambda_n)$, then X is orthogonal (i.e., $X^{-1} = X^T$) and

$$A = XBX^T. \tag{1.5}$$

Using the eigenvalue equation, we get

$$|\lambda||x_i| \le \sum_{j \sim i} |x_j| \le \Delta|x_i|. \tag{1.6}$$

In other words, $\lambda \le \Delta$, for any eigenvalue λ of any graph G. It can be verified by direct computation that a graph is r-regular if and only if $\mathbf{j}$ is an eigenvector of G corresponding to the largest eigenvalue equal to r.

Since the adjacency matrix of a connected graph is irreducible, the next result is an immediate consequence of Theorem 1.1.

Theorem 1.4 (cf. [98, Theorem 0.7] or [102, Corollary 1.3.8]) *A graph is connected if and only if its spectral radius is a simple eigenvalue with an eigenvector whose coordinates are non-zero and of the same sign.*

Any positive eigenvector corresponding to the spectral radius is called a *Perron eigenvector*. If this vector is considered as a unit, then it is called a *principal eigenvector*. We have the following corollary.

Corollary 1.5 (cf. [98, Theorem 0.7] or [102, Proposition 1.3.10]) *For any vertex u or any edge e of a connected graph G, $\lambda_1(G-u) < \lambda_1(G)$ and $\lambda_1(G-e) < \lambda_1(G)$.*

Proof Both inequalities are considered using the Rayleigh principle and the above theorem. We prove only the first one. Let $A = \begin{pmatrix} A' & \mathbf{r} \\ \mathbf{r}^T & O \end{pmatrix}$ be the adjacency matrix of G, where A' is the adjacency matrix of $G - u$. If $\mathbf{y}$ is the principal eigenvector of $G - u$, then by setting $\mathbf{x} = \begin{pmatrix} \mathbf{y} \\ 0 \end{pmatrix}$ we get $\lambda_1(G-u) = \mathbf{x}^T A \mathbf{x} \le \lambda_1(G)$, with equality if and only if $\mathbf{x}$ is the principal eigenvector of G, which is not possible since it has a zero coordinate. $\square$

We now consider eigenvalue interlacing.

Theorem 1.6 (Interlacing Theorem, cf. [98, Theorem 0.10] or [102, Corollary 1.3.12]) *Let G be a graph with n vertices and eigenvalues $\lambda_1(G) \ge \lambda_2(G) \ge \cdots \ge \lambda_n(G)$, and let H be an induced subgraph of G with n' vertices and eigenvalues $\lambda_1(H) \ge \lambda_2(H) \ge \cdots \ge \lambda_{n'}(H)$. Then*

$$\lambda_{n-n'+i}(G) \leq \lambda_i(H) \leq \lambda_i(G) \ (1 \leq i \leq n').$$

We usually say that the eigenvalues of H interlace those of G. The corresponding inequality is known as the *Cauchy inequality*, and the theorem itself is frequently used in the theory of graph spectra.

Remark 1.7 It can be proved that some of the previous statements hold for wider classes of matrices. For some later situations it is convenient to say that Theorem 1.6 holds for any Hermitian matrix [98, Theorem 0.10].   □

An *internal path* of a graph G is a sequence of vertices $u_1, u_2, \ldots, u_k$ such that all u_i are distinct (except possibly $u_1 = u_k$), the vertex degrees satisfy $d_{u_1}, d_{u_k} \geq 3$, $d_{u_2} = d_{u_3} = \cdots = d_{u_{k-1}} = 2$ (unless $k = 2$), and u_i is adjacent to u_{i+1} ($1 \leq i \leq k-1$).

Theorem 1.8 (Hoffman and Smith [213]) *Let $G_{u,v}$ be the graph obtained from a graph G by inserting a new vertex into the edge uv of G. If uv is an edge on an internal path of G and $G \not\cong T_n^{2,n-3}$ ($n \geq 6$), then $\lambda_1(G_{u,v}) > \lambda_1(G)$.*

For bipartite graphs we have the following result, rediscovered many times in the literature.

Theorem 1.9 (Cvetković, cf. [98, Theorem 3.11] or [102, Theorems 3.2.3 and 3.2.4]) *A graph G is bipartite if and only if its spectrum is symmetric with respect to zero. If G is connected then it is bipartite if and only if $\lambda_1 = -\lambda_n$.*

By the following theorem the spectrum of a regular graph contains full information about the spectrum of its complement.

Theorem 1.10 (Sachs, cf. [98, Theorem 2.6]) *If an r-regular graph has eigenvalues $r = \lambda_1 \geq \lambda_2 \geq \cdots \geq \lambda_n$, then the eigenvalues of $\overline{G}$ are $n-1-r \geq -\lambda_n - 1 \geq \cdots \geq -\lambda_2 - 1$.*

1.2.2 Laplacian spectrum of a graph

Let D denote the diagonal matrix of vertex degrees in a graph G, that is, $D = \text{diag}(d_1, d_2, \ldots, d_n)$. The *Laplacian matrix* L ($= L(G)$) of G is defined as

$$L = D - A.$$

This is named after P.-S. Laplace, since by discretizing the Laplace equation $\Delta \phi = \mu \phi$ one gets the Laplacian matrix of the discretized space [325]. In the literature, L is also referred to as the *Kirchhoff matrix* or *admittance matrix*. Following [98], we write $C_G(x)$ for $\det(xI - L)$, called the *Laplacian characteristic polynomial* of a graph G. It can be seen that L is positive semidefinite, and so its eigenvalues are non-negative. The eigenvalues of L are referred to as the *Laplacian eigenvalues* or *L-eigenvalues* of G. They are denoted by

$$\mu_1 \; (= \mu_1(G)), \mu_2 \; (= \mu_2(G)), \ldots, \mu_n \; (= \mu_n(G)),$$

with the assumption that $\mu_i \geq \mu_j$, whenever $i < j$. μ_1 is referred to as the *L-spectral radius* of a graph.

The spectrum, spectral radius, and other quantities defined prior to the eigenvalues of the adjacency matrix are defined analogously, along with the adaptation of the corresponding terminology. For example, the collection of all L-eigenvalues is called the *Laplacian spectrum* or *L-spectrum* of G.

Note that L is a singular matrix with all-1 eigenvector $\mathbf{j} \in \mathbb{R}^n$ corresponding to μ_n. We can deal simultaneously with the L-eigenvalues of a graph and its complement. Namely, if $\{\mathbf{x_1}, \mathbf{x_2}, \ldots, \mathbf{x_n}\}$ is an orthogonal basis of $\mathbb{R}^n$ such that $L(G)\mathbf{x_i} = \mu_i \mathbf{x_i}$ then, since $L(\overline{G}) = nI - L(G) - J$, we have $L(\overline{G})\mathbf{x_n} = 0$ and $L(\overline{G})\mathbf{x_i} = (n - \mu_i)\mathbf{x_i}$ $(1 \leq i \leq n - 1)$, which gives the following theorem.

Theorem 1.11 *If the L-eigenvalues of a graph G are $\mu_1 \geq \mu_2 \geq \cdots \geq \mu_n = 0$ then the L-eigenvalues of its complement $\overline{G}$ are $n - \mu_{n-1} \geq n - \mu_{n-2} \geq \cdots \geq n - \mu_1$, and 0.*

Some additional properties of L-eigenvalues are given in the following theorems.

Theorem 1.12 ([315]) *For a graph G, $\mu_1(G) \leq n$. The multiplicity of n in the L-spectrum of G equals the number of components of $\overline{G}$ minus one.*

Using the complementary graph we get the following result.

Theorem 1.13 (cf. [102, Theorem 7.1.2]) *The multiplicity of zero in the L-spectrum of a graph is equal to the number of its components.*

And there is an analogue to the interlacing theorem.

Theorem 1.14 (*L*-Interlacing Theorem, cf. [170, Theorem 13.6.2]) *Let G be a graph whose L-eigenvalues are $\mu_1(G) \geq \mu_2(G) \geq \cdots \geq \mu_n(G)$, and let H be a graph obtained by removing k $(k < n)$ edges from G. The eigenvalues $\mu_1(H) \geq \mu_2(H) \geq \cdots \geq \mu_n(H)$ of H satisfy*

$$\mu_i(G) \geq \mu_i(H) \geq \mu_{i+k}(G) \; (1 \leq i \leq n - k).$$

In particular, $\mu_1(G) > \mu_1(H)$ if G is a connected bipartite graph.

1.2.3 Signless Laplacian spectrum of a graph

The *signless Laplacian matrix Q $(= Q(G))$* is defined as

$$Q = D + A.$$

In other words, this matrix is obtained by removing the minus sign from the Laplacian matrix. This matrix is sometimes called the *co-Laplacian matrix*. We also write $Q_G(x)$ $(= \det(xI - Q))$ for the corresponding *signless Laplacian characteristic polynomial*. The matrix Q is also positive semidefinite. The eigenvalues of Q are called the *signless Laplacian eigenvalues* or *Q-eigenvalues* of G. They are denoted by

$$\kappa_1 \ (= \kappa_1(G)), \kappa_2 \ (= \kappa_2(G)), \ldots, \kappa_n \ (= \kappa_n(G)),$$

with the assumption that $\kappa_i \geq \kappa_j$, whenever $i < j$. κ_1 is referred to as the *Q-spectral radius* of a graph.

The remaining terminology is analogous to that for the (usual) spectrum of a graph. For example, the collection of all Q-eigenvalues is called the *signless Laplacian spectrum* or *Q-spectrum* of G.

Since the matrix Q of a connected graph is irreducible, the following two results are analogous to Theorem 1.4 and Corollary 1.5.

Theorem 1.15 *A graph is connected if and only if its Q-spectral radius is a simple eigenvalue with an eigenvector whose coordinates are non-zero and of the same sign.*

Any positive eigenvector corresponding to the Q-spectral radius is called a *Perron eigenvector*. If this vector is considered as a unit, then it is called a *principal eigenvector*.

Corollary 1.16 *For any vertex u or any edge e of a connected graph G, $\kappa_1(G-u) < \kappa_1(G)$ and $\kappa_1(G-e) < \kappa_1(G)$.*

Using the vertex–edge incidence matrix R, we get a connection between the Q-spectrum of a graph G and the spectrum of its line graph. Since

$$RR^T = Q \quad \text{and} \quad R^T R = A(\text{Line}(G)) + 2I,$$

we have

$$P_{\text{Line}(G)}(x) = (x+2)^{m-n} Q_G(x+2).\tag{1.7}$$

There is another connection. The *subdivision graph* $S(G)$ of a graph G is the graph obtained by inserting a new vertex into every edge of G.

Theorem 1.17 ([418] or [98, Theorem 2.17]) *Let G be a graph with n vertices and m edges, and let Q_G denote the characteristic polynomial of its signless Laplacian. Then*

$$P_{S(G)}(x) = x^{m-n} Q_G(x^2).\tag{1.8}$$

Proof Since each inserted vertex can be identified with the corresponding edge, we get that the subdivision graph is a bipartite graph whose adjacency matrix is of the form

$$\begin{pmatrix} O_m & R^T \\ R & O_n \end{pmatrix},$$

where R is the vertex–edge incidence matrix of $S(G)$. Thus, we get

$$P_{S(G)}(x) = \begin{vmatrix} xI_m & -R^T \\ -R & xI_n \end{vmatrix} = x^m \left| xI_n - R\frac{I_m}{x}R^T \right|$$
$$= x^{m-n} \left| x^2 I_n - RR^T \right| = x^{m-n} Q_G(x^2).$$

$\square$

If G is a non-trivial connected graph, for a vector $\mathbf{x} = (x_1, x_2, \ldots, x_n)^T$ we have $Q\mathbf{x} = 0$ if and only if $R^T \mathbf{x} = 0$, which holds if and only if $x_i = x_j$ whenever $i \sim j$. It follows that if zero is an eigenvalue of Q, then the corresponding eigenvector is an $(1, -1)$-vector where the sign determines the part in a bipartite graph. Thus, we have the following theorem.

Theorem 1.18 *The multiplicity of zero in the Q-spectrum of a graph is equal to the number of its bipartite components.*

Since the adjacency matrix of a bipartite graph can be written in the form

$$A = \begin{pmatrix} O & B \\ B^T & O \end{pmatrix}\tag{1.9}$$

we get $D + A = P^{-1}(D - A)P$, where $P = \begin{pmatrix} I & O \\ O & I \end{pmatrix}$, and thus we get the next result.

Theorem 1.19 *For any bipartite graph, the Q-spectrum coincides with the L-spectrum.*

The next theorem is quite similar to the L-interlacing theorem.

Theorem 1.20 (Q-Interlacing Theorem, cf. [104]) *Let G be a graph whose Q-eigenvalues are $\kappa_1(G) \geq \kappa_2(G) \geq \cdots \geq \kappa_n(G)$, and let H be a graph obtained by removing k ($k < n$) edges from G. The eigenvalues $\kappa_1(H) \geq \kappa_2(H) \geq \cdots \geq \kappa_n(H)$ of H satisfy*

$$\kappa_i(G) \geq \kappa_i(H) \geq \kappa_{i+k}(G) \quad (1 \leq i \leq n - k).$$

1.2.4 Relations between A, L, and Q

There are two specific relations. The first considers the largest eigenvalues of all three matrices, and the second considers all eigenvalues of these matrices in the case of a regular graph.

Theorem 1.21 (cf. [102, p. 38] or [315]) *For a graph G,*

$$\mu_1(G) \leq \kappa_1(G) = 2 + \lambda_1(\text{Line}(G)). \tag{1.10}$$

Equality holds if and only if G is bipartite.

The spectrum of a regular graph gives full information about the L-spectrum or the Q-spectrum of the same graph, as follows.

Theorem 1.22 (cf. [102, p. 5]) *If $r = \lambda_1 \geq \lambda_2 \geq \cdots \geq \lambda_n$ is the spectrum of an r-regular graph, then the L-spectrum and the Q-spectrum are $r - \lambda_n \geq r - \lambda_{n-1} \geq \cdots \geq r - \lambda_1 = 0$ and $r + \lambda_1 \geq r + \lambda_2 \geq \cdots \geq r + \lambda_n$, respectively.*

In the following example we give the eigenvalues of some particular types of graph.

Example 1.23 The spectrum of the totally disconnected graph $\overline{nK_1}$ consists of n zeros.

The spectrum of the complete graph K_n consists of $n - 1$, and $n - 1$ numbers all equal to -1.

The cocktail party $CP(n)$ is the complement of nK_2, and therefore, by Theorem 1.10, its spectrum consists of $2n - 2$, n numbers all equal to 0, and $n - 1$ numbers all equal to -2.

The spectrum of the complete bipartite graph K_{n_1,n_2} consists of $\sqrt{n_1 n_2}$, $-\sqrt{n_1 n_2}$, and $n_1 + n_2 - 2$ numbers all equal to 0. The L-spectrum of the same graph consists of $n_1 + n_2$, $n_2 - 1$ numbers all equal to n_1, $n_1 - 1$ numbers all equal to n_2, and 0.

The spectrum of the cycle C_n consists of the numbers

$$2\cos\left(\frac{2\pi}{n}i\right) \quad (1 \leq i \leq n).$$

Using the Chebyshev polynomial of the first kind (cf. [98, p. 73]), we get that the spectrum of the path consists of the numbers

$$2\cos\left(\frac{\pi}{n+1}i\right) \quad (1 \leq i \leq n).$$

Except for the path P_n, the L-eigenvalues and the Q-eigenvalues of the above graphs can be computed using Theorems 1.19 and 1.22.

For $n \geq 2$, both the L-spectrum and the Q-spectrum of P_n consist of all distinct L-eigenvalues of C_{2n} except the largest one. □

Remark 1.24 The spectrum gives information on whether a graph is bipartite (Theorem 1.9) but not whether it is connected. In contrast, the L-spectrum provides information on whether a graph is connected (Theorem 1.13) but not whether it is bipartite. If we know that a graph is connected then, by Theorem 1.18, the Q-spectrum gives information on whether it is bipartite and vice versa, if we know that a graph is bipartite then the Q-spectrum gives information on whether it is connected. □

The spectrum of the adjacency matrix has been investigated much more thoroughly than the spectrum of any other matrix associated with graphs, and naturally most of the results in the theory of graph spectra are related to this spectrum.

On the contrary, the L-spectrum is much more informative in some respects. For example, by Theorem 1.11, it contains full information about the L-spectrum of the complementary graph.

Only recently (say in the last 10 years) has the signless Laplacian attracted the attention of researches. From [114], a strong reason for believing that studying graphs by their Q-spectra is more efficient than studying them by their spectra or L-spectra lies in the fact that computational results show the tendency for a smaller number of graphs sharing the same Q-spectrum than the number of graphs sharing the same spectrum or L-spectrum.

Some similarities between the Q-spectrum and the spectrum follow from Theorem 1.4 and Corollary 1.5 (considering the spectral radius), Theorem 1.15 and Corollary 1.16 (considering the Q-spectral radius) or from the relations (1.7) and (1.8). The Q-spectrum and the L-spectrum are both non-negative and they coincide for bipartite graphs. Based on this discussion, we can say that, in an informal sense, the Q-spectrum is closer to the spectrum than the L-spectrum, and also closer to the L-spectrum than the spectrum.

1.3 Some more specific elements of the theory of graph spectra

The purpose of this section is to focus on some specific elements of the theory of graph spectra that will frequently be used.

1.3.1 Eigenvalue interlacing

We first consider more eigenvalue interlacing. In [102] the interlacing theorem is given in its general form, and the following result has arisen as a consequence.

Theorem 1.25   (cf. [102, Corollary 1.3.13] or [201]) *Let M be a real symmetric matrix with eigenvalues $v_1 \geq v_2 \geq \cdots \geq v_n$. Given a partition $\{1, 2, \ldots, n\} = P_1 \sqcup P_2 \sqcup \cdots \sqcup P_k$, with $|P_i| = n_i > 0$, consider the corresponding blocking $M = (M_{ij})$, where M_{ij} is an $n_i \times n_j$ block. Let e_{ij} be the sum of entries in M_{ij}, and set $B = \left(\frac{e_{ij}}{n_i} \right)$. Then the eigenvalues of B interlace those of M.*

In relation to the previous theorem, B is called the *quotient matrix*, and the corresponding vertex partition is said to be *equitable* if each block M_{ij} has constant row sum.

Given a graph G, let Π denote an equitable partition $V(G) = P_1 \sqcup P_2 \sqcup \cdots \sqcup P_k$ such that each vertex in P_i has b_{ij} neighbours in P_j, then the directed multi-graph[3] with vertices $P_1, P_2, \ldots, P_k$ and b_{ij} arcs from P_i to P_j is called the *divisor* of G (with respect to Π). The matrix $B = (b_{ij})$ is called the *divisor matrix*. We have the following result.

---

[3] In a directed (multi)graph each edge (called an arc) has a direction associated with it.

Theorem 1.26 (Petersdorf and Sachs, cf. [98, Theorem 4.5] or [102, Theorem 3.9.5]) *The characteristic polynomial of any divisor of a graph divides its characteristic polynomial.*

An eigenvalue of a graph G is called a *main eigenvalue* provided that the corresponding eigenvector is not orthogonal to the vector $\mathbf{j}$. Otherwise, the eigenvalue is non-main. The *main part* of the spectrum of G consists of its main eigenvalues.

Theorem 1.27 (Cvetković, cf. [103, Theorem 2.4.5]) *The spectrum of any divisor of a graph G includes the main part of the spectrum of G.*

1.3.2 Small perturbations

In what follows we need two simple graph perturbations. Let u, v, u', and v' be the distinct vertices of a graph G, and assume that u and v are adjacent, while u' is non-adjacent to u and v'. A *relocation* of type

- $\mathscr{R}$ $(= \mathscr{R}(u, v, u'))$ (usually called a *rotation*) consists of a deletion of the edge uv and an addition of the edge uu',
- $\mathscr{S}$ $(= \mathscr{S}(u, v, u', v'))$ consists of a deletion of the edge uv and an addition of the edge $u'v'$.

The next lemma sublimates the results for λ_1 and κ_1. The proof for λ_1 is taken from [409], but its essential part was given earlier in [407]. The same result for κ_1 can be found in [106].

Lemma 1.28 ([106, 409]) *Let* $\mathbf{x} = (x_1, x_2, \ldots, x_n)^T$ *be the Perron eigenvector of a connected graph G corresponding to* ν_1 *(where* ν_1 *stands for either* λ_1 *or* κ_1*). If G' is obtained from G by one of the relocations* $\mathscr{R}$ *or* $\mathscr{S}$*, then*

  (i) for $\mathscr{R}$*, if* $x_{u'} \geq x_v$ *then* $\nu_1(G') > \nu_1(G)$*,*
  (ii) for $\mathscr{S}$*, if* $x_{u'}x_{v'} \geq x_ux_v$ *then* $\nu_1(G') > \nu_1(G)$*.*

Proof   Consider $\mathscr{R}$ for λ_1. We may assume $\|\mathbf{x}\| = 1$, and then we have

$$\lambda_1(G') - \lambda_1(G) = \sup_{\mathbf{y} \in \mathbb{R}^n, \|\mathbf{y}\| = 1} \mathbf{y}^T A' \mathbf{y} - \mathbf{x}^T A \mathbf{x} \geq \mathbf{x}^T A' \mathbf{x} - \mathbf{x}^T A \mathbf{x} = \mathbf{x}^T (A' - A) \mathbf{x}.$$

We get $\mathbf{x}^T(A' - A)\mathbf{x} = 2x_u(x_{u'} - x_v)$, and so $\lambda_1(G') \geq \lambda_1(G)$, where equality holds if and only if $2x_u(x_{u'} - x_v) = 0$ and $\mathbf{x}$ is an eigenvector of G', but it can easily be seen that then the eigenvalue equation does not hold for the vertices v and u'.

The proof for κ_1 is obtained by replacing A with Q. The relocation $\mathscr{S}$ is considered in a very similar way.   □

In the following lemma we consider another graph perturbation preserving order and size.

Lemma 1.29 (cf. [103, Theorem 6.2.2] or [183, 184]) *Let $G(k,l)$ be the graph obtained from a non-trivial connected graph G by attaching at a vertex u two hanging paths whose lengths are k and l. If $k \geq l \geq 1$, then*

> *(i) $\lambda_1(G(k,l)) > \lambda_1(G(k+1,l-1))$,*
> *(ii) $\lambda_n(G(k,l)) \leq \lambda_n(G(k+1,l-1))$,*
> *(iii) $\mu_1(G(k,l)) \geq \mu_1(G(k+1,l-1))$,*
> *(iv) $\mu_{n-1}(G(k,l)) \geq \mu_{n-1}(G(k+1,l-1))$,*
> *(v) $\kappa_1(G(k,l)) > \kappa_1(G(k+1,l-1))$,*
> *(vi) $\kappa_n(G(k,l)) \geq \kappa_1(G(k+1,l-1))$.*

The proof of (i) can be found in [103]. An elegant proof of (ii) is given later in Corollary 3.24, while (vi) can be proved in a similar way by using Theorem 7.18. The inequalities (iii) and (iv) are similar to (i); their explicit proofs are given in [184] and [183], respectively. Finally, (v) follows from (i) and (1.7).

1.3.3 Hoffman program

Let $M \in \{A,L,Q\}$ and let $v_1(G) \geq v_2(G) \geq \cdots \geq v_n(G)$ be the eigenvalues of $M (= M(G))$. A real number t is said to be the *limit point* of the kth largest eigenvalue of M if there exist a sequence G_i of graphs such that $v_k(G_i) \neq v_k(G_j)$, $i \neq j$ and $\lim_{i \to \infty} v_k(G_i) = t$.

According to Hoffman [209], the smallest limit point for the spectral radius is 2 and

$$2 = \alpha_1 < \alpha_2 < \cdots$$

are all limit points for the spectral radius smaller than $\tau = \sqrt{2 + \sqrt{5}} \approx 2.0582$, where $\alpha_k = \sqrt{\beta_k} + \frac{1}{\sqrt{\beta_k}}$ and β_k is the positive root of $x^{k+1} - (1 + x + x^2 + \cdots + x^{k-1}) = 0$. In addition, $\lim_{k \to \infty} \alpha_k = \tau$. In the same paper, Hoffman proved that τ is the limit point for the spectral radius of closed caterpillars with exactly one pendant vertex, and posed the question of determining all graphs whose spectral radius is less than τ. This number is known as the *Hoffman limit point*.

Later, the same investigation was extended to the remaining two matrices (L and Q). Guo [178] determined the Laplacian Hoffman limit point (i.e., the limit point of the L-spectral radius of the mentioned closed caterpillars) and the signless Laplacian Hoffman limit point by showing that both are equal to $2 + \varepsilon$, where ε is the real root of $x^3 - 4x - 4 = 0$. These limit points are close to 4.3830.

The problem of determining all graphs whose largest eigenvalue (of any of the three matrices A, L or Q) is less than the corresponding Hoffman limit point is known as the *Hoffman program*. In other words, the Hoffman program considers graphs with very small largest eigenvalue. As we will see in Subsection 2.3.1, Section 6.3, and Section 7.3, this program is realized for all three matrices.

1.3.4 Star complement technique

The *star complement technique* is a spectral tool for constructing graphs from their induced subgraphs called star complements. We deal with this technique in Section 3.2 and Subsection 4.2.3. Here we explain its main concepts, while for more details the reader can consult the books [102, 105] or papers [392, 419, 424].

Following the above references, if λ is an eigenvalue of a connected graph G of multiplicity k, then the *star set* for λ in G is a set X of k vertices taken from G such that λ is not an eigenvalue of $G - X$. The graph $H = G - X$ is then called the *star complement* for λ in G. (Star sets and star complements exist for any graph and any of its eigenvalues, and they need not be unique.) If G has the star complement H for λ, and G is not a proper induced subgraph of some other graph with the same star complement for λ, then G is a *maximal graph* with the star complement H for λ.

Given a graph H, a subset U of $V(H)$, and an isolated vertex u, denote by $H(U)$ the graph obtained from H by joining u to all vertices of U. We say that u is the good vertex and $H(U)$ is the good extension for λ and H if λ is an eigenvalue of $H(U)$, but is not an eigenvalue of H. Assume now that U_1 and U_2 are the subsets of $V(H)$ corresponding to the vertices u_1 and u_2, and let $H(U_1, U_2; 0)$ and $H(U_1, U_2; 1)$ be the graphs obtained from H by adding both vertices u_1 and u_2, so they are non-adjacent in the former graph, while they are adjacent in the latter graph. We say that u_1 and u_2 are good partners if λ is an eigenvalue of multiplicity two either in $H(U_1, U_2; 0)$ or in $H(U_1, U_2; 1)$. Note that if u_1 and u_2 are good partners and $\lambda \notin \{-1, 0\}$, then the sets U_1 and U_2 are non-empty and distinct (cf. [105, Proposition 7.2.1]). It also follows that any vertex set X in which all vertices are good, both individually and in pairs, gives rise to a graph G in which X can be viewed as a star set for λ with H as the corresponding star complement.

The above consideration shows us how to introduce the star complement technique: in order to find the maximal extensions for $\lambda \notin \{-1, 0\}$ of a given star complement H, we form an *extendability graph* whose vertices are good vertices for λ in H, and add an edge between two good vertices whenever they are good partners. It is easy to see that the search for maximal extensions is reduced to the search for maximal cliques in the extendability graph.

The implementation of the star complement technique requires the use of a computer. Some useful software called the SCL (Star Complement Library) has been developed by Stanić and Stefanović [424, 426].

1.4 A few more words

In this section we focus on relevant applications. We also say something about spectral inequalities, extremal graph theory, and computer support in graph theory research.

1.4.1 Selected applications

In chemical graph theory, a molecular graph is a graph whose vertices correspond to the atoms of a compound and whose edges correspond to chemical bonds. Its vertices are labelled with the kinds of corresponding atom and its edges are labelled with the types of bond. Therefore, there is a natural and tight connection between graph theory (in particular, the theory of graph spectra) and chemistry.

A more specific chemical theory is the *Hükel molecular orbital theory*. Following [102, Section 9.2.1], we recall that this theory applies to conjugate hydrocarbons – chemical compounds composed of carbons and hydrogens with a single bond between a hydrogen atom and a carbon atom and a single or double bond between two carbon atoms. It is also assumed that all carbons have valency 4 and all hydrogens have valency 1. The corresponding *Hükel graph* has carbon atoms as its vertices which are connected by an edge if and only if there is a single or double bond between the corresponding carbon atoms. According to [118], the total π-electron energy of a hydrocarbon is computed as $nc_1 + 2c_2 \sum_{i=1}^{\frac{n}{2}} \lambda_i$, where c_1 and c_2 are constants and the summation is taken over the first $\frac{n}{2}$ eigenvalues of the adjacency matrix of the corresponding Hükel graph. In the special case when the Hükel graph is bipartite, we have $2 \sum_{i=1}^{\frac{n}{2}} = \sum_{i=1}^{n} |\lambda_i|$. The last quantity is called the graph energy and molecular graphs of maximal energy are of great importance. This spectral invariant is considered in Section 8.5.

Another chemical parameter is the HOMO–LUMO separation. Its definition and connection with graph eigenvalues are given in Remark 2.78.

We continue with applications in physics. From [102, Section 9.1], suppose we have a membrane held fixed along its boundary in the (x,y)-plane with displacement $f(x,y,t)$ orthogonal to the (x,y)-plane in time t described by the partial differential equation $\frac{\partial f}{\partial t} = c^2 \left(\frac{\partial^2 f}{\partial x^2} + \frac{\partial^2 f}{\partial y^2} \right)$, where c is a constant depending on the properties of the membrane. The solutions $f(x,y,t)$ subject to the boundary condition $f(x,y,t) = 0$ are called the eigenfunctions and they correspond to the infinite sequence of numbers called the eigenvalues. In practice, eigenfunctions and eigenvalues are approximated by discrete functions and a finite number of the corresponding eigenvalues, computed by means of numerical mathematics. A *membrane graph* is a planar graph whose boundary vertices lie on the boundary of a membrane. The membrane graph is not unique and is constructed in such a way that the allocation of its vertices and their connectedness by edges give a good approximation of the membrane; usually each non-boundary vertex has degree four. Then there is a connection (which depends on the function $f = f(x,y,t)$) between the eigenvalues of the membrane and those of the corresponding graph, and so the solutions to the problem can be approximated by the eigenvalues of the membrane graph.

Another physical application to the so-called *dimer problem*, which arises in the investigation of the thermodynamic properties of a system of diatomic molecules, is based on computing the eigenvalues of a lattice graph. Details can be found in [102, Section 9.1].

In biology, the *networks of protein interactions* are represented by graphs whose vertices are adjacent if and only if the corresponding proteins interact. One problem considered is *network alignment*, which deals with comparing two or more protein networks by taking into consideration their parts that are mutually similar. Several algorithms for network alignment are based on spectra of small subgraphs [99]. Another problem deals with locating the smaller groups of proteins (in a given protein network) within which the interaction between proteins is stronger than outside. The authors of [328] propose a so-called *lock-and-key* method for detecting such groups of proteins using the eigenvectors of the adjacency matrix. In [171], the authors propose another method based on the eigenvectors of the signless Laplacian matrix.

Following [110], we reinterpret some applications in computer science.

A *complex network* is the common name for a number of real networks which are represented by graphs of enormously large order. One of the main characteristics of complex networks is the *eigenvalue distribution*, that is, the distribution of the eigenvalues of the adjacency matrix associated with the network. This distribution describes many properties of the network, including its connectivity, vertex degrees, their pairwise distances, etc. [81, 318].

A graph is said to be an *expander* if each subset S of its vertex set is well connected with the vertices outside S. In other words, there is a large number of edges with exactly one end in S. Expanders are significant in constructing communication networks, error-correcting codes, sorting algorithms [113] or combinatorial optimization [326]. Such graphs can be constructed from regular graphs with a small second largest eigenvalue λ_2, graphs with a large second smallest L-eigenvalue μ_{n-1} or graphs with a large ratio $\frac{\mu_{n-1}}{\mu_1}$ [110]. Expanders are considered in Subsections 4.1.1 and 6.6.3.

The spectral radius λ_1 plays an important role in modelling virus propagation in computer networks. The smaller the spectral radius, the larger the robustness of a network against the spread of viruses [119]. More details are given in Subsection 2.3.3. In research and development networks it is desirable that knowledge is spread throughout the network as widely as possible. To achieve this, networks are constructed in such a way that the spectral radius of the corresponding graph is as large as possible [250, 251]. Graphs with large spectral radius are considered in Subsections 2.3.2, 2.3.4, and 2.3.5.

In multiprocessor systems, a task is divided into smaller parts (elementary items) assigned to given processors. Since many tasks are usually considered at the same time, each processor receives many elementary items to be handled. The distribution of elementary tasks among processors can be represented by a non-negative vector $\mathbf{x}$ whose coordinates are associated with graph vertices which are adjacent if there is a direct communication between the corresponding processors. It is natural to expect that the distribution of elementary items among processors be as uniform as possible. To achieve this, processors with large ballast send some items to adjacent processors. Therefore, the vector $\mathbf{x}$, which controls the whole process, usually changes during the time and it is of great importance to know how it looks at every moment. By [110], if $\mathbf{x}$ is spent by the eigenvectors corresponding to distinct L-eigenvalues of the associated graph, then $\mathbf{x}$ is determined iteratively by $\mathbf{x}^{(i)} = \left(I - \frac{1}{\mu_i}L\right)\mathbf{x}^{(i-1)}$.

Statistical databases (such as social networks) are those bases that allow only statistical access to their data. It is usual that some data in such bases is confidential (i.e., non-visible to the user). If a user is able to reveal confidential data, the statistical base is called compromised. Otherwise, it is said to be secure. The results of [54, 55] give a connection between secure statistical bases and graphs whose least eigenvalue is equal to -2. These graphs are considered in Section 3.2.

Following [108], we recall an application in control theory. The following differential equation is a standard model for physical systems:

$$\frac{d\mathbf{x}}{dt} = A\mathbf{x} + \mathbf{b}u. \tag{1.11}$$

Here $\mathbf{x} = \mathbf{x}(t)$ is called the state vector, with given $\mathbf{x}(0)$, and the scalar $u = u(t)$ is the control input. The matrix A has size $n \times n$, while both $\mathbf{x}$ and $\mathbf{b}$ have size $n \times 1$.

The above system is called *controllable* if the following is true: given any vector $\mathbf{x}^*$ and time t^*, there always exists a control function $u(t)$, $0 < t < t^*$, such that the solution of (1.11) gives $\mathbf{x}(t^*) = \mathbf{x}^*$ irrespective of $x(0)$. That is, the state can be steered to any point of n-dimensional vector space arbitrarily quickly. It is well known in control theory (see [78]) that this system is controllable if and only if the *controllability matrix*

$$[\mathbf{b}|A\mathbf{b}|A^2\mathbf{b}|\cdots|A^{n-1}\mathbf{b}] \tag{1.12}$$

has full rank n.

In some situations the matrix A can be read as an adjacency matrix of a graph G whose vertices are integrators (agents) and whose edges denote signal exchanges between agents. If all agents have the same sensitivity to a common external signal u, then we can take $\mathbf{b} = \mathbf{j}$. In this case, the matrix (1.12) is just the walk matrix of G, that is, the $n \times n$ matrix $W = (w_k(i))$, where $w_k(i)$ denotes the number of k-walks starting at the ith vertex of G.

In this context, the controllability is related to the main eigenvalues of the graph G, because W has full rank n if and only if G has n distinct main eigenvalues [443]. A connected graph in which all eigenvalues are mutually distinct and main is called *controllable*.

Any controllable graph is asymmetric [108], K_1 is controllable, while there are no other connected controllable graphs on fewer than six vertices. On the contrary, it is conjectured in [108] that almost all graphs are controllable, and for example there are more than 5.5 million connected controllable graphs with 10 vertices [416]. More results on these graphs can be found in [108, 109, 416, 479]. In particular, these graphs are considered in Subsection 2.3.1 and Section 3.2.

1.4.2 Spectral inequalities and extremal graph theory

The main idea of the theory of graph spectra is to exploit relations between graphs and matrices in order to study graphs by means of eigenvalues of some matrices associated with graphs. Since there are several matrices which can be used for this purpose, we can talk about the theory of graph spectra based on a matrix M, or M-theory. As we have seen, M can be any of A, L or Q.

In spectral graph theory (or in particular M-theory), *spectral inequalities* are the inequalities that include one or more spectral invariants, usually eigenvalues, of a graph. Such inequalities help in describing the structure of a graph and, as we have just seen, they have a wide range of applications. Spectral inequalities are derived using different techniques, some of which are based on the Rayleigh principle, eigenvalue interlacing, spectral moments, inspection of suitable eigenvectors or more specific connections between spectral and structural invariants.

Special cases of spectral inequalities are bounds for spectral invariants. A (lower or upper) bound for a spectral invariant usually applies to all graphs or a fixed set of graphs. In both cases, *extremal graphs* for the same invariant in the same set of graphs are just graphs that attain a given bound (if any). In this book we consider extremal graphs and also graphs that are 'close' to extremal ones. For example, it is known that $\lambda_1(G) \geq 0$ with equality if and only if G is totally disconnected, but in Subsection 2.3.1 we consider the problem of determining all graphs whose spectral radius is close to zero. More precisely, the graphs with $\lambda_1 \leq \frac{3\sqrt{2}}{2}$. Once such graphs are determined, the next natural problem is to arrange them in order with respect to the fixed invariant. In 1981, Cvetković [95] listed 12 directions for further investigation of graph spectra, and one of them was to classify and order the graphs with respect to their spectra. This problem is very difficult, even if it is considered in the restricted classes of graphs. The ordering of graphs in the context of extremal graph theory seems to be more practical.

It seems impossible to cover all the interesting results in one book, even if we limit ourselves to the last decade. We have tried to include the most powerful and representative spectral inequalities, which should be interesting to those who deal with the theory of graph spectra or its applications. We also present various proof techniques, examples, open problems, and directions for further investigation.

Bearing in mind the current expansion of studies related to extremal graphs, it is natural to expect that many new results will appear after the publication of this book. In any case, we believe that this book is a separate entity dealing with all major directions in these studies. For example, we have considered graphs with maximal spectral radius within the set of graphs of fixed order and size, graphs with minimal spectral radius within the set of graphs of fixed diameter, graphs with small second largest eigenvalue, graphs with large or small algebraic connectivity or graphs whose largest eigenvalue (of any spectrum introduced in the previous section) is comparatively small. Consequently, we have presented the realization of the Hoffman program for the most widely investigated graph matrices.

1.4.3 Computer help

As we have seen, there are a number of applications of the theory of graph spectra and many of them are in computer science. However, many researches in graph theory are to a great extent computer aided. Several software systems have been developed to support research in graph theory, among other things, by helping to prove theorems or to pose and verify conjectures.

The expert system GRAPH [111] includes programs for solving problems on particular graphs and proving theorems in graph theory. The user communicates with this system by a simplified and formalized sublanguage of English. Its user-friendly extension newGRAPH [434] allows for interactive work via automated computing of a large number of graph invariants.

The library of programs nauty, [310] written in a portable subset of C, includes, for example, programs for generating graphs with prescribed properties and computing the automorphism groups of graphs.

Among the software specialized for producing conjectures, we mention Graffiti [146], GraPHedron [311], and AutoGraphiX [68]. The former two software packages are supplied with the bases of graphs and are capable of evaluating mathematical formulae formed from graph invariants. If none of the graphs in the considered base is a counterexample to the particular formula, then the formula is considered as a conjecture. The third software package uses the variable neighbourhood search metaheuristic in the search for graphs that attain the extremal value of functions in various sets of graphs. Some results presented in this book are assumed with the support of these software packages (mostly AutoGraphiX). We mention in particular only those presented in Subsection 2.3.2 and Section 8.2.

2

Spectral radius

The most frequently investigated graph eigenvalue is the largest eigenvalue of its adjacency matrix. In Section 2.1 we demonstrate the interplay between the spectral radius and some structural invariants of a graph. Specific classes of graphs are considered in Section 2.2. Section 2.3 deals with graphs that have comparatively small spectral radius, extremal graphs, and the ordering of graphs with respect to their spectral radius.

In the whole chapter we recall some classical results obtained at the very beginning of the theory of graph spectra, together with later developments.

2.1 General inequalities

In [89] Collatz and Sinogowitz gave the following lower and upper bound for the spectral radius of a connected graph with n vertices:

$$2\cos\frac{\pi}{n+1} \leq \lambda_1 \leq n-1. \tag{2.1}$$

The upper bound was proved immediately by the same authors, while the lower bound was proved by Lovás and Pelikán [295]. The equalities are attained for the path P_n and the complete graph K_n, respectively. If a graph has no edges then $\lambda_1 = 0$, while otherwise $\lambda_1 \geq 1$.

There is another range for λ_1 given in [89],

$$\delta \leq \overline{d} \leq \lambda_1 \leq \Delta, \tag{2.2}$$

where $\overline{d} = \frac{2m}{n}$ is the average vertex degree in a graph. The first inequality is trivial, the second is proved by using the Rayleigh principle for $\mathbf{y} = \frac{1}{\sqrt{n}}\mathbf{j}$, and the third follows from (1.6). Moreover, $\delta = \bar{d} = \lambda_1$ if and only if G is regular. If G is connected then $\lambda_1 = \Delta$ if and only if G is regular.

In what follows we consider the inequalities for the spectral radius that can be applied to an arbitrary graph. The exposition is separated into several subsections concerning a number of structural invariants of a graph.

2.1.1 Walks in graphs

Here we give the lower and upper bounds for the spectral radius expressed in terms of walks in graphs, together with their consequences that bound the spectral radius in terms of vertex degrees. There are also inequalities involving the independence number or the clique number that are derived by considering walks in graphs.

The main connection between the walks in a graph and its eigenvalues is given in the following theorem.

Theorem 2.1 (cf. [98, Theorem 1.9]) *Given a graph G with eigenvalues λ_1, $\lambda_2, \ldots, \lambda_n$ and the corresponding orthogonal unit eigenvectors $\mathbf{x_1}, \mathbf{x_2}, \ldots, \mathbf{x_n}$, let $\mathbf{x_i} = (x_{i1}, x_{i2}, \ldots, x_{in})^T$ and set $c_i = (\sum_{j=1}^{n} x_{ij})^2$. Then, for every $k \geq 1$,*

$$w_k = \sum_{i=1}^{n} c_i \lambda_i^{k-1}.$$

Proof If $X = (x_{ij})$ is the orthogonal matrix of the eigenvectors of the adjacency matrix $A = (a_{ij})$, and $A^{k-1} = \left(a_{ij}^{(k-1)}\right)$ then, according to (1.5),

$$a_{ij}^{(k-1)} = \sum_{l=1}^{n} x_{il} x_{jl} \lambda_l^{k-1}.$$

Since $w_k = \sum_{1 \leq i,j \leq n} a_{ij}^{(k-1)}$, we get the assertion. $\qquad \square$

Theorem 2.2 and its consequences

The main result here is as follows.

Theorem 2.2 (Nikiforov Inequalities [346]) *For a graph G with n vertices, positive integer r, and positive odd integer q,*

$$\frac{w_{q+r}}{w_q} \leq \lambda_1^r \leq \frac{\omega - 1}{\omega} w_r, \tag{2.3}$$

where ω and w_k are the clique number and the number of k-walks in G, respectively.

In order to prove this theorem we need two additional statements.

Lemma 2.3 ([346]) *Given a graph G then, for every positive integer k,*

$$w_{2k} \leq \frac{\omega - 1}{\omega} w_k^2.$$

Proof We compute

$$w_{2k} = \sum_{i \sim j} w_k(i) w_k(j) \leq \frac{\omega - 1}{\omega} \left(\sum_{i=1}^{n} w_k(i) \right)^2 = \frac{\omega - 1}{\omega} w_k^2.$$

$\square$

Multiple application of the previous lemma gives the following corollary.

Corollary 2.4 ([346]) *For a graph G and positive integers k and l,*

$$\frac{\omega - 1}{\omega} w_{2^l k} \leq \left(\frac{\omega - 1}{\omega} w_k \right)^{2^l}.$$

We are ready to prove the initial theorem.

Proof of Theorem 2.2 Using Theorem 2.1, we get the left inequality in (2.3) by

$$\frac{w_{q+r}}{w_q} = \frac{\sum_{i=1}^{n} c_i \lambda_i^{q+r-1}}{\sum_{i=1}^{n} c_i \lambda_i^{q-1}} = \lambda_1^r \frac{\sum_{i=1}^{n} c_i \left(\frac{\lambda_i}{\lambda_1} \right)^{q+r-1}}{\sum_{i=1}^{n} c_i \left(\frac{\lambda_i}{\lambda_1} \right)^{q-1}} \leq \lambda_1^r.$$

To prove the right inequality, we first assume that G is non-bipartite, and contrary to the statement let

$$\lambda_1^r > (c+1) \frac{\omega - 1}{\omega} w_r,$$

for some $c > 0$. By Corollary 2.4, we get

$$\lambda_1^{2^l r} > \left((c+1)\frac{\omega-1}{\omega}w_r \right)^{2^l} \geq (c+1)^{2^l}\frac{\omega-1}{\omega}w_{2^l r}.$$

Next, by Theorem 2.1,

$$c_1\lambda_1^{q-1} < (\varepsilon+1)w_q,$$

for any $\varepsilon > 0$ and sufficiently large q. Taking $q = 2^l r$ and l sufficiently large, by the inequality (2.88) we get

$$\lambda_1^{2^l r} \leq \frac{\omega-1}{\omega}c_1\lambda_1^{2^l r-1} < (\varepsilon+1)\frac{\omega-1}{\omega}w_{2^l r}.$$

A contradiction.

Assume now that G is bipartite. Here we have $w = 2$, and the corresponding inequality reduces to $\lambda_1^r \leq \frac{w_r}{2}$, for every $r \geq 2$. If r is odd, (2.88) and Theorem 2.1 imply

$$\lambda_1^r \leq \frac{1}{2}c_1\lambda_1^{r-1} \leq \frac{1}{2}w_r.$$

If r is even, write $\overline{w}_k$ for the number of closed walks on k vertices in G. Bearing in mind that $\overline{w}_k = tr(A^k) = \sum_{i=1}^n \lambda_i$ and the fact that the spectrum of a bipartite graph is symmetric about zero, we get $2\lambda_1^r \leq \overline{w}_{r+1} \leq w_r$, and the inequality follows. $\qquad\square$

Remark 2.5 By Nikiforov, if $q = 1$ the left equality in (2.3) holds if and only if each component of G has the spectral radius λ_1 and is regular or, if r is even, semiregular bipartite. For $q > 1$ and r odd, equality holds if and only if each component of G has the spectral radius λ_1 and its vertices have the same average 2-degree. For $q > 1$ and r even, equality holds if and only if each component of G is bipartite with the spectral radius λ_1 and its vertices have the same average 2-degree whenever they belong to the same part.

Next, if $r = 1$ and the right equality in (2.3) holds then G is a regular complete ω-partite graph, while if $r > 1$ then G has a dominant component H described as follows. For $\omega = 2$, H is complete bipartite, and if r is odd then H is a regular graph. For $\omega > 2$, H is a regular complete ω-partite graph. $\qquad\square$

Theorem 2.2 is a strong result, and a number of known lower or upper bounds for λ_1 arise from (2.3). We start with lower ones. For $r = q = 1$, we get $\overline{d} \leq \lambda_1$ (see (2.2)). For $r = 2, q = 1$, we get the inequality of Hofmeister [214]

$$\lambda_1 \geq \sqrt{\frac{1}{n} \sum_{i=1}^{n} d_i^2}. \tag{2.4}$$

Concerning the same inequality, Hofmeister proved that if G is non-regular, there is a unique real number p such that $\lambda_1 = \sqrt[p]{\frac{1}{n} \sum_{i=1}^{n} d_i^p}$.

Setting $r = 2, q = 3$, we get the inequality obtained by Yu et al. [481]

$$\lambda_1 \geq \sqrt{\frac{\sum_{i=1}^{n} (d_i m_i)^2}{\sum_{i=1}^{n} d_i^2}}, \tag{2.5}$$

together with its consequence

$$\lambda_1 \geq \frac{\sqrt{\sum_{i=1}^{n} (d_i m_i)^2}}{\sum_{i=1}^{n} \sqrt{d_i}}. \tag{2.6}$$

Setting $r = 2, q = 5$, we get the inequality of Hong and Zhang [221]

$$\lambda_1 \geq \sqrt{\frac{\sum_{i=1}^{n} \left(\sum_{j \sim i} d_j m_j \right)^2}{\sum_{i=1}^{n} (d_i m_i)^2}}. \tag{2.7}$$

Consider now the upper bound of (2.3). Setting $r = 1$, we get the inequality obtained by Wilf [462] in 1967.

Corollary 2.6 *For the clique number ω of a graph G,*

$$\omega \geq \frac{n}{n - \lambda_1}. \tag{2.8}$$

Since for the coordinates of the principal eigenvector $\sum_{i=1}^{n} x_i \leq \sqrt{n}$, the last corollary follows from Exercise 2.4, as well. The inequality (2.8) obviously implies the inequality of Edwards and Elphick [139] obtained in 1982,

$$\omega \geq \frac{n}{n - \lambda_1} - \frac{1}{3}. \tag{2.9}$$

Since the chromatic number χ satisfies $\chi \geq \omega$, we deduce that (2.8) implies the inequality of Cvetković [93] obtained in 1972,

$$\chi \geq \frac{n}{n - \lambda_1}. \tag{2.10}$$

Setting next $r = 2$, we get

$$\lambda_1^2 \leq \frac{\omega - 1}{\omega} 2m, \tag{2.11}$$

or equivalently $\omega \geq \frac{2m}{2m - \lambda_1^2}$, which, using the fact that $\lambda_1(G) + \lambda_1(\overline{G}) \geq n - 1$ (see the Courant–Weyl inequalities), gives the upper bound for the independence number α in terms of the order, size, and spectral radius [339]:

$$\alpha \geq \frac{n(n-1) - 2m}{n(n-1) - 2m - (n - 1 - \lambda_1)^2}. \tag{2.12}$$

Other walks-based bounds

For $p \geq 0, r \geq 1$, by setting $y_i = \sqrt{\frac{w_p(i)}{w_p}}$ for all coordinates y_i $(1 \leq i \leq n)$ of the vector $\mathbf{y}$ and using the Rayleigh principle, we get the following general lower bound [346]:

$$\lambda_1^r \geq \mathbf{y}^T A^r \mathbf{y} = \frac{1}{w_p} \sum_{i,j \in V(G)} w_{r+1}(i,j) \sqrt{w_p(i)w_p(j)}.$$

By the Cauchy–Schwarz inequality,

$$\sum_{i,j \in V(G)} w_{r+1}(i,j) \sqrt{w_p(i)w_p(j)} \sum_{i,j \in V(G)} \frac{w_{r+1}(i,j)}{\sqrt{w_p(i)w_p(j)}} \geq \left(\sum_{i,j \in V(G)} w_{r+1}(i,j) \right)^2$$

$$= w_{r+1}^2,$$

which yields

$$\lambda_1^r \sum_{i,j \in V(G)} \frac{w_{r+1}(i,j)}{\sqrt{w_p(i)w_p(j)}} \geq \frac{w_{r+1}^2}{w_p}. \tag{2.13}$$

Setting $p = 2, r = 1$ in the above inequality, we get the following bounds obtained by Favaron et al. [157]:

$$\lambda_1 \geq \frac{1}{2m} \sum_{i \sim j} \sqrt{d_i d_j} \qquad (2.14)$$

and

$$\lambda_1 \geq \frac{2m}{\sum_{i \sim j} \frac{1}{\sqrt{d_i d_j}}}. \qquad (2.15)$$

Since λ_1^r is equal to the largest eigenvalue of $B^{-1} A^r B$, where B is the diagonal matrix with entries of the main diagonal $b_{ii} = w_p(i)$ $(1 \leq i \leq n)$, it is less than or equal to the largest row sum in $B^{-1} A^r B$. In this way we get the following result.

Theorem 2.7 (Nikiforov [346]) *For a graph G and integers $p, r \geq 1$,*

$$\lambda_1^r \leq \max \left\{ \frac{w_{p+r}(i)}{w_p(i)} : 1 \leq i \leq n \right\}. \qquad (2.16)$$

This upper bound also improves some results. For example, by setting $p = 1, r = 2$ or $p = 2, r = 1$, we get the upper bounds obtained in [157]:

$$\lambda_1 \leq \max\{\sqrt{d_i m_i} : 1 \leq i \leq n\} \qquad (2.17)$$

and

$$\lambda_1 \leq \max\{m_i : 1 \leq i \leq n\}. \qquad (2.18)$$

Concerning the last two bounds, it is worth mentioning that similar lower bounds can be derived [400]:

$$\lambda_1 \geq \min\{\sqrt{d_i m_i} : 1 \leq i \leq n\} \quad \text{and} \quad \lambda_1 \geq \min\{m_i : 1 \leq i \leq n\}. \qquad (2.19)$$

An easy consequence of (2.17) is

$$\lambda_1 \leq \max\{\sqrt{d_i d_j} : 1 \leq i, j \leq n\}, \qquad (2.20)$$

which can also be verified by considering the Perron eigenvector $\mathbf{x}$ of G and setting $x_s = \max\{x_i \; : \; i \in V(G)\}$, $x_t = \max\{x_i \; : \; ij \in E(G)\}$. Then, from the eigenvalue equation, we get

$$\lambda_1 x_s = \sum_{i \sim s} x_i \leq d_s x_t \quad \text{and} \quad \lambda_1 x_t = \sum_{i \sim t} x_i \leq d_t x_s.$$

Thus, $\lambda_1^2 x_s x_t \leq d_s d_t x_s x_t$, yielding (2.20).

Finally, setting $p = 2, r = 2$ in (2.16) and considering walks in graphs, we get

$$\lambda_1^2 \leq \max_{i \in V(G)} \frac{w_4(i)}{d_i} = \max_{i \in V(G)} \frac{w_3(i)}{d_i} \frac{w_4(i)}{w_3(i)} \leq \max_{i \in V(G)} \frac{w_3(i)}{d_i} \left(\frac{1}{d_i} \sum_{j \sim i} \frac{w_3(i)}{d_j} \right)$$
$$\leq \max_{i \sim j} \frac{w_3(i)}{d_i} \frac{w_3(j)}{d_j} = \max_{i \sim j} m_i m_j, \tag{2.21}$$

which is the upper bound obtained by Das and Kumar [128].

The next inequality concerns the interaction between the number of cliques in a graph and its spectral radius.

Theorem 2.8 (Bollobás and Nikiforov [48]) *If k_s is the number of cliques with s vertices in a graph G then, for any $r \geq 2$,*

$$\lambda_1^{r+1} \leq (r+1)k_{r+1} + \sum_{s=2}^{r} (s-1)k_s \lambda_1^{r+1-s}. \tag{2.22}$$

Proof We need two results taken from [341]:

$$\lambda_1^{\omega} \leq \sum_{s=2}^{\omega} (s-1)k_s \lambda_1^{\omega-s} \tag{2.23}$$

and

$$\sum_{i=1}^{n} \left(k_s(i)w_{l+1}(i) - k_{s+1}(i)w_l(i) \right) \leq (s-1)k_s w_l, \tag{2.24}$$

for all $2 \leq s \leq \omega$ and $l \geq 2$, where $k_s(i)$ denotes the number of cliques with s vertices containing i.

Following [48], it is sufficient to prove the inequality for $2 \leq r > \omega$, since the case $r \geq \omega$ follows from (2.23). Summing the inequalities (2.24) for $2 \leq s \leq r$, we get

$$\sum_{i=1}^{n} \left(k_2(i)w_{l+r-1}(i) - k_{r+1}(i)w_l(i) \right) \leq \sum_{s=2}^{r} (s-1)k_s w_{l+r-s}$$

or

$$w_{l+r} - \sum_{s=2}^{r} (s-1)k_s w_{l+r-s} \leq \sum_{i=1}^{n} k_{r+1}(i)w_l(i).$$

Observing that $w_l(i) \leq w_{l-1}(i)$, we get

$$\sum_{i=1}^{n} k_{r+1}(i)w_l(i) \leq w_{l-1} \sum_{i=1}^{n} k_{r+1}(i) = (r+1)k_{r+1}w_{l-1}$$

and thus

$$\frac{w_{l+r}}{w_{l-1}} - \sum_{s=2}^{r} (s-1)k_s \frac{w_{l+r-s}}{w_{l-1}} \leq (r+1)k_{r+1}.$$

Now, since $\omega > 2$, G is non-bipartite, and therefore $\lambda_1 > -\lambda_n$. Using the equality of Theorem 2.1, we get

$$\lim_{l \to \infty} \frac{w_{l+r}}{w_{l-1}} = \lambda_1^{r+1},$$

and the assertion follows. $\qquad\square$

The following upper bound for the number of cliques can be derived in a similar way [48]:

$$k_{s+1} \geq \left(\frac{\lambda_1}{n} - 1 + \frac{1}{s} \right) \frac{s(s-1)}{s+1} \left(\frac{n}{s} \right)^{s+1}, \qquad (2.25)$$

which is left as an exercise.

We illustrate some of the previous bounds.

Example 2.9 Consider the graph obtained from the well-known Petersen graph by adding an edge (see Fig. 2.1(a)) and the graph on 20 vertices depicted in Fig. 2.1(b). The latter graph is considered in Subsection 4.2.3 in the context

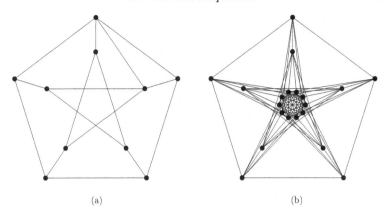

(a)   (b)

Figure 2.1 Graphs for Example 2.9.

of graphs with small second largest eigenvalue. It is a bidegreed graph containing two disjoint induced subgraphs: the Petersen graph and K_{10} (removal of one gives the other), with the spectrum $[11, 1^{10}, -1^5, -4^4]$.

The results of numerical computation are given below.

Lower bounds

Graph	(2.2)	(2.4)	(2.5)	(2.7)	λ_1
Fig. 2.1(a)	3.200	3.225	3.256	3.261	3.262
Fig. 2.1(b)	10	10.440	10.996	10.999	11

Upper bounds

Graph	(2.3), $r = 20$	(2.3), $r = 5$	(2.3), $r = 3$	(2.17)	(2.18)	(2.20)
Fig. 2.1(a)	3.376	3.743	4.108	3.317	3.667	4
Fig. 2.1(b)	11.216	11.886	12.519	12.042	11.154	13

The lower bounds obtained from the left inequality of (2.3) give a good estimate for λ_1, even for small values of the parameters r and q (see the results for (2.5) and (2.7)). For r sufficiently large, the same conclusion holds for the corresponding upper bounds. □

Example 2.10 In this example we compute the same bounds for the pineapple graphs $P(k, l)$.

Lower bounds

(k, l)	(2.2)	(2.4)	(2.5)	(2.7)	λ_1
$(5,5)$	3	3.873	4.171	4.247	4.284
$(5,50)$	2.182	7.422	7.437	7.442	7.527
$(50,5)$	44.727	46.821	48.997	49.002	49.002

Upper bounds

(k, l)	(2.3), $r = 20$	(2.3), $r = 5$	(2.3), $r = 3$	(2.17)	(2.18)	(2.20)
$(5,5)$	4.358	4.613	4.932	4.583	9	6
$(5,50)$	7.900	10.604	13.433	8.124	54	14.697
$(50,5)$	49.012	49.041	49.070	49.051	54	51.439

The exact value of (2.7) computed for $P(50,5)$ differs from $\lambda_1(P(50,5))$ in its fifth decimal. Observe that (2.18) gives the same estimate for $\lambda_1(P(5,50))$ and $\lambda_1(P(50,5))$, much closer to the latter spectral radius.

Nikiforov [346] asked if the lower bound of (2.3) holds for any even $q \geq 2$ and any $r \geq 2$. We get a counterexample by taking $P(5,5)$ and setting $r = 2, q = 4$. Then we compute $\lambda_1 \approx 4.284 < 4.335 \approx \frac{w_6}{w_4}$. $\qquad\square$

2.1.2 Graph diameter

Here we consider the bounds for the spectral radius that include the graph diameter. Connections between the diameter and the coordinates of the principal eigenvector, and consequently between the diameter and the spectral radius of a graph, are demonstrated in Theorems 2.12 and 2.16.

Upper bounds

Besides the graph diameter, all upper bounds exposed in this subsection include the maximal vertex degree. Since the spectral radius of any regular graph coincides with the vertex degree, only non-regular graphs are of interest. Before we start, we mention the inequality obtained by Stevanović [430] in 2004 that bounds the spectral radius of a non-regular graph in terms of its order and maximal vertex degree:

$$\lambda_1 < \Delta - \frac{1}{2n(n\Delta - 1)\Delta^2}. \tag{2.26}$$

In the following few years a number of similar inequalities (in addition, including the graph diameter) appeared. Following the idea explained in the Preface, we first give the last improvement and then the inequalities which can be derived from it.

Theorem 2.11 (Cioabă [85]) *For a connected non-regular graph G with diameter D,*

$$\lambda_1 < \Delta - \frac{1}{nD}. \tag{2.27}$$

If $\mathbf{x} = (x_1, x_2, \ldots, x_n)^T$ denotes the principal eigenvector of G and s is a vertex of G such that $x_s = \max\{x_i \ : \ i \in V(G)\}$ then, since G is non-regular, it follows that $x_s > \frac{1}{\sqrt{n}}$. If now $d_s < \Delta$, the above inequality follows from

$$\lambda_1 x_s = \sum_{j \sim s} x_j \leq (\Delta - 1)x_s.$$

The case $d_s = \Delta$ is considered in a long proof which will not be interpreted here. We note that the bound (2.27) was conjectured in [88], where the authors obtained a similar inequality. Here we give the proof of this result, with a comment that the main idea and some parts of the proof of Theorem 2.11 are similar.

Theorem 2.12 (Cioabă et al. [88]) *For a connected non-regular graph G with diameter D,*

$$\lambda_1 < \Delta - \frac{n\Delta - 2m}{n(D(n\Delta - 2m) + 1)}. \tag{2.28}$$

Proof If $\mathbf{x}$ is the principal eigenvector of the adjacency matrix A of G, then

$$\lambda_1 = \mathbf{x}^T A \mathbf{x} = 2 \sum_{i \sim j} x_i x_j = \sum_{i \sim j} (x_i^2 + x_j^2) - \sum_{i \sim j} (x_i - x_j)^2$$
$$= \sum_{i=1}^n d_i x_i^2 - \sum_{i \sim j} (x_i - x_j)^2.$$

This yields $\Delta - \lambda_1 = \sum_{i \sim j} (x_i - x_j)^2 + \sum_{i=1}^n (\Delta - d_i)x_i^2$.

Denote $x_s = \max\{x_i \ : \ i \in V(G)\}$ and $x_t = \min\{x_i \ : \ i \in V(G)\}$. Since G is non-regular, $s \neq t$. Let $s = i_0, i_1, \ldots, i_k = t$ be the vertices in the shortest path from s to t (given in natural order). By the Cauchy–Schwarz inequality, we have

$$\sum_{j=0}^{k-1}(x_{i_j} - x_{i_{j+1}})^2 \geq \frac{1}{D}(x_s - x_t)^2,$$

which, together with the above equality, gives

$$\Delta - \lambda_1 \geq \frac{1}{D}(x_s - x_t)^2 + (n\Delta - 2m)x_t^2,$$

where the right-hand side attains its minimum when $x_t = \frac{x_s}{D(n\Delta - 2m)+1}$. Putting back this expression for x_t, and substituting x_s^2 by $\frac{1}{n}$ (since $||\mathbf{x}|| = 1$, we have $x_s^2 > \frac{1}{n}$), we get the assertion. $\square$

Remark 2.13   Note that (2.27) is an obvious improvement of (2.28). Moreover, both bounds improve the result of Zhang [506],

$$\lambda_1 < \Delta - \frac{(\sqrt{\Delta} - \sqrt{\delta})^2}{nD\Delta}, \tag{2.29}$$

which is, in fact, a finer bound than (2.26).

Concerning the expressions of all these bounds, it is noteworthy to add that the last improvement (2.27) implies the existence of a constant $c \geq 1$ such that

$$\lambda_1 < \Delta - \frac{c}{nD},$$

for all connected non-regular graphs.

Finally, considering the graphs with maximal spectral radius within the set of connected graphs of fixed order with maximal vertex degree, Liu and Li [278] proved that

$$\lambda_1 < \Delta - \frac{\Delta + 1}{n(3n + \Delta - 8)} \tag{2.30}$$

holds for any of these graphs. It is also verified that for any of these graphs $D \leq \frac{3n+\Delta-8}{\Delta+1}$ holds, unless we have two non-adjacent vertices of degree $\Delta - 1$. Setting $D = \frac{3n+\Delta-8}{\Delta+1}$ in the upper bound (2.27), we get that this bound is better than (2.30) for all graphs whose diameter satisfies the mentioned inequality. $\square$

Figure 2.2 Graph H_1 for Example 2.14.

Example 2.14   It is not difficult to obtain a graph for which the bound (2.27) (along with its consequences (2.28), (2.29), and (2.26)) does not give a good estimate. Namely, it is sufficient to choose a graph with large gap between its spectral radius and maximal vertex degree. For example, if we take the star $K_{1,n-1}$ then we get $\lambda_1(K_{n-1,1}) = \sqrt{n-1}$, while the right-hand side of (2.27) gives $n - 1 - \frac{1}{2n}$.

In other situations, all these bounds can give much better results. We illustrate this for the following two graphs.

Let H_1 denote the graph depicted in Fig. 2.2. It is obtained by deleting an edge from a 3-regular graph on 12 vertices, and its diameter is 6. Let H_2 denote the graph obtained from two copies of K_{100} by deleting an edge in each copy and joining a vertex of degree 98 in one copy with a vertex of the same degree in the other. The diameter of H_2 is 5, this graph is non-regular, and so the results of Theorems 2.11 and 2.12, and those of Remark 2.13 can be applied. The computational results follow.

Graph	λ_1	(2.27)	(2.28)	(2.29)	(2.26)
H_1	2.9134	2.9861	2.9872	2.9995	2.9999
H_2	98.9901	98.9990	98.9991	99.0000	99.0000

□

A lower bound

To prove Theorem 2.16, we need a simple combinatorial result.

Lemma 2.15   ([86]) *For a vertex i of a graph with n vertices and diameter D,*

$$w_D(i) \geq n - 1.$$

Moreover, $w_D(i) \geq n - 1 - d_i + d_i^2$, *whenever* $D \geq 3$.

Theorem 2.16 (Cioabă et al. [86]) *For a graph G with n vertices and diameter D,*

$$\lambda_1 \geq (n-1)^{\frac{1}{D}}. \tag{2.31}$$

Proof By the previous lemma, the number of D-walks starting at a vertex i is at least $n - 1$. Next, the number of D-walks from i to j is equal to a_{ij}^D (the (i,j)-entry of A^D), and therefore the total number of D-walks starting at i is $A_i^D \mathbf{j}$, where A_i^D denotes the ith row in A^D. Using the Rayleigh principle, we get

$$n - 1 \leq \frac{\mathbf{j}^T A^D \mathbf{j}}{n} = \frac{\mathbf{j}^T A^D \mathbf{j}}{\mathbf{j}\mathbf{j}^T} \leq \lambda_1(A^D) = \lambda_1^D,$$

and the inequality follows. $\square$

Remark 2.17 A *Moore graph* is a graph with diameter D and girth $2D + 1$, for some $D > 1$. Moore graphs were named after E. F. Moore, who posed the question of describing the graphs of fixed degree with maximal number of vertices. It is known that any Moore graph must be regular (see, e.g., [102]), and then the following bound for n in terms of diameter D and vertex degree r holds:

$$n \leq \begin{cases} r\frac{(r-1)^D - 1}{r-2} + 1, & \text{if } D > 2, \\ 2D + 1, & \text{if } D = 2. \end{cases} \tag{2.32}$$

So, alternatively, a Moore graph can be defined as a regular graph that attains the above bound for n. Next, the Moore graph can exist only if $r = 2$ (the graphs being the cycles C_{2D+1}) or if $D = 2$ and $r \in \{3, 7, 57\}$. Here, for $r = 3$ and $r = 7$ there exists a unique Moore graph (the Petersen graph and the Hoffman–Singleton graph,[1] respectively); for $r = 57$ the answer is not known [43, 212].

From Lemma 2.15, equality in (2.31) can only hold for $D \leq 2$, and it indeed holds for $D = 1$, while for $D = 2$, by [119], equality holds if and only if G is the star $K_{1,n-1}$ or a Moore graph.

For $D = 3$ the inequality is slightly improved to

---

[1] The Hoffman–Singleton graph is a 7-regular graph with 50 vertices. It is the unique strongly regular graph with parameters $(50, 7, 0, 1)$. For more details, see [102, p. 170].

$$\lambda_1^3 - \lambda_1^2 \geq n - 1 - \frac{2m}{n}, \tag{2.33}$$

with equality if and only if the corresponding graph is the cycle C_7 [86]. Using $\lambda_1 \geq \frac{2m}{n}$, we get

$$\lambda_1^3 - \lambda_1^2 + \lambda_1 \geq n - 1. \tag{2.34}$$

$\square$

Example 2.18 In relation to the previous remark, except for the star $K_{1,n-1}$, any graph of diameter 2 that attains the lower bound of (2.31) must be a Moore graph. Since there is no Moore graph for $r \geq 3$ and $D \geq 3$, in the following two situations we consider the graphs with maximal number of vertices which do not attain the upper bound of (2.32).

It has been shown in [41] that a cubic graph of diameter 3 can have at most 20 vertices and there is a unique such graph with exactly 20 vertices. Similarly, by [64], a cubic graph of diameter 4 can have at most 38 vertices.

Computing the bound (2.31) for these graphs, we get

$$3 = \lambda_1 \geq \sqrt[3]{19} \approx 2.6684 \quad \text{and} \quad 3 = \lambda_1 \geq \sqrt[4]{37} \approx 2.4663,$$

respectively. For the former graph, the inequality (2.33) gives

$$18 = \lambda_1^3 - \lambda_1^2 \geq n - 1 - \frac{2m}{n} = 16.$$

$\square$

2.1.3 Other inequalities

Here we give more bounds for λ_1 that are mainly expressed in terms of the order, size, and (some) vertex degrees. The first is a frequently used upper bound considered independently by several authors. Subsequently, it will be compared with an upper bound of Shu and Wu.

Theorem 2.19 and its consequences

Theorem 2.19 (Hong et al. [220], Nikiforov [341], Zhou and Cho [522]) *For a graph G,*

$$\lambda_1 \leq \frac{\delta - 1 + \sqrt{(\delta + 1)^2 + 4(2m - \delta n)}}{2}. \tag{2.35}$$

Equality holds if and only if G is regular or G has a component of order $\Delta + 1$ in which every vertex is of degree δ or Δ, and all other components are δ-regular.

Proof Let $s_i(M)$ denote the row sum in a matrix M with n rows which corresponds to its ith row ($1 \leq i \leq n$). If A is the adjacency matrix of G and i is an arbitrary vertex, then we have

$$
\begin{aligned}
s_i(A^2) = \sum_{j \sim i} d_j &= 2m - d_i - \sum_{d(i,j) \geq 2} d_j \\
&\leq 2m - d_i - (n - d_i - 1)\delta \\
&= 2m + (\delta - 1)d_j - \delta(n - 1).
\end{aligned}
$$

In other words, $s_i(A^2 - (\delta - 1)A) \leq 2m - \delta(n-1)$ and then $\lambda_1^2 - (\delta - 1)\lambda_1 \leq 2m - \delta(n-1)$. Solving this inequality, we get (2.35).

If the equality holds, then

$$\sum_{d(i,j) \geq 2} d_j = (n - d_i - 1)\delta$$

must hold for all i ($1 \leq i \leq n$), which gives the assertion. The other implication is proved in a similar way.   □

Remark 2.20 It can be checked that the upper bound (2.35) is sharp for regular graphs, and also for the graphs that maximize the *degree variance*

$$\mathrm{var}(G) = \frac{1}{n} \sum_{i=1}^{n} \left(d_i - \frac{2m}{n} \right)^2. \tag{2.36}$$

The degree variance is the quantity measuring how much a graph differs from a regular graph. Other similar measures are

$$\varepsilon(G) = \lambda_1 - \frac{2m}{n}$$

or

$$s(G) = \sum_{i=1}^{n} \left| d_i - \frac{2m}{n} \right|. \tag{2.37}$$

Bell [36] compared $\varepsilon(G)$ and $\mathrm{var}(G)$ and proved that neither of them could be preferred to the other as a measure of irregularity. An inequality that involves $\mathrm{var}(G)$ and $s(G)$ is given later in Theorem 2.30. □

Some classical results of the theory of graph spectra are, in fact, particular cases of Theorem 2.19.

Corollary 2.21 (Stanley Inequality [427]) *The spectral radius of a graph G with m edges satisfies*

$$\lambda_1 \le \frac{1}{2}(\sqrt{8m+1} - 1). \tag{2.38}$$

Equality occurs if and only if $m = \binom{k}{2}$ *and G is a disjoint union of the complete graph* K_k *and isolated vertices.*

The result of the previous corollary extends the result of Brualdi and Hoffman [60] claiming that if $m = \binom{k}{2}$ then $\lambda_1 \le k - 1$.

Corollary 2.22 (Schwenk Inequality [397]) *For any connected graph G,*

$$\lambda_1 \le \sqrt{2m - n + 1}. \tag{2.39}$$

Equality holds if and only if G is a star or a complete graph.

The last result is also known as the *Hong inequality* [216]. Moreover, Hong noticed that (2.39) holds for any graph with minimal vertex degree at least 1. Observe that this inequality is a finer result than the *Wilf inequality* [463]

$$\lambda_1 \le \sqrt{2m\left(1 - \frac{1}{n}\right)}. \tag{2.40}$$

There is another corollary.

Corollary 2.23 ([69, 128]) *For a graph G,*

$$\lambda_1 \le \sqrt{2m - (n-1)\delta + (\delta - 1)\Delta}. \tag{2.41}$$

Equality holds if and only if G is either a regular graph or a star.

Proof The result follows from (2.35) by

$$\lambda_1^2 \leq 2m - (n-1)\delta + (\delta - 1)\lambda_1 \leq 2m - (n-1)\delta + (\delta - 1)\Delta,$$

where the last equality holds either if $\delta = \Delta$ or $\delta = 1$. □

A similar lower bound is given in Exercise 2.8.

More bounds in terms of vertex degrees

In the next result we deal with all vertex degrees d_i $(1 \leq i \leq n)$.

Theorem 2.24 (Shu and Wu [403]) *For a connected graph with vertex degrees $d_1 \geq d_2 \geq \cdots \geq d_n$,*

$$\lambda_1 \leq \frac{d_i - 1 + \sqrt{(d_i + 1)^2 + 4(i-1)(d_1 - d_i)}}{2}, \qquad (2.42)$$

where $1 \leq i \leq n$. If $i = 1$, equality holds if and only if G is regular. If $2 \leq i \leq n$, equality holds if and only if G is either a regular graph or a bidegreed graph with $\Delta = d_1 = d_2 = \cdots = d_{i-1} = n - 1$, and $d_i = d_{i+1} = \cdots = d_n = \delta$.

As a consequence we get the bound

$$\lambda_1 \leq \frac{\Delta' - 1 + \sqrt{(\Delta' + 1)^2 + 4p(\Delta - \Delta')}}{2}, \qquad (2.43)$$

where p and Δ' stand for the number of vertices of degree Δ and the second largest degree, respectively.

Remark 2.25 It can be deduced from Theorem 2.24 that equality holds in (2.43) if and only if G is regular or isomorphic to $K_p \nabla H$, where H is a $(\Delta' - p)$-regular graph on $n - p$ vertices.

If G is non-regular and $4p \leq \Delta + \Delta' + 2$, we have a simple bound

$$\lambda_1 \leq \frac{\Delta + \Delta'}{2}. \qquad (2.44)$$

In particular, it holds whenever $\Delta > \Delta'$.

Finally, concerning the efficiency of (2.43), it can be verified that it gives an estimate at least equal to (2.35) whenever $p + q \geq \Delta + 1$ (where q is the number of vertices of the second largest degree) [403]. □

There are three similar inequalities. The proof is left as an exercise for the reader.

Theorem 2.26 (Feng et al. [158]) *For a connected graph G,*

$$\lambda_1 \leq \max\left\{\sqrt{\frac{d_i^2 + d_i m_i}{2}} : 1 \leq i \leq n\right\}, \qquad (2.45)$$

$$\lambda_1 \leq \max\left\{\sqrt{\frac{d_i(d_i + m_i) + d_j(d_j + m_j)}{2}} : i \sim j\right\}, \qquad (2.46)$$

and

$$\lambda_1 \leq \max\left\{\frac{\Delta + \sqrt{\Delta^2 + 8(d_i m_i)}}{4} : 1 \leq i \leq n\right\}. \qquad (2.47)$$

In all cases, equality holds if and only if G is regular.

We continue with two lower bounds of Kumar [254] obtained by using a slightly different arguments. Any principal submatrix of order two of A^2 is

$$\begin{pmatrix} A_i^T A_i & A_i^T A_j \\ A_j^T A_i & A_j^T A_j \end{pmatrix},$$

where A_i is the ith column of A. The square root of the largest eigenvalue of the above matrix is

$$\frac{1}{\sqrt{2}}\sqrt{d_i + d_j + \sqrt{(d_i - d_j)^2 + 4c_{ij}^2}},$$

where c_{ij} is the number of common neighbours of i and j. Since $\lambda_1(G)$ is not less than the largest eigenvalue of any principal submatrix of A, we get the following result.

Theorem 2.27 ([254]) *For a graph G on $n \geq 2$ vertices labelled $1, 2, \ldots, n$,*

$$\lambda_1 \geq \frac{1}{\sqrt{2}} \max_{j<i} \sqrt{d_i + d_j + \sqrt{(d_i - d_j)^2 + 4c_{ij}^2}}, \qquad (2.48)$$

where c_{ij} is the number of common neighbours of i and j.

The next theorem follows by the Rayleigh principle applied to A^2 for $\mathbf{y} = \mathbf{j}$.

Theorem 2.28 ([254]) *For a graph G on n vertices labelled $1, 2, \ldots, n$,*

$$\lambda_1 \geq \sqrt[4]{\frac{1}{n} \sum_{i=1}^n \left(d_i + 2 \sum_{j>i} c_{ij} \right)^2}, \qquad (2.49)$$

where c_{ij} is the number of common neighbours of i and j.

We point out another similar result which is proved by considering the possible common neighbourhoods of two vertices i and j specified in the theorem.

Theorem 2.29 (Das and Kumar [128]) *For a non-trivial graph G,*

$$\lambda_1 \geq \sqrt{\frac{(\Delta + \tilde{d} - 1) + \sqrt{(\Delta + \tilde{d} - 1)^2 - 4(\Delta - 1)(\tilde{d} - 1) + 4\tilde{c}^2 + 8\tilde{c}\sqrt{\Delta}}}{2}}, \qquad (2.50)$$

where if i is a vertex with maximal degree Δ in G and j is a vertex with maximal degree $\tilde{d}$ in $N(i)$, then $\tilde{c}$ is the number of common neighbours of i and j.

In the next theorem we bound λ_1 by the quantities (2.36) and (2.37).

Theorem 2.30 (Nikiforov [337]) *For a graph G,*

$$\frac{\mathrm{var}(G)}{2\sqrt{m}} \leq \lambda_1 - \frac{2m}{n} \leq \sqrt{s(G)}, \qquad (2.51)$$

where $\mathrm{var}(G)$ and $s(G)$ are defined by (2.36) and (2.37), respectively.

The left inequality of (2.51) follows by

$$2\sqrt{2m}\left(\lambda_1 - \frac{2m}{n}\right) \geq 2\lambda_1\left(\lambda_1 - \frac{2m}{n}\right) \geq \lambda_1^2 - \left(\frac{2m}{n}\right)^2$$

$$\geq \frac{1}{n}\sum_{i=1}^{n} d_i^2 - \left(\frac{2m}{n}\right)^2 = \mathrm{var}(G),$$

where the first inequality in the above chain follows from (2.38), and the third follows from (2.4).

The proof of the right inequality of (2.51) given in [337] is based on two supplementary statements, and will not be repeated here.

Using $\frac{s^2(G)}{n^2} \leq \mathrm{var}(G) \leq s(G)$, we get

$$\frac{s^2(G)}{2n^2\sqrt{2m}} \leq \lambda_1 - \frac{2m}{n} \leq \sqrt[4]{n^2\mathrm{var}(G)}.$$

Concerning the sharpness of both inequalities of (2.51), Nikiforov [337] conjectured that an improvement

$$\frac{s^2(G)}{2n^2\sqrt{m}} \leq \lambda_1 - \frac{2m}{n} \leq \sqrt{\frac{s(G)}{2}},$$

may hold for sufficiently large n and m.

Another similar result for non-regular graphs is proved in [332]:

$$\lambda_1 - \frac{2m}{n} > \begin{cases} \frac{1}{n(\Delta+2)}, & \text{if } \Delta - \delta = 1 \text{ and all but one vertices have equal degree,} \\ \frac{1}{2(m+n)}, & \text{otherwise.} \end{cases} \tag{2.52}$$

The lower bound of (2.51) and the lower bound of (2.52) improve the bound $\overline{d} = \frac{2m}{n} \leq \lambda_1$ (given in (2.2)).

Example 2.31 We compute some lower and upper bounds obtained in this subsection for the graphs considered in Example 2.9.

| **Lower bounds** | | | | | |
Graph	(2.48)	(2.50)	(2.51)	(2.52)	λ_1
Fig. 2.1(a)	2.358	2.524	3.220	3.226	3.262
Fig. 2.1(b)	4.690	5.008	10.450	10.025	11

| **Upper bounds** | | | | | |
Graph	(2.35)	(2.39)	(2.38)	(2.43)	(2.45)	(2.51)
Fig. 2.1(a)	3.450	4.796	5.179	3.450	3.674	4.989
Fig. 2.1(b)	11.718	13.454	13.651	11.718	12.787	17.746

For these graphs, the lower bounds of (2.51) and (2.52) give significantly better results than those of (2.48) and (2.50). In addition, the lower bound of (2.51) has less deviation from λ_1 than the corresponding upper bound. Concerning the remaining upper bounds, recall that (2.39) and (2.38) are both improved by (2.35). For both graphs we have $p + q \geq \Delta + 1$, and so, by Remark 2.25, (2.35) should not be better than (2.43), but they are in fact equal. $\square$

Example 2.32   The lower bound of (2.51) and the lower bound of (2.52) can be used together with the upper bound of (2.27) in order to obtain the admissible magnitude for λ_1.

For example, for the graph H_1 (resp. H_2) of Example 2.14, the lower bound of (2.51) gives $\lambda_1 \geq 2.8824$ (resp. $\lambda_1 \geq 98.9901$) and the lower bound of (2.52) gives $\lambda_1 > 2.8438$ (resp. $\lambda_1 > 98.9914$). Combining the better of these results with those obtained by (2.27) (see Example 2.14), we get

$$\lambda_1(H_1) \in [2.8824, \ 2.9861) \quad \text{and} \quad \lambda_1(H_2) \in (98.9914, \ 98.9990).$$

The lengths of these intervals are close to 0.1037 and 0.0076. $\square$

2.2 Inequalities for spectral radius of particular types of graph

As we have seen in the previous section, general inequalities concerning the interaction between the spectral radius and other graph invariants are widely studied. Apart from these, there are more results related to some particular classes of graph. Here we consider bipartite graphs, graphs which do not contain a specified induced subgraph, some graphs related to regular ones, and nested graphs.

2.2.1 Bipartite graphs

Recall that some general bounds for λ_1 attain equality exactly for specific bipartite graphs (e.g., this occurs for the lower bound of (2.3) when n is even, and consequently for all derived lower bounds given in (2.5)–(2.7)). On the contrary, in some cases, we can get a refinement or simple expression for bipartite graphs. Say we have $\tilde{c} = 0$ in (2.50).

The examination of other general bounds is left to the reader. Rather than that, in what follows we consider some bounds which are not related directly to the previous ones.

In the following theorem we use a more general bound for matrix eigenvalues.

Theorem 2.33   (Kumar [254]) *For a bipartite graph G with adjacency matrix A,*

$$\frac{m}{p} + \frac{s}{\sqrt{p-1}} \leq \lambda_1^2 \leq \frac{m}{p} + s\sqrt{p-1}, \qquad (2.53)$$

where $p = \lfloor \frac{n}{2} \rfloor$, $s^2 = \frac{M_4}{2p} - \left(\frac{m}{p}\right)^2$, and M_4 is the fourth spectral moment or

$$\frac{M_{2l-1}}{2p} + \frac{s_l}{\sqrt{p-1}} \leq \lambda_1^{2^{l-1}} \leq \frac{M_{2l-1}}{2p} + s_l\sqrt{p-1}, \qquad (2.54)$$

where p is the same as above, and $s_l^2 = \frac{M_{2l}}{2p} - \left(\frac{M_{2l-1}}{2p}\right)^2$.

Proof   We use the result

$$t + \frac{s}{\sqrt{n-1}} \leq \lambda_1 \leq t + s\sqrt{n-1}, \qquad (2.55)$$

which holds for all $n \times n$ complex or real matrices B with real eigenvalues (cf. [465]), with $t = \frac{\text{tr}(B)}{n}$ and $s^2 = \frac{\text{tr}(B^2)}{n} - t^2$. To get (2.53) it is sufficient to set $B = A^2$. Similarly, by setting $B = A^{2^{l-1}}$, we get (2.54). $\square$

If G is a bipartite graph with n_1 (resp. n_2) vertices in the first (resp. second) part, then its adjacency matrix can be written in the form (1.9) where B is an $n_1 \times n_2$ submatrix. The singular eigenvalues of such a matrix B are the square roots of eigenvalues of BB^T, and therefore we can obtain an upper bound for λ_1 by replacing B with BB^T in (2.55). In this way we get

$$\lambda_1^2 \le \frac{\text{tr}(BB^T)}{n_1} + \sqrt{\frac{n_1-1}{n_1}\left(\text{tr}((BB^T)^2) - \frac{\text{tr}(BB^T)^2}{n_1}\right)}. \qquad (2.56)$$

Using this result we prove the next theorem.

Theorem 2.34 (Merikoski et al. [312]) *Let G be a bipartite graph with parts $X = \{u_1, u_2, \ldots, u_{n_1}\}$ and $Y = \{v_1, v_2, \ldots, v_{n_2}\}$, $n_1 \le n_2$. Let the vertices of X have degrees $d_1 \ge d_2 \ge \cdots \ge d_{n_1}$, and let d_{ij} stand for the number of common neighbours of the vertices u_i and u_j $(1 \le i,j \le n_1)$. Denote $e = \sum_{i=1}^{n_1} d_i$, $g = \sum_{i,j} d_{ij}^2$, and $f = d_1^2 + 3d_2^2 + \cdots + (2n_1 - 1)d_{n_1}^2$. Then we have*

$$\lambda_1 \le \sqrt{\frac{e}{n_1} + \sqrt{\frac{n_1-1}{n_1}\left(g - \frac{e^2}{n_1}\right)}} \le \sqrt{\frac{e}{n_1} + \sqrt{\frac{n_1-1}{n_1}\left(f - \frac{e^2}{n_1}\right)}} \le \sqrt{e}. \qquad (2.57)$$

Proof The first inequality follows from (2.56), since $\text{tr}(BB^T) = e$ and

$$\text{tr}\left((BB^T)^2\right) = g.$$

To get the second or third inequality it is sufficient to show $g \le f$ (or equivalently, $\sum_{i \ne j} d_{ij}^2 \le d_1^2 + 3d_2^2 + \cdots + (2n_1 - 1)d_{n_1}^2$) or $f \le e^2$ (or equivalently, $d_1^2 + 3d_2^2 + \cdots + (2n_1 - 1)d_{n_1}^2 \le \left(\sum_{i=1}^{n_1} d_i\right)^2$), respectively. Both claims follow easily by the induction argument, and we get the assertion. $\qquad\square$

Remark 2.35 Note that for $n_1 = 1$, the equalities hold in (2.57). For $n_1 \ge 2$, by [312], the first equality holds if and only if $n_1 - 1$ smallest eigenvalues of BB^T are mutually equal, while the remaining two equalities hold if and only if G is complete bipartite.

Using similar reasoning, it can be proved that

$$\lambda_1 \le \sqrt{\frac{M+1}{2} + \frac{M-1}{2}\sqrt{\frac{2M^2 + 3M + 1}{3}}} \le \sqrt{\frac{M(M+1)}{2}} \qquad (2.58)$$

holds whenever $d_i \le d_1 - i + 1$ $(2 \le i \le n_1)$, where $M = \max\{d_1, n_1\}$. $\qquad\square$

Example 2.36 Let $K(n_1, n_2, n_3)$ denote the graph obtained from the complete bipartite graph K_{n_1, n_2} by joining a vertex that belongs to the part of size n_1 with

n_3 isolated vertices. Further, let $C(k,l)$ denote the comet defined on page 5. Computational results for the inequalities (2.53) and (2.56) are given below.

Graph	(2.53), lower bound	λ_1	(2.56)	(2.53), upper bound
$K(5,10,5)$	3.253	7.147	7.155	7.157
$K(5,10,50)$	2.349	8.507	8.750	8.795
$C(5,5)$	1.732	2.499	2.584	2.569
$C(20,5)$	1.914	4.588	4.681	5.115

□

Trees

If T is a tree with n vertices, the minimal value of $\lambda_1(T)$ is given in (2.1) and it is attained if and only if T is the path P_n. In contrast, one of the oldest results concerning extremal eigenvalues of trees is the corresponding upper bound obtained by Lovász and Pelikán [295]:

$$\lambda_1(T) \leq \sqrt{n-1}, \tag{2.59}$$

where equality occurs if and only if T is the star $K_{1,n-1}$.[2] Later, it was proved by Shao [401] that this bound is part of a more general result concerning a larger number of eigenvalues (see page 148).

Recently, Song et al. [415] gave a valuable discussion concerning the bounds for λ_1 with respect to the order and maximal vertex degree of a tree. Their results are summarized as follows. If the maximal degree of an arbitrary tree lies between 3 and $n-2$, then

$$\lambda_1 \begin{cases} \leq \sqrt{\frac{n-1+\sqrt{(n-2\Delta)^2+2n-3}}{2}}, & \text{if } n \leq 2\Delta, \\ \leq \sqrt{2\Delta-1}, & \text{if } 2\Delta < n \leq \Delta^2+1, \\ < 2\sqrt{\Delta-1}\cos\frac{\pi}{2k+1}, & \text{if } \Delta^2+1 < n, \end{cases} \tag{2.60}$$

where $k = \left\lceil \log_{\Delta-1}\left(\frac{(\Delta-2)(n-1)}{\Delta}+1\right)\right\rceil + 1$.

Trees that attain the first two bounds are given in [415]. This result extends the result of Stevanović [429]

---

[2] We provide an elegant proof of $\lambda_1(P_n) \leq \lambda_1(T) \leq \lambda_1(K_{1,n-1})$ in Theorem 2.67 (Subsection 2.3.4).

$$\lambda_1(T) < 2\sqrt{\Delta - 1}. \tag{2.61}$$

The last inequality is proved by considering the *rooted Bethe trees* defined as follows. The (rooted) Bethe tree $T_{\Delta,1}$ is a single vertex which is simultaneously a root. For $k \geq 2$, the Bethe tree $T_{\Delta,k}$ consists of a single root u and $\Delta - 1$ copies of $T_{\Delta,k-1}$, along with $\Delta - 1$ edges joining u with roots of all $T_{\Delta,k-1}$.

Rojo and Robbiano [384] have shown that

$$\lambda_1(T_{\Delta,k}) = 2\sqrt{\Delta - 1}\cos\frac{\pi}{k+1}.$$

They also determined the full spectrum of $T_{\Delta,k}$.

Example 2.37   Considering the inequality (2.61) for trees $T_{\Delta,k}$, it can be observed that it does not depend on k.

If we take the family of binary trees $T_{3,k}$ ($k \geq 2$), we get $\lambda_1(T_{3,k}) \leq 2\sqrt{2} \approx 2.8284$. On the contrary, the spectral radius of these trees increases as k increases. Setting $k = 10$, we get the tree on $2^{10} - 1 = 1023$ vertices whose spectral radius is close to 2.7139.

For the family of ternary trees $T_{4,k}$ ($k \geq 2$), we get $\lambda_1(T_{4,k}) \leq 2\sqrt{3} \approx 3.4641$, while for example $\lambda_1(T_{4,7}) \approx 3.2004$ (this tree has 1093 vertices).   □

We list without proof two more upper bounds. The first is part of the research of Das and Kumar [128], who produced several eigenvalue bounds including (2.50). The second was obtained by Ming and Wang [322].

$$\lambda_1(T) \leq \sqrt{n - 1 - \frac{(\Delta + \tilde{d} - 1) - \sqrt{(\Delta + \tilde{d} - 1)^2 - 4(\Delta - 1)(\tilde{d} - 1)}}{2}}, \tag{2.62}$$

where $\tilde{d}$ denotes the maximal vertex degree within the set of neighbours of a vertex with degree Δ.

$$\lambda_1(T) \leq \sqrt{\frac{1}{2}\left(n - \mu + 1 + \sqrt{(n - \mu + 1)^2 - 4(n - 2\mu + 1)}\right)}, \tag{2.63}$$

where μ is the size in a maximal matching of T.

2.2.2 Forbidden induced subgraphs

In 1970, Nosal [351] obtained the following result.

Theorem 2.38 ([351]) *If a graph G is triangle-free, then*

$$\lambda_1 \leq \sqrt{m}. \tag{2.64}$$

This result was extended slightly in [339] by inserting the strict inequality which holds for all graphs except a graph whose unique non-trivial component is complete bipartite. Furthermore, a similar result concerning quadrangle-free graphs was given in [342]. If $m \geq 9$ and G is a quadrangle-free graph distinct from the star $K_{1,9}$, then $\lambda_1 < \sqrt{m}$.

We now turn on to the triangle and quadrangle-free graphs, that is, the graphs of girth at least five.

Theorem 2.39 (Nikiforov [331]) *For a graph G of girth at least five,*

$$\lambda_1 \leq \min\{\Delta, \sqrt{n-1}\}. \tag{2.65}$$

Proof $\lambda_1 \leq \Delta$ follows from (2.2), so it remains to prove $\lambda_1 \leq \sqrt{n-1}$.

By (2.17), we have $\lambda_1^2 \leq \max_{i \in V(G)} \sum_{j \sim i} d_j$. Next, if i is an arbitrary vertex of G, since the girth of G is at least five, we have that any two neighbours of i are non-adjacent and their unique common neighbour is i. Hence, if $m(N(i), V(G) \backslash N(i))$ denotes the number of edges between the set of neighbours of i and the remaining vertices of G, we have

$$\sum_{j \sim i} d_j = m(N(i), V(G) \backslash N(i)) \leq d_i + n - d_i - 1 = n - 1,$$

and the proof follows. $\qquad\square$

Remark 2.40 Equality holds in (2.65) if and only if G is either the star $K_{1,n-1}$, a Δ-regular Moore graph of diameter 2, or contains two components of girth at least 5 where one is Δ-regular and the maximal vertex degree in the other is at most Δ (cf. [331]).

The bound (2.65) is always finer than the upper bound of Lu et al. [297] (which concerns connected graphs of girth at least five)

$$\lambda_1 \leq \frac{\sqrt{4n + 4\Delta - 3} - 1}{2}. \tag{2.66}$$

Namely, the right-hand side of the above inequality is never less than $\min\{\Delta, n - 1\}$. $\qquad\square$

The proof of the next inequality is very similar to that of the previous theorem.

Theorem 2.41 (Nikiforov [332]) *For a graph G with no induced subgraphs isomorphic to $K_2 \cup \overline{K}_{k+1}$ or $K_{2,l+1}$ $(0 \le k \le l)$,*

$$\lambda_1 \le \min\left\{\Delta, \frac{k-l+1+\sqrt{(k-l+1)^2+4l(n-1)}}{2}\right\}. \qquad (2.67)$$

If G is connected, equality holds if and only if either $\Delta^2 - \Delta(k-l+1) \le l(n-1)$ and G is Δ-regular or $\Delta^2 - \Delta(k-l+1) > l(n-1)$ and every two vertices have k common neighbours if they are adjacent, and l common neighbours otherwise.

2.2.3 Nearly regular graphs

In the following two theorems we consider the effects of removing or adding an edge to an r-regular graph on a spectral radius.

Theorem 2.42 (Cioabă [85]) *Let G be an r-regular graph and e be an edge of G such that $H = G - e$ is connected. Then*

$$\frac{1}{nD} < r - \lambda_1(H) < \frac{2}{n}, \qquad (2.68)$$

where D is the diameter of H.

Proof The proof is a direct consequence of Theorem 2.11. $\qquad\square$

The left inequality was improved by Nikiforov [344], who showed that this inequality holds for any proper subgraph H when D stands for the diameter of G. Both results improve the inequality of [340]: $\frac{1}{n(D+1)} < r - \lambda_1(H)$.

For $f, g : \mathbb{N} \mapsto [0, \infty)$, we write $f(n) = O(g(n))$ if there are $c, n_0 > 0$ such that $f(n) \le cg(n)$, for $n \ge n_0$. We also write $f(n) = \Theta(g(n))$ if $f(n) = O(g(n))$ and $g(n) = O(f(n))$.

If $r \ge 3$ is fixed then the diameter of H is at least $\log_{r-1} n + O(1)$, and in this case we have the following estimate [85]:

$$O\left(\frac{1}{n\log_{r-1} n}\right) < r - \lambda_1(H) < \frac{2}{n}.$$

Example 2.43 The lower bound obtained in Theorem 2.42 is the same bound given in Theorem 2.11, already tested for a graph obtained by removing an edge from a regular graph (see graph H_1 in Example 2.14). The upper bound can be tested for the same graph, and in this case we get

$$3 - \lambda_1(H_1) \approx 0.0866 < 0.1667 \approx \frac{2}{n}.$$

$\square$

In the next theorem we modify the result of Cioabă [85] by changing the left inequality in (2.69) together with the corresponding part of the proof.

Theorem 2.44 (cf. [85]) *Let G be a connected r-regular graph and H be obtained by adding an edge e to G. In addition, assume that $r - \lambda_2(G) > 1$. Then*

$$\frac{2}{n} + \frac{1}{n(r+2)+2} < \lambda_1(H) - r < \frac{2}{n}\left(\frac{1}{r - \lambda_2(G) - 1} + 1\right). \qquad (2.69)$$

Proof Consider the left inequality. By (2.52) we have

$$\lambda_1(H) > \frac{2m(H)}{n} + \frac{1}{2(m(H)+n)}.$$

Since G is r-regular, we have $r = \frac{2(m(H)-1)}{n}$, or equivalently $m(H) = \frac{nr+2}{2}$. Setting this in the above inequality, we get the assertion.

The right inequality is obtained in [85] by straightforward computation based on the result of Maas [303], claiming that if i and j are non-adjacent vertices of G then $\lambda_1(G+ij) - \lambda_1(G) < 1 + \xi - \lambda_1(G) + \lambda_2(G)$, where ξ satisfies the equation $\lambda_1(G) - \lambda_2(G) = \frac{\xi(\xi+1)(\xi+2)}{(x_i+x_j)^2+\xi(2x_ix_j+\xi+2)}$ and x_i, x_j are the corresponding coordinates of the principal eigenvector. $\square$

Example 2.45 For the graph H of Theorem 2.44 we can take the graph depicted in Fig. 2.1(a). In this case we compute

$$3.2588 < 3.2618 \approx \lambda_1(H) < 3.5030.$$

☐

Remark 2.46   For fixed $r \geq 3$ and $\varepsilon > 0$, Friedland [164] proved that the most r-regular graphs satisfy $\lambda_2 \leq 2\sqrt{r-1} + \varepsilon$. This implies that for the most r-regular graphs G,

$$\lambda_1(G+e) - \lambda_1(G) = \Theta\left(\frac{1}{n}\right).$$

☐

2.2.4 Nested graphs

We give some lower and upper bounds for the spectral radius of nested graphs. All results exposed here are proved by considering the coordinates of the principal eigenvector. The notation introduced on page 6 is used throughout this subsection.

Nested split graphs

Following [408], for an NSG we define the quantities

$$e_s = \sum_{i=1}^{s} m_i N_i$$

(the number of edges with one end in $U_1 \cup U_2 \cup \cdots \cup U_s$),

$$\bar{e}_s = M_s N_s - e_s = \sum_{j=1}^{s} n_j M_{j-1}$$

(the number of non-edges between $U_1 \cup U_2 \cup \cdots \cup U_s$ and $V_1 \cup V_2 \cup \cdots \cup V_s$),

$$\widehat{e}_h = \sum_{i=1}^{h} m_i \frac{N_i^2}{N_h}$$

(since $m_i N_i$ is the number of edges with one end in U_i, $\widehat{e}_h$ is the weighted mean of these quantities).

We are ready to present one lower and two upper bounds for λ_1. All three results were obtained by Simić et al. [408].

Theorem 2.47 ([408]) *For a connected NSG,*

$$\lambda_1 \geq \frac{1}{2}\left(N_h - 1 + \sqrt{(N_h - 1)^2 + 4\widehat{e}_h}\right). \tag{2.70}$$

Proof Let $\mathbf{y} = (y_1, y_2, \ldots, y_n)^T$, where $y_u = N_i$ if $u \in U_i$ for some i $(1 \leq i \leq h)$ and $y_v = \lambda_1$ if $v \in V_j$ $(1 \leq j \leq h)$. The above inequality is now easily obtained by using the Rayleigh principle. $\square$

In order to give two upper bounds we need to consider the principal eigenvector $\mathbf{x} = (x_1, x_2, \ldots, x_n)^T$ of an NSG. Observe that all vertices within the sets U_s or V_t $(1 \leq s, t \leq h)$ correspond to mutually equal coordinates of $\mathbf{x}$ (since they belong to the same orbit). Let $x_u = a_s$ if $u \in U_s$ and $x_v = b_t$ if $v \in V_t$. We prove a lemma.

Lemma 2.48 (cf. [408]) *For any s, t $(1 \leq s, t \leq h)$, we have*

$$\frac{1}{\lambda_1 + 1}(1 - a_t M_{t-1}) \leq b_t \leq \frac{1}{\lambda_1 + 1}(1 - a_1 M_{t-1}) \tag{2.71}$$

and

$$a_s \leq \frac{1}{\lambda_1^2 + \lambda_1}(N_s - a_1 \overline{e}_s). \tag{2.72}$$

Proof We first derive some (in)equalities on the coordinates of the principal eigenvector $\mathbf{x}$. From the eigenvalue equation for λ_1, we get

$$\lambda_1 a_s = \sum_{i=1}^{s} n_i b_i \quad (1 \leq s \leq h), \tag{2.73}$$

$$\lambda_1 b_t = \sum_{i=t}^{h} m_i a_i + \sum_{j=1}^{h} n_j b_j - b_t \quad (1 \leq t \leq h).$$

By normalization, we get $\sum_{i=1}^{h} m_i a_i + \sum_{j=1}^{h} n_j b_j = 1$, and thus we have

$$a_s = \frac{1}{\lambda_1}\sum_{i=1}^{s} n_i b_i \quad (1 \leq s \leq h), \tag{2.74}$$

$$b_t = \frac{1}{\lambda_1 + 1}\left(1 - \sum_{i=1}^{t-1} m_i a_i\right) \quad (1 \leq t \leq h). \tag{2.75}$$

Setting $a_0 = b_0 = 0$, we next get

$$\lambda_1(a_{s+1} - a_s) = n_{s+1}b_{s+1} \quad (0 \leq s \leq h-1), \qquad (2.76)$$
$$(\lambda_1 + 1)(b_1 - b_0) = 1 \quad (t = 0),$$
$$(\lambda_1 + 1)(b_{t+1} - b_t) = -m_t a_t \quad (1 \leq t \leq h-1).$$

Since all coordinates of **x** are positive, we have that

$$a_{s+1} > a_s \quad (1 \leq s \leq h-1), \qquad (2.77)$$
$$b_{t+1} < b_t \quad (1 \leq t \leq h-1).$$

Now, the inequality (2.71) follows from (2.75).

From (2.74) and (2.71), we get

$$a_s = \frac{1}{\lambda_1} \sum_{i=1}^{s} n_i b_i \leq \sum_{i=1}^{s} n_i \frac{1}{\lambda_1 + 1} (1 - a_1 M_{i-1})$$
$$= \frac{1}{\lambda_1^2 + \lambda_1} \left(N_s - a_1 \sum_{i=1}^{s} n_i M_{i-1} \right) = \frac{1}{\lambda_1^2 + \lambda_1} (N_s - a_1 \bar{e}_s),$$

and the proof follows. $\square$

Theorem 2.49   ([408]) *For a connected NSG,*

$$\lambda_1 \leq \frac{1}{2} \left(N_h - 1 + \sqrt{(N_h - 1)^2 + 4e'_h} \right), \qquad (2.78)$$

where $e'_h = e_h - n_1 \left(\frac{\bar{e}_h}{n} + \frac{\sum_{i=1}^{h} m_i \bar{e}_i}{2m} \right)$.

Proof   Setting $s = h$ in (2.73), we get $\sum_{s=1}^{h} m_s a_s + \lambda_1 a_h = 1$. Using (2.71), we get

$$1 \leq \sum_{s=1}^{h} m_s \frac{1}{\lambda_1^2 + \lambda_1} (N_s - a_1 \bar{e}_s) + \frac{1}{\lambda_1 + 1} (N_h - a_1 \bar{e}_h),$$

which gives

$$\lambda_1^2 + \lambda_1 \leq \sum_{s=1}^{h} m_s (N_s - a_1 \bar{e}_s) + \lambda_1 (N_h - a_1 \bar{e}_h).$$

The above inequality can be rewritten as

$$\lambda_1^2 - (N_h - 1)\lambda_1 - \left(e_h - a_1\left(\lambda_1\bar{e}_h + \sum_{s=1}^{h} m_s\bar{e}_s\right)\right) \leq 0.$$

Computing $a_1 = \frac{N_1}{\lambda_1^2 + \lambda_1}$ (see (2.76)) and taking into account that $\lambda_1 \leq n - 1$ and $\lambda_1^2 + \lambda_1 \leq 2m$ (follows from (2.38)), we get the assertion.   □

The proof of the following theorem is similar.

Theorem 2.50   (Simić et al. [408]) *For a connected NSG,*

$$\lambda_1 \leq \frac{1}{2}\left(N_h' - 1 + \sqrt{(N_h' - 1)^2 + 4e_h''}\right),   (2.79)$$

where $N_h' = N_h - \frac{n_1\bar{e}_h}{2m}$ *and* $e_h'' = \frac{n_1 \sum_{i=1}^{h} m_i\bar{e}_i}{2m}$.

We now test the previous bounds. The next example and Example 2.55 are respectively taken from [408] and [20].

Example 2.51   For an NSG with parameters $(40, 20, 10, 10; 20, 10, 50, 40)$, we have the following results.

(2.70)	λ_1	(2.79)	(2.78)
134.045	134.266	137.183	138.377

For an NSG with parameters $(400, 200, 100, 100; 200, 100, 500, 400)$, we have the following results.

(2.70)	λ_1	(2.79)	(2.78)
1348.54	1350.74	1379.78	1391.70

□

Double nested graphs

DNGs are considered by Anđelić et al. [20].

For a connected DNG, denote the degree of a vertex $u \in U_s$ (resp. $v \in V_t$) by $d_s' = N_{h+1-s}$ (resp. $d_t'' = M_{h+1-t}$). Setting $y_u = d_s'$ if $u \in U_s$ $(1 \leq s \leq h)$ and $y_v = \lambda_1$ for $v \in V_t$ $(1 \leq t \leq h)$, and applying the Rayleigh principle to the vector $\mathbf{y} = (y_1, y_2, \ldots, y_n)$, we get the theorem.

Theorem 2.52   ([20]) *For a connected DNG,*

$$\lambda_1 \geq \max \left\{ \sqrt{\sum_{k=1}^{h} \frac{m_k}{N_h}(d_k')^2}, \sqrt{\sum_{k=1}^{h} \frac{n_k}{M_h}(d_k'')^2} \right\}. \tag{2.80}$$

The next two statements are analogous to Theorems 2.49 and 2.50.

Theorem 2.53   ([20]) *For a connected DNG,*

$$\lambda_1 \geq \min \left\{ \sqrt{m - n_1 \frac{\sum_{s=1}^{h} m_s \overline{f}_{h+1-s}}{m}}, \sqrt{m - m_1 \frac{\sum_{s=1}^{h} n_s \overline{e}_{h+1-s}}{m}} \right\}, \tag{2.81}$$

where $\overline{e}_s = \sum_{i=1}^{s} m_i(N_h - N_{h+1-i})$ and $\overline{f}_s = \sum_{j=1}^{s} n_j(M_h - M_{h+1-j})$.

Theorem 2.54   ([20]) *For a connected DNG,*

$$\sqrt{\frac{1}{2}\left(m - \sqrt{m^2 - 4g}\right)} \leq \lambda_1 \leq \sqrt{\frac{1}{2}\left(m + \sqrt{m^2 - 4g}\right)}, \tag{2.82}$$

where $g = \max\{m_1 \sum_{s=1}^{h} n_s \overline{e}_{h+1-s}, n_1 \sum_{s=1}^{h} m_s \overline{f}_{h+1-s}\}$.

Example 2.55   For a DNG with parameters $(1,2,3,2;2,1,3000,1)$, we have the following results.

(2.80)	λ_1	(2.82), upper bound	(2.81)
94.9105	94.9176	94.9369	94.9369

For a DNG with parameters $(1,2,3,2;2,1,3,1000)$, we have the following results.

(2.80)	λ_1	(2.82), upper bound	(2.81)
31.7192	31.7192	31.7192	31.7287

$\square$

2.3 Extremal graphs

According to (2.1), the spectral radius of any connected graph lies in the segment $[2\cos\frac{\pi}{n+1}, n-1]$. Concerning possible values of λ_1 that are close to $n-1$ we have that for any even n, the spectral radius of the cocktail party graph $CP(\frac{n}{2})$ is equal to $n-2$, and, by Corollary 1.5, adding k ($0 \le k \le \frac{n}{2}$) edges to $CP(\frac{n}{2})$ gives a graph G_k whose spectral radius belongs to $[n-2, n-1]$. In particular, $G_0 = CP(\frac{n}{2})$ and $G_{\frac{n}{2}} = K_n$. Moreover, for any G_k,

$$\lambda_1(G_k) = \frac{n-3+\sqrt{n^2-2n+8k+1}}{2}$$

(see [234]), and therefore all graphs G_k ($0 \le k \le \frac{n}{2}$) may be ordered by the spectral radius, that is, $i \le j$ yields $\lambda_1(G_i) < \lambda_1(G_j)$.

The previous example illustrates the subject of this section. For a given set of graphs, we consider those with minimal or maximal spectral radius or, in some cases, we consider the determining or the ordering of all graphs within the given set.

In the following subsection we consider the determining of graphs with comparatively small spectral radius. In the sequel, we consider two frequently investigated problems concerning graphs of fixed order and size with maximal spectral radius, and graphs of fixed order and diameter with minimal or maximal spectral radius. Trees and some other graphs with extremal spectral radius are considered in Subsections 2.3.4 and 2.3.5, while the ordering of graphs is considered in Subsection 2.3.6.

2.3.1 Graphs whose spectral radius does not exceed $\frac{3\sqrt{2}}{2}$

We start with the Hoffman program (see page 21) with respect to the adjacency matrix. The first step in its realization was carried out in [413] (see also [102, Theorem 3.11.1]), where Smith determined all connected graphs with $\lambda_1 = 2$. These graphs are illustrated in Fig. 2.3, and they are known as the Smith graphs. Apart from the first four graphs, there are two infinite families of Smith graphs. The first consists of T-shape trees which, according to the notation described on page 5, can be denoted by $T_n^{2,n-3}$ (with $n \ge 5$), and the second consists of cycles C_n ($n \ge 3$). It is easy to see that any connected graph is either a proper induced subgraph of a Smith graph, a Smith graph or a proper supergraph of a Smith graph, and any connected graph whose spectral radius is less than 2 is an induced subgraph of some Smith graph.

Further, in [97] Cvetković et al. gave a nearly complete determination of

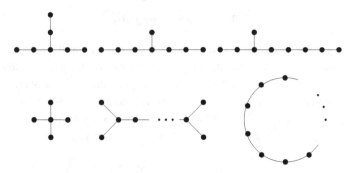

Figure 2.3 Smith graphs.

all graphs with $2 < \lambda_1 \leq \tau = \sqrt{2 + \sqrt{5}}$. This determination was completed by Brower and Neumaier [59]. All the resulting graphs are limited to being part of two very specific families of T-shape trees $S_{i,j,k}$ and caterpillars $T_n^{i,j}$. Next, the exact values of the corresponding parameters for which $2 < \lambda_1(S_{i,j,k}), \lambda_1(T_n^{i,j}) \leq \tau$ can be determined by computing the corresponding characteristic polynomial, which is done for $S_{i,j,k}$ in [97]. For $T_n^{i,j}$, the authors of [59] used the concept of exit values described in [329]. The result follows.

Theorem 2.56 (Brower and Neumaier [59], Cvetković et al. [97]) *If G is connected and $2 < \lambda_1(G) \leq \tau = \sqrt{2 + \sqrt{5}}$, then G is one of the following graphs:*

(i) $S_{1,2,k}$ $(k \geq 6)$,

(ii) $S_{1,3,k}$ $(k \geq 4)$,

(iii) $S_{1,j,k}$ $(k \geq j \geq 4)$,

(iv) $S_{2,2,k}$ $(k \geq 3)$,

(v) $S_{2,3,3}$,

(vi) $T_n^{i,j}$ $(i < j, \; n \leq 2j - 2i + 1)$,

(vii) $T_n^{3,j}$ $(i < j, \; n \leq 2j - 4)$,

(viii) $T_n^{2,j}$ $(i < j, \; n \leq 2j)$,

(ix) $T_7^{2,3}$,

(x) $T_{11}^{3,7}$,

(xi) $T_{13}^{3,8}$,

(xii) $T_{16}^{4,11}$,

(xiii) $T_{18}^{4,12}$.

Note that that none of the graphs listed in the previous theorem attains $\lambda_1 = \tau$.

Remark 2.57 On page 25 we gave a definition of controllable graphs together with their application in control theory. In the context of this subsection, we recall that Stanić [416] gave an exact description of the graphs whose spectral radius lies in the subinterval $(2, \zeta)$, where

$$\zeta = \sqrt{\frac{14 + \sqrt[3]{188 + 12\sqrt{93}} + \sqrt[3]{188 - 12\sqrt{93}}}{6}} \approx 2.0366.$$

These graphs are

(i) T_n^3 ($n \geq 10$), T_n^4 ($n \geq 9$), T_n^5 ($10 \leq n \leq 11$),

(ii) $T_n^{2,n-4}$ ($n \geq 7$), $T_n^{3,n-4}$ ($n \geq 15$),

(iii) $S_{2,2,3}$.

Next, considering the controllability of the graphs obtained, we get that a controllable graph whose spectral radius does not exceed ζ must be either K_1, T_n^3 or T_n^4 ($n \geq 7$).

Controllable graphs whose least eigenvalue is bounded from below by -2 are considered in Section 3.2. □

Collecting the results above, we get the following theorem.

Theorem 2.58 (Hoffman Program) *If G is a connected graph whose spectral radius does not exceed the Hoffman limit point* $\tau = \sqrt{2 + \sqrt{5}}$ *then G is either an induced subgraph of some Smith graph, a Smith graph or one of the graphs listed in Theorem 2.56.*

Once the Hoffman program was realized, Woo and Neumaier increased the upper bound for the spectral radius to $\frac{3\sqrt{2}}{2} \approx 2.1312$ and gave the following result.

Theorem 2.59 (Woo and Neumaier [467]) *A connected graph whose spectral radius satisfies*

$$2 < \lambda_1 \leq \frac{3\sqrt{2}}{2} \tag{2.83}$$

is $T_n^{2,2}$ *with* $n \geq 6$, *an open quipu or a closed quipu.*

Conversely, $2 < \lambda_1(T_n^{2,2}) \leq \frac{3\sqrt{2}}{2}$ whenever $n \geq 6$. By [467], all open quipus with (in a precise sense) a sufficiently long gap between vertices of degree 3 satisfy (2.83). The set of forbidden subgraphs for the same spectral condition is infinite.

The last theorem is generalized in [86].

2.3.2 Order, size, and maximal spectral radius

The search for graphs of fixed order and size with maximal spectral radius is one of the oldest problems in spectral graph theory, according to [165], started by Schwarz in 1965. Brualdi and Hoffman [60] proved that such graphs admit a stepwise form of the adjacency matrix. It has also been observed by Hansen that any graph with such an adjacency matrix is a split graph with a nesting property, called NSG. Later on, it was proved that each NSG is $\{P_4, C_4, 2K_2\}$-free and vice versa, and so this is usually taken to be their shortest definition (see page 6). Hence, the solution to the problem is some NSG (the elegant proof proposed by Simić et al. will be given right after this discussion).

The general solution was provided by Rowlinson [391] in 1988: the graph of fixed order and size with maximal spectral radius is formed from a single clique of maximal allowed size and a single vertex adjacent to at least one vertex of the clique, along with the set of isolated vertices. On the contrary, for connected graphs the problem has not been solved yet, but there are many specific results.

The following result is crucial in this investigation.

Theorem 2.60 ([409]) *Let G be a connected graph of fixed order and size with maximal spectral radius. Then G does not contain as an induced subgraph any of the graphs $P_4, C_4,$ and $2K_2$.*

Proof Let $\mathbf{x}$ denote the principal eigenvector of G, and let $H \in \{P_4, C_4, 2K_2\}$ be the induced subgraph of G. Let v be a vertex of H whose coordinate (in $\mathbf{x}$) is minimal. Let u, u' be non-adjacent vertices of H such that the first of them is adjacent to v. Considering the possible candidates for H, we get that such a triplet of vertices always exists.

Now, applying the relocation $\mathscr{R}$ (with the same notation as in Lemma 1.28), we get the graph G' satisfying $\lambda_1(G') > \lambda_1(G)$. If in addition G' is connected, we are done. If G' is disconnected then $G - uv$ consists of two components, say G_1 and G_2, such that u and u' are in G_1 and v is in G_2. Since G has a maximal spectral radius, $x_u > x_s$ holds for any $s \in V(G_1)$ other than u (otherwise the deletion of uv and the addition of vs gives a graph with greater spectral radius). If w is a vertex adjacent to u' and belonging to a path between u and u' then the relocation $\mathscr{R}$ applied to $u', t,$ and u (i.e., deletion of $u'w$ and addition of $u'u$) necessarily gives a graph with a greater spectral radius. A contradiction. $\square$

In what follows we denote by $\mathscr{H}_{n,m}$ the set of connected graphs of fixed order n and size m. The graph with maximal spectral radius within $\mathscr{H}_{n,m}$ will be denoted by $H_{n,m}$. We need the following notation (taken from [35]).

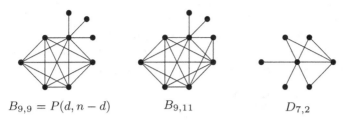

$$B_{9,9} = P(d, n - d) \qquad\qquad B_{9,11} \qquad\qquad D_{7,2}$$

Figure 2.4 Graphs $B_{9,9}, B_{9,11}$, and $D_{7,2}$.

Let t be a fixed non-negative integer and let d be the largest integer such that $\binom{d-1}{2} \leq t + 1$. For $\binom{d-1}{2} = t + 1$, by $P(d, n - d)$ we denote the pineapple with $n - d$ pendant vertices. In this case we have $3 \leq d \leq n$. For $\binom{d-1}{2} < t + 1$, there is a unique way to write $\binom{d-1}{2} + r = t + 1$, where $1 \leq r \leq d - 2$, and then, following [35], we denote by $B_{n,t}$ the graph on n vertices obtained from $P(d, n - d)$ by joining one pendant vertex to r vertices of degree $d - 1$. Here we have $3 \leq d \leq n - 1$. Finally, for $t \leq n - 3$ we denote by $D_{n,t}$ the graph obtained from the star $K_{1,n-1}$ by joining one pendant vertex to $t + 1$ other pendant vertices. The size of all graphs defined is $m = n + t$, and all of them are NSGs. See the examples illustrated in Fig. 2.4.

The graph $H_{n,m}$ is identified in some cases that depend on m. First, if $\binom{n-1}{2} \leq m \leq \binom{n}{2}$, this is the graph obtained by Rowlinson (see above) with isolated vertices removed and it coincides with $B_{d+1,t}$.

Due to Brualdi and Solheid [61], the problem is solved for $t \leq 2$. For $t = 1$, the resulting graph is $B_{n,t}$ $(= D_{n,t})$ and for $t = 2$, this is $B_{n,t}$. It is assumed in the same paper that, for small values of n, $H_{n,m}$ is $B_{n,t}$ while for large values of n, it is $D_{n,t}$. This assumption is confirmed for $3 \leq t \leq 5$.

Cvetković and Rowlinson [100] extended this result by showing that, for any fixed $t \geq 6$ and sufficiently large values of n, $H_{n,m}$ can only be $D_{n,t}$.

Finally, there is the following theorem.

Theorem 2.61 (Bell [35]) *Let* $5 \leq d \leq n - 1$, $t = \binom{d-1}{2} - 1$, *and*

$$g(d) = \frac{d(d+5)}{2} + 7 + \frac{32}{d-4} + \frac{16}{(d-4)^4}.$$

If $H_{n,m}$ *is the graph with maximal spectral radius within the set of connected graphs of fixed order n and size m, then*

  (i) $H_{n,m} = B_{n,t}$, *if* $n < g(d)$,
  (ii) $H_{n,m} = B_{n,t}$ *or* $H_{n,m} = D_{n,t}$ *and* $\lambda_1(B_{n,t}) = \lambda_1(D_{n,t})$, *if* $n = g(d)$,
  (iii) $H_{n,m} = D_{n,t}$, *if* $n > g(d)$.

From the above theorem we can deduce that $H_{n,m}$ is not necessarily unique. Until now only the graphs $B_{n,t}$ and $D_{n,t}$ appear as solutions to the problem. Concerning their structure, Simić et al. [409] posed two conjectures (see the notation for NSGs – page 6):

- For any $H_{n,m}$, $h \leq 3$.
- If $m \leq \binom{n-1}{2} + 1$, then $n_1 = 1$.

The analogous problem of determining graphs with maximal spectral radius within the set of connected bipartite graphs of fixed number of vertices in each part and size was introduced by Cvetković et al. [117]. It is proved in [37, 42] that the resulting graph must be a DNG. The problem is resolved in certain particular cases, and the result appears to be a graph obtained from a complete bipartite graph by joining some vertices that belong to one part with an isolated vertex.

Both problems (concerning non-bipartite and bipartite graphs) are still open. Recall that some bounds for the spectral radius of NSGs and DNGs are given in Subsection 2.2.4, but nevertheless it seems that (as noted in [409]) computer interaction will be of great importance in future research.

2.3.3 Diameter and extremal spectral radius

We consider the problem of determining the graphs of fixed order and diameter with minimal or maximal spectral radius. Although some similar results appeared before 2007, it seems that both problems were posed in that year, the first by van Dam and Kooij [119] and the second by van Dam only [120].

The second problem was completely resolved in the same paper, and the resulting graph is obtained from the complete graph K_{n-D+2} by deleting an edge uv and attaching a path $P_{\lfloor \frac{D}{2} \rfloor - 1}$ at u, and $P_{\lceil \frac{D}{2} \rceil - 1}$ at v.

The motivation for the research on graphs of fixed order and diameter with minimal spectral radius has been found in a real application. Namely, according to [119] (see also [110]), in the case of the Susceptible–Infected–Susceptible (SIS) infection model (covering a number of viruses), for effective spreading rates (of network viruses) below $\frac{1}{\lambda_1}$ the virus contamination in the network dies out, and for effective spreading rates above the same value the virus is prevalent. This means precisely that a smaller spectral radius of the graph corresponding to a given network yields better virus protection. The networks can be designed in various ways; for example, the communication networks often have high connectivity and consequently a small diameter. So, except for the order, it is natural to consider the diameter as an additional invariant.

In this subsection a graph that attains the minimal spectral radius within the set of all graphs of order n and finite diameter D will be called a *minimizer*.

Since the complete graph K_n (resp. path P_n) is the unique graph (with n vertices) of diameter $D = 1$ (resp. $D = n - 1$), it follows that this graph is the corresponding minimizer for this diameter.

In the sequel we give some results of van Dam and Kooij [119], who determined all the minimizers for $D \in \{2, \lfloor \frac{n}{2} \rfloor, n - 3, n - 2\}$. By using a computer, the authors also determined almost all the minimizers of order at most 20 (in almost all cases the corresponding spectral radius is around 2).

The minimizers of diameter $n - 2$ or $n - 3$ are determined in the following theorem.

Theorem 2.62 ([119]) *Within the set of graphs on $n \geq 4$ vertices and with diameter $D = n - 2$, the graph T_n^2 has the minimal spectral radius. Within the set of graphs on $n \geq 5$ vertices and with diameter $D = n - 3$, the minimal spectral radius is equal to 2. The minimizers are the Smith graphs (Fig. 2.3) of the corresponding diameter.*

Proof By computing the spectral radius of T_n^2, we get $\lambda_1(T_n^2) = 2\cos\frac{\pi}{2n-2}$, so it is less than 2. Therefore, to prove the first part of the theorem it is sufficient to consider the possible candidates (of diameter $n - 2$) whose spectral radius is also less than 2. Considering the graphs with small spectral radius (Subsection 2.3.1), we find only three such graphs T_5^2, T_6^3, and T_7^3. The result follows by direct computation.

The second part of the theorem follows from the result of Smith (mentioned in the same subsection, and obtained in [413]). □

The next theorem is also a consequence of the same result of Smith.

Theorem 2.63 ([119]) *The spectral radius of graphs on $n \geq 5$ vertices and with diameter $D = \lfloor \frac{n}{2} \rfloor$ is at least 2. The graphs attaining this value are the Smith graphs (Fig. 2.3) of the corresponding diameter.*

Finally, the case $D = 2$ is resolved in the same paper, but here it is already mentioned in Remark 2.17 (page 42).

The investigation started in [119] is continued in the subsequent period, and we give a brief review of selected results. Let $P_{n_1,n_2,\ldots,n_t,p}^{m_1,m_2,\ldots,m_t}$ denote a tree obtained from the path P_p (with vertices labelled $1, 2, \ldots, p$)[3] by attaching one pendant path P_{n_i} at vertex m_i, for each $i \in \{1, 2, \ldots, t\}$.

Since the spectral radius of $P_{1,2,n-3}^{2,n-5}$ decreases as n increases, we have

$$\lambda_1 \left(P_{1,2,n-3}^{2,n-5} \right) \leq \lambda_1 \left(P_{1,2,8}^{2,6} \right) \approx 2.0684,$$

and thus by considering the possible candidates for the minimizer of diameter $n - 4$ (within the set of graphs with small spectral radius), we get the following result.

Theorem 2.64 (Cioabă et al. [87], Yuan et al. [492]) *For $n \geq 11$, the graph $P_{1,2,n-3}^{2,n-5}$ is the unique minimizer with n vertices and diameter $n - 4$.*

We set $D = n - e \, (e \geq 3)$ and

$$\mathscr{P}_{n,e} = \left\{ P_{2,1,\ldots,1,2,n-e+1}^{3,m_2,\ldots,m_{e-4},n-e-1} : 2 < m_2 < \cdots < m_{e-4} < n - e - 1 \right\}.$$

The next result is conjectured in [87] and positively solved in [256].

Theorem 2.65 (Lan et al. [256]) *For fixed $e \geq 5$, a minimizer with n vertices and diameter $D = n - e$ is in the family $\mathscr{P}_{n,e}$, for n sufficiently large.*

The unique minimizers for $5 \leq e \leq 8$ are determined, the first in [87] (see also Exercise 2.20) and the remaining three in [256] (not to be listed here).

Concerning the minimizers of diameter near $\frac{n}{2}$, the authors of [87] conjectured that for n sufficiently large and such that $n - e$ is even, the unique minimizer with n vertices and diameter $D = \frac{n+e}{2}$ is obtained from the cycle C_{n-e} by taking two vertices at maximal distance and attaching the path $P_{\lfloor \frac{e}{2} \rfloor}$ at one and $P_{\lceil \frac{e}{2} \rceil}$ at the other. The conjecture is confirmed for $e \leq 4$.

We conclude with bipartite graphs of given diameter that maximize the spectral radius. The co-clique chain graphs defined on page 6 are needed.

Theorem 2.66 (Zhai et al. [496]) *The graph $G_*^D \left(\lceil \frac{D-1}{2} \rceil \cdot 1, \lceil \frac{n-D+1}{2} \rceil, \lfloor \frac{n-D+1}{2} \rfloor, \lfloor \frac{D-1}{2} \rfloor \cdot 1 \right)$ is the unique graph with maximal spectral radius within the set of bipartite graphs of fixed order and diameter (where $\lceil \frac{D-1}{2} \rceil \cdot 1$ denotes that there are $\lceil \frac{D-1}{2} \rceil$ consecutive V_i's with one vertex).*

Moreover, $\lambda_1 \left(G_*^D \left(\lceil \frac{D-1}{2} \rceil \cdot 1, \lceil \frac{n-D+1}{2} \rceil, \lfloor \frac{n-D+1}{2} \rfloor, \lfloor \frac{D-1}{2} \rfloor \cdot 1 \right) \right)$ is an increasing function on $D \, (2 \leq D \leq n - 1)$.

---

[3] In [87, 492] the vertex labelling starts with 0, and so our presentation of the corresponding results differs from the original one.

2.3.4 Trees

Among all trees of fixed order, those with minimal or maximal spectral radius are already identified: they are paths (cf. (2.1)) or stars (cf. (2.59)). Although the proofs of these results can be found in many places (say in [295]), we offer another elegant proof based on Lemma 1.29.

Theorem 2.67 *For any tree T with n vertices, $\lambda_1(P_n) \leq \lambda_1(T) \leq \lambda_1(K_{1,n-1})$. The first (resp. second) equality holds if and only if $T \cong P_n$ (resp. $T \cong K_{1,n-1}$).*

Proof Consider the first inequality and assume that T is not a path. Since T is a tree there is at least one vertex with hanging paths of non-zero lengths $k_1, k_2, \ldots, k_t$ ($t \geq 3$) attached. If we denote $T = T(k_1, k_2, \ldots, k_t)$, by Lemma 1.29 we get $\lambda_1(T) > \lambda_1(T(k_1 + k_2 + \cdots + k_t, 0, 0, \ldots, 0))$. If $T(k_1 + k_2 + \cdots + k_t, 0, 0, \ldots, 0)$ is a path we are done, otherwise multiple application of Lemma 1.29 gives the result.

Consider the second inequality and assume that T is a tree with maximal spectral radius. Then T must be a DNG (as mentioned on page 68), which implies that its diameter is at most 3. If it is equal to 2, T is the star $K_{1,n-1}$ and we are done. Otherwise, it contains two adjacent vertices, say u and v, of degree at least 3. Let u' be a pendant vertex adjacent to v. If the corresponding coordinates of the principal eigenvector satisfy $x_u \geq x_{u'}$, by Lemma 1.28 the relocation of the edge vu' to the non-edge uu' gives a tree with larger spectral radius. If $x_u < x_{u'}$, we get the same result by relocating vu to uu'. A contradiction.

Since the inequalities for λ_1 in Lemmas 1.28 and 1.29 are strict, we get $\lambda_1(T) = \lambda_1(P_n)$ if and only if $T \cong P_n$ and $\lambda_1(T) = \lambda_1(K_{1,n-1})$ if and only if $T \cong K_{1,n-1}$. □

The remaining investigation on trees with extremal spectral radius is focused on particular types of tree.

Trees of fixed order and maximal vertex degree with maximal spectral radius

These trees are considered by Simić and Tošić [411]. To formulate the main result we need the following graphs. A rooted tree is said to be *balanced* if all vertices at the same distance from the root have the same degree. Such a tree can be described by, say, h parameters $p_0, p_1, \ldots, p_{h-1}$, where p_i is the degree of any vertex at distance i from the root. If T is a balanced tree with the above parameters, we write $T = T(p_0, p_1, \ldots, p_{h-1})$. For $p_0 = \Delta - 1, p_1 =$

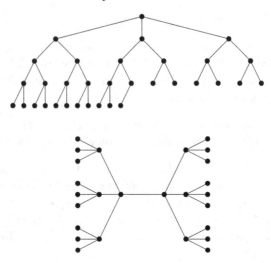

Figure 2.5 Examples of trees of fixed order and maximal vertex degree with maximal spectral radius.

$p_2 = \cdots = p_h - 1 = \Delta$, we get the Bethe trees introduced on page 54. For $p_0 = p_1 = \cdots = p_{h-1} = \Delta$, we write $T = T(h; \Delta)$.

The depth-first search (DFS) [379] of rooted trees gives the ordering of the vertices at the same distance from the root such that the vertices sharing the same (first) ancestor are ordered arbitrarily, while the remaining vertices are ordered according to the ordering of their ancestors. We briefly call this ordering a *natural ordering*.

Theorem 2.68 ([411]) *If $\widehat{T}$ is a tree with maximal spectral radius within the set of trees of fixed order n and maximal vertex degree Δ, then $\widehat{T}$ is obtained from $T(h - 1; \Delta)$ by attaching the maximal number (which does not exceed $\Delta - 1$) of pendant vertices at vertices at distance $h - 1$ from the root in natural order until the total number of vertices is n.*

In Fig. 2.5 we give two examples: the first has $n = 33$ and $\Delta = 3$, while the second has $n = 26$ and $\Delta = 4$.

Bidegreed trees of fixed order with minimal spectral radius

Consider the set of trees with n vertices, all of whose vertices other than the pendant ones are of degree Δ. It is easy to see that this set is non-empty only if $n \equiv 2 \mod(\Delta - 1)$, and then each tree has $\frac{n-2}{\Delta-1}$ vertices of degree Δ.

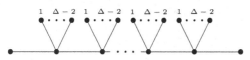

Figure 2.6 Bidegreed tree of fixed order with minimal spectral radius.

Theorem 2.69 (Belardo et al. [28]) *If $\widehat{T}$ is a tree with minimal spectral radius within the set of bidegreed trees of fixed order n, then $\widehat{T}$ is a tree as illustrated in Fig. 2.6.*

The importance of caterpillars in studying trees with maximal spectral radius is also revealed in Theorem 2.71.

Some caterpillars with maximal spectral radius

We first prove a general result and then consider two types of caterpillar graph. The following lemma is needed.

Lemma 2.70 *Let S be a star with at least 3 vertices whose central vertex is s, and t any of its pendant vertices. For a non-trivial rooted graph G, denote by G_s (resp. G_t) a graph obtained from G by identifying its root with s (resp. t) of S. Then $\lambda_1(G_s) > \lambda_1(G_t)$.*

Proof The proof is a direct consequence of Lemma 1.29. □

Theorem 2.71 (Simić et al. [410]) *For any tree T of order n and diameter D other than a caterpillar, there exists a caterpillar T' of the same order and diameter such that $\lambda_1(T) < \lambda_1(T')$.*

Proof Let u and v be the vertices of T at the largest distance, and let P denote the unique path between them. Let w be a vertex of T at the largest distance from P. (This distance is at least 2 and w is a pendant vertex.) Let $s \sim w$ and $t \sim s$, such that t belongs to the path between s and P. If S_s is a star induced by s and all vertices adjacent to s, then by applying the previous lemma we get a tree with a greater spectral radius and greater number of pendant vertices. Repeating this procedure, we arrive at a caterpillar. □

In other words, among all trees of fixed order and diameter, a tree with maximal spectral radius is a caterpillar. We now consider the caterpillars with these invariants fixed. Recall that any caterpillar of diameter D may be denoted by $T(m_2, \ldots, m_D)$, where m_i is the number of pendant vertices attached at the vertex i ($2 \le i \le D$) of the diametral path P_{D+1}.

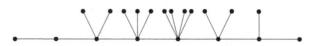

Figure 2.7 A caterpillar with maximal spectral radius in $\mathscr{T}(4,3,2,2,1,0)$.

Theorem 2.72 ([410]) *Let T be a tree with n vertices and diameter D. Then $\lambda_1(T) < \lambda_1(T(0,0,\ldots,m_{\lceil\frac{D+1}{2}\rceil},0,\ldots,0,0))$, where $m_{\lceil\frac{D+1}{2}\rceil} = n - D - 1$.*

Proof By the previous theorem, T is a caterpillar, and so we may write $T = T(m_2,\ldots,m_D)$. Let u and v be two vertices (if any) in a diametral path whose degree is at least three. Without loss of generality, we can assume $x_v \geq x_u$ for the corresponding coordinates of the principal eigenvector. But then, relocating a pendant edge from u to v gives a caterpillar with greater spectral radius. So, $m_i \neq 0$ just for one i.

The remainder of the proof follows from Lemma 1.29. □

A more difficult question arises if we consider the caterpillars with fixed vertex degrees. Note that in this situation the order and diameter are determined, so fixed, and therefore only the possible degrees for non-terminal vertices of the diametral path are of interest. Denote by $\mathscr{T}(m_2,m_3,\ldots,m_D)$ the set of caterpillars which differ from the caterpillar $T(m_2,m_3,\ldots,m_D)$ only up to a permutation of m_i's ($2 \leq i \leq D$).

Under this assumption, we can ask for those caterpillars with the largest spectral radius, and the answer is given in [410]. If the quantities $m_2,m_3,\ldots$, m_D are arranged in non-increasing order as $u_1,u_2,u_3,\ldots,u_{D-1}$, then the corresponding vertices should be arranged as $u_{D+1},\ldots,u_2,u_1,u_3,\ldots,u_D$ if D is even or $u_D,\ldots,u_2,u_1,u_3,\ldots,u_{D+1}$ if D is odd. An example is given in Fig. 2.7.

2.3.5 Various results

We give a brief proofless review of selected results concerning maximal or minimal spectral radius within a particular set of graphs. All graphs considered have fixed order n and some additional prescribed properties. We start with the following results, and then give some comments.

The graphs with exactly k cut vertices with maximal spectral radius are determined by Berman and Zhang [40]. They are obtained from the complete graph K_{n-k} by attaching the paths of almost equal lengths at distinct vertices of K_{n-k} such that the total number of vertices in the resulting graph is n. The result is followed by an upper bound for λ_1 (of the described graphs):

$$\lambda_1 < \begin{cases} n - k - 1 + \frac{k}{(n-k-1)^2}, & \text{for } k \leq \frac{n}{2}, \\ n - k - 1 + \frac{1}{n-k-1}, & \text{for } \frac{n}{2} < k \leq n - 3. \end{cases} \tag{2.84}$$

Similarly, from Liu and Liu [287], the graph with exactly k cut edges with maximal spectral radius is the pineapple $P(n - k, k)$.

Concerning the graphs with (vertex or edge) connectivity at most k, Li et al. [262] identified the unique graph with maximal spectral radius. It is obtained from K_{n-1} by joining its k vertices to an isolated vertex.

The above graphs have simple and somewhat expected structure, although the results are not easily proved. The situation is similar for many other graphs with extremal spectral radius. For example, a unicyclic graph with fixed diameter attaining maximal spectral radius can be obtained by adding an edge to a tree which is described in Theorem 2.72 [284]. Similarly for a bicyclic graph with the same properties.

Many singular results concerning graphs with extremal spectral radius are improved by some more general result. For example, Huang et al. [228] identified the unique graph with maximal spectral radius within the set of all connected cacti on $2n$ vertices, k cycles, and with perfect matching. This graph is given in Fig. 2.8(b), and the result itself improves two results, obtained in [473] and [75], respectively, concerning trees and unicyclic graphs with the same properties. It is also proved in [76] that a bicyclic graph with perfect matching and maximal spectral radius must be a cactus, and therefore bicyclic graphs are covered by the above result as well. If we remove the 'perfect matching' condition, we get a slightly different graph obtained by Borovćanin and Petrović [51] that maximizes the spectral radius within the set of all connected cacti with n vertices and k cycles (Fig. 2.8(a)), and this result again improves similar results concerning trees or unicyclic graphs.

We mention a few more results related to graphs with given domination number, clique number or matching number.

Recall from [395] that the *surjective split graph* $SSG(n, k; a_1, a_2, \ldots, a_k)$ defined for positive integers $n, k, a_1, a_2, \ldots, a_k$, $n \geq k \geq 3$, satisfying $\sum_{i=1}^{k} a_i = n - k, a_1 \geq a_2 \geq \cdots \geq a_k$, is a split graph on n vertices formed from the complete graph K_{n-k} and an independent set S with k vertices, in such a way that the ith vertex of S is adjacent to a_i vertices of K_{n-k} and no two vertices of S have a common neighbour in K_{n-k}. It is known that surjective split graphs have the maximal number of edges among graphs with no isolated vertices and a given domination number $\varphi \geq 3$.

From Stevanović et al. [433], if G is a graph with n vertices and domination

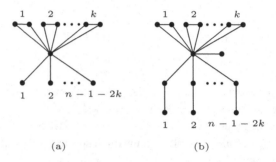

(a)   (b)

Figure 2.8 Cacti with maximal spectral radius.

number φ, then $\lambda_1(G) \leq n - \varphi$. Equality holds if and only if $G \cong K_{n-\varphi+1} \cup (\varphi - 1)K_1$ or, when $n - \varphi$ is even, $G \cong \frac{n-\varphi+2}{2}K_2 \cup (\varphi - 2)K_1$.

If G has no isolated vertices then we have the following situation:

- if $\varphi = 1$, then $\lambda_1(G) \leq n - 1$ with equality if and only if $G \cong K_n$;
- if $\varphi = 2$ and n is even, then $\lambda_1(G) \leq \lambda_1\left(\frac{n}{2}K_2\right)$ with equality if and only if $G \cong \frac{n}{2}K_2$;
- if $\varphi = 2$ and n is odd, then $\lambda_1(G) \leq \lambda_1\left(\overline{\left(\frac{n}{2} - 1\right)K_2 \cup P_3}\right)$ with equality if and only if $G \cong \overline{\left(\frac{n}{2} - 1\right)K_2 \cup P_3}$;
- if $3 \leq \varphi \leq \frac{n}{2}$, then $\lambda_1(G) \leq \lambda_1(SSG(n, \varphi; n - 2\varphi + 1, 1, 1, \dots, 1))$ with equality if and only if $G \cong SSG(n, \varphi; n - 2\varphi + 1, 1, 1, \dots, 1)$.

Finally, from Tan and Fan [440], the complete bipartite graph $K_{\mu,n-\mu}$ is the unique graph with maximal spectral radius among all bipartite graphs of order n with matching number μ.

Nikiforov [332] proved that a graph with maximal spectral radius within the set of graphs of fixed order and clique number ω must be the complete ω-partite graph in which the numbers of vertices in each part are as equal as possible. Stevanović and Hansen [435] proved that the graph with minimal spectral radius within the same set is the kite $K(\omega, n - \omega)$.

2.3.6 Ordering graphs

The study of ordering graphs by the spectral radius was started by Collatz and Sinogowitz [89] in 1957. In his list of directions in further investigations of graph spectra, Cvetković [95] suggested the lexicographical ordering (of, say, connected graphs) according to the following graph invariants: number of vertices, number of edges (this invariant may be omitted for some classes), and eigenvalues in non-increasing order. For co-spectral graphs, an idea is to use some quantities based on the eigenvectors. If we consider a set of graphs in which no two graphs have equal spectral radius then the problem becomes easier: instead of the whole spectrum it is sufficient to consider the spectral radius.

Since there are only a few connected graphs of fixed number of vertices with the property $\lambda_1 \leq 2$ (see page 63), all of them can easily be ordered with respect to their spectra. In contrast, by [399], any real number greater than or equal to $\sqrt{2 + \sqrt{5}}$ is a limit point for the spectral radius of graphs. Thus, it seems that the ordering of graphs with eigenvalues greater than $\sqrt{2 + \sqrt{5}}$ makes a difficult task, and so it is natural to consider the graphs whose spectral radius is in the interval $(2, \sqrt{2 + \sqrt{5}})$. These graphs are characterized and listed in Theorem 2.56. Note that all of them are trees of diameter $n - 3$ or $n - 2$. As we will see, there are many results concerning their ordering, but until now the problem has not been completely resolved.

We start with the three results of Zhang and Chen [499]. The T-shape trees described in Theorem 2.56(i)–(v) are considered in the following theorem.

Theorem 2.73 ([499]) *For $n \geq 10$,*

$$\lambda_1(S_{1,2,n-4}) < \lambda_1(S_{1,3,n-5}) < \cdots < \lambda_1\left(S_{1,\lfloor \frac{n-2}{2} \rfloor, \lceil \frac{n-2}{2} \rceil}\right) < \lambda_1(S_{2,2,n-5}).$$

Proof All inequalities except the last one follow immediately from Lemma 1.29. The last inequality is proved by direct computation of the characteristic polynomials of the corresponding graphs. $\qquad\square$

There is a similar result concerning the remaining graphs of Theorem 2.56. It is proved using of Lemma 1.28.

Theorem 2.74 ([499]) *For $n \geq 3k$ and $k \geq 2$,*

$$\lambda_1(T_n^{2,n-k-1}) < \lambda_1(T_n^{3,n-k-1}) < \cdots < \lambda_1(T_n^{k,n-2k+1,k}).$$

Combining the previous two theorems and using the result of Hoffman (originally proved in [210]; see also [102, Theorem 2.2.2] and Exercise 2.15) concerning the upper bounds for the spectral radii of the above trees, we get the next result.

Theorem 2.75 ([499]) *For $k \geq 3$, there is an $n_0 \in \mathbb{N}$ such that for any $n > n_0$, we have*

$$\lambda_1(S_{1,2,n-4}) < \lambda_1(T_n^{2,n-4}) < \lambda_1(T_n^{3,n-4}) < \lambda_1(S_{1,3,n-5})$$
$$< \lambda_1(T_n^{2,n-5}) < \lambda_1(T_n^{3,n-5}) < \lambda_1(T_n^{4,n-6})$$
$$< \lambda_1(S_{1,4,n-6}) < \cdots < \lambda_1(T_n^{k,n-k-1}) < \lambda_1(S_{1,k,n-k-2})$$
$$< \lambda_1(S_{1,k+1,n-k-3}) < \lambda_1(S_{1,k+2,n-k-4})$$
$$< \cdots < \lambda_1\left(S_{1,\lfloor\frac{n-2}{2}\rfloor,\lceil\frac{n-2}{2}\rceil}\right) < \lambda_1(S_{2,2,n-5}).$$

The previous theorem, providing only asymptotic ordering of the graphs considered, was extended by Belardo et al. in their papers [29, 30, 31]. In Table 2.1 we give the trees of diameter $D = n - 3$, that is, the caterpillars $T_n^{i,j}$.[4] If we omit n in the notation and denote $k = j - i > 0$, then for every k ($1 \le k \le D-2$) we have a row of Table 2.1 which contains all the trees $T^{i,i+k}$ for $2 \le i \le \frac{D+2-k}{2}$. A column contains all the trees $T^{i,j}$ in which the position of the vertex i is fixed.

The key result in [30] is the following lemma.

Lemma 2.76 ([30]) *Let $G_i = T^{i,i+k}$. Then, for any fixed D,*

> (i) *if $D > 2k - 1$ then $\lambda_1(G_i) < \lambda_1(G_j)$, for all $2 \le i < j \le \lfloor\frac{D+2-k}{2}\rfloor$,*
> (ii) *if $D < 2k - 1$ then $\lambda_1(G_i) > \lambda_1(G_j)$, for all $2 \le i < j \le \lfloor\frac{D+2-k}{2}\rfloor$,*
> (iii) *if $D = 2k - 1$ then $\lambda_1(G_i) = \lambda_1(G_j)$, for all $2 \le i < j \le \lfloor\frac{D+2-k}{2}\rfloor$.*

As a consequence, we have that the trees in any fixed row of Table 2.1 are totally ordered with respect to their spectral radii. To consider the trees from different rows we need the following result.

Lemma 2.77 ([29]) *For $D \ge 5$,*

> (i) $\lambda_1(T^{i-1,j}) < \lambda_1(T^{i,j})$, *for all $3 \le i < j$,*
> (ii) $\lambda_1(T^{i,j}) < \lambda_1(T^{i,j-1})$, *for all $2 \le i < j - 1$.*

Proof Consider the tree $T_{n-1}^{i-1,j-1}$. By inserting a vertex between the vertices $i - 1$ and $j - 1$, we get a tree $T_n^{i-1,j}$. By Theorem 1.8, $\lambda_1(T_n^{i-1,j}) < \lambda_1(T_{n-1}^{i-1,j-1})$. Similarly, by inserting a vertex between 1 and $i - 1$, we get a tree $T_n^{i,j}$. By Corollary 1.5, $\lambda_1(T_n^{i,j}) > \lambda_1(T_{n-1}^{i-1,j-1})$ and so (i) follows. In a similar way we prove (ii). $\square$

It follows immediately from the previous two lemmas that Table 2.1 is partially ordered. We illustrate this ordering in Table 2.2 by taking $D \equiv 2 \pmod 4$. The graphs whose diameter is greater than $D - 3$ are added to the upper left

---

[4] The vertices of the diametral path of any such caterpillar are labelled according to our notation introduced on page 5, so $1, 2, \ldots, D + 1$. This labelling is not the same as in [29, 31], and thus our presentation of the corresponding results differs from the original one.

Table 2.1 *Caterpillars with two vertices of degree 3. Adapted from [29] with permission from Elsevier.*

$k=D-2$	$T^{2,D}$			
$k=D-3$	$T^{2,D-1}$			
$k=D-4$	$T^{2,D-2}$	$T^{3,D-1}$		
$k=D-5$	$T^{2,D-3}$	$T^{3,D-2}$		
$\ldots$				
k	$T^{2,k+2}$	$T^{3,k+3}$	$\ldots$	$T^{\lfloor(D+2-k)/2\rfloor,\lfloor(D+2+k)/2\rfloor}$
$\ldots$			$\ldots$	$\ldots$
$k=1$	$T^{2,3}$	$T^{3,4}$	$\ldots$	$T^{\lfloor(D+1)/2\rfloor,\lfloor(D+3)/2\rfloor}$

Table 2.2 *Table 2.1 with the ordering given in Lemmas 2.76 and 2.77. Adapted from [29] with permission from Elsevier.*

$$P_n$$
$$<$$
$$\vdots$$
$$<$$
$$T^{k+2} \;>\; \cdots \;>\; \cdots \;>\; T^{2,k+2} \;>\; \cdots \;>\; T^{\lfloor(D+2-k)/2\rfloor,\lfloor(D+2+k)/2\rfloor} \qquad \left(k>\tfrac{D+2}{2}\right)$$
$$< \qquad\quad < \qquad\quad < \qquad\quad < \qquad\quad <$$
$$\vdots \qquad\quad \vdots \qquad\quad \vdots$$
$$< \qquad\quad < \qquad\quad <$$
$$T^{(D+6)/2} \;=\; \cdots \;=\; T^{2,(D+6)/2} \;=\; \cdots \;=\; T^{(D+2)/4,(3D+6)/4} \qquad \left(k=\tfrac{D+2}{2}\right)$$
$$< \qquad\quad < \qquad\quad < \qquad\quad <$$
$$T^{(D+4)/2} \;<\; \cdots \;<\; T^{2,(D+4)/2} \;<\; \cdots \;<\; T^{(D+2)/4,(3D+2)/4} \qquad \left(k=\tfrac{D}{2}\right)$$
$$< \qquad\quad < \qquad\quad < \qquad\quad <$$
$$\vdots \qquad\quad \vdots \qquad\quad \vdots$$
$$< \qquad\quad < \qquad\quad <$$
$$T^{2,k+2} \;<\; \cdots \;<\; T^{\lfloor(D+2-k)/2\rfloor,\lfloor(D+2+k)/2\rfloor} \qquad \left(k<\tfrac{D}{2}\right)$$
$$< \qquad\quad < \qquad\quad <$$
$$\vdots \qquad\quad \vdots \qquad\quad \vdots$$
$$< \qquad\quad < \qquad\quad <$$
$$T^{2,3} \;<\; \cdots \;<\; T^{D/2,(D+2)/2}$$

part of this table. We also denote $G < G'$ if $\lambda_1(G) < \lambda_1(G')$ and $G = G'$ if $\lambda_1(G) = \lambda_1(G')$.

The ordering of graphs at diagonal positions of Table 2.1 is considered in [29]. In this way a significant portion of the graphs whose spectral radius is close to the endpoints of the interval $(2, \sqrt{2+\sqrt{5}})$ is ordered.

Remark 2.78   A possible application of graph ordering in chemistry was pointed out by Zhang and Chen [499]. Namely, the difference between the highest occupied molecular orbit and the lowest unoccupied molecular orbit is called the *HOMO–LUMO separation*, and it is well known to be an important parameter in chemistry. For the majority of molecular graphs the number of vertices is even and the condition $\lambda_{\frac{n}{2}} \geq 0 \geq \lambda_{\frac{n}{2}+1}$ is obeyed [271]. In this case the HOMO–LUMO separation is equal to the difference between these two eigenvalues, that is, the difference between the smallest positive and the largest negative eigenvalues, and the molecules with large HOMO–LUMO separation are of special interest.

It turns out that in some cases the ordering of graphs by the spectral radius can provide the ordering of molecular graphs by the HOMO–LUMO separation. First, for any bipartite graph, the HOMO–LUMO separation is just twice the smallest positive eigenvalue. Second, if T is a tree with n vertices, let $\widetilde{T}$ denote the tree obtained from T by attaching a new pendant edge at each vertex of T. Clearly, $\widetilde{T}$ has $2n$ vertices. By the result of Zhang and Chang [501], the smallest positive eigenvalue of $\widetilde{T}$ is $\frac{\sqrt{\lambda_1^2(T)+4}-\lambda_1(T)}{2}$. Hence, the ordering of any set of trees T (and, for example, any connected graph whose spectral radius is in the interval $(2, \sqrt{2+\sqrt{5}})$ is a tree) by the spectral radius produces the ordering of the corresponding trees $\widetilde{T}$ by the HOMO–LUMO separation. $\square$

Similar orderings can be obtained by fixing some invariants or graph properties (see the example at the beginning of this section concerning graphs obtained from a cocktail party graph). In general, a wider class of graphs considered implies a more difficult ordering. Therefore, we can also consider a partial (instead of total) ordering. Such orderings produce a (fixed) set of graphs (belonging to the considered class) ordered by the spectral radius. We give an example.

For positive integers $p, q, r, s,$ and t satisfying $rt \leq p$, $st \leq q$, let $G(p,q;r,s;t)$ be the bipartite graph with parts $\{u_1, u_2, \ldots, u_p\}$ and $\{v_1, v_2, \ldots, v_q\}$, such that u_i and v_j are not adjacent if and only if there exists a positive integer k ($1 \leq k \leq t$) such that $(k-1)r+1 \leq i \leq kr$ and $(k-1)s+1 \leq j \leq ks$. It is clear that these graphs are missing at most two edges from the complete bipartite graph $K_{p,q}$, and therefore it is natural to expect the spectral radius of any such graph to be close to $\lambda_1(K_{p,q}) = \sqrt{pq}$. The spectral radius of $G(p,q;r,s;t)$ can be computed in various ways, say by considering the adjacency matrix partitioned into the corresponding blocks. This is done explicitly in [73], and the result is

$$\lambda_1(G(p,q;r,s;t)) = \sqrt{\frac{f + \sqrt{f^2 - 4rs(p-rt)(q-st)}}{2}}, \qquad (2.85)$$

where $f = f(p,q,r,s,t) = pq - 2rst + rs$. If we denote by $G^{(i)}$ the bipartite graph with the ith largest spectral radius among all bipartite graphs with $2n$ vertices, then we have the following result.

Theorem 2.79 (Chen et al. [73]) *Using the notation from the above, for $n \geq 3$, $G^{(1)} = K_{n,n}$, $G^{(2)} = K_{n-1,n+1}$, $G^{(3)} = G(n,n;1,1;1)$, $G^{(4)} = G(n-1,n+1;1,1;1)$, $G^{(5)} = G(n,n;2,1;1)$, $G^{(6)} = G(n,n;1,1;2)$, $G^{(7)} = K_{n-2,n+2}$, $G^{(8)} = G(n-1,n+1;2,1;1)$, $G^{(9)} = G(n-1,n+1;1,2;1)$, and $G^{(10)} = G(n-1,n+1;1,1;2)$.*

Proof Let $1 \leq p \leq q$ and $p+q = 2n$. Clearly, $G^{(1)}$ is the complete bipartite graph $K_{p,q}$ for which $\sqrt{pq}$ is maximal, that is, $G^{(1)} = K_{n,n}$.

Following [73], by Corollary 1.5 and the fact that $\sqrt{pq}$ (with $p+q = 2n$) is increasing as the value of p grows from 1 to n, it is sufficient to consider $G(n,n;1,1;1)$ and $K_{n-1,n+1}$ for $G^{(2)}$. By (2.85), we get

$$\lambda_1(G^{(2)}) = \max\left\{ \sqrt{\frac{n^2 - 1 + \sqrt{n^4 - 6n^2 + 8n - 3}}{2}}, \sqrt{n^2 - 1} \right\}.$$

Since $\sqrt{\frac{n^2-1+\sqrt{n^4-6n^2+8n-3}}{2}} \leq \sqrt{\frac{n^2-1+\sqrt{n^4-6n^2+8n-3+4n^2-8n+4}}{2}} \leq \sqrt{n^2-1}$, we get $G^{(2)} = K_{n-1,n+1}$.

To find $G^{(3)}$, it is sufficient to consider $G(n,n;1,1;1)$, $G(n-1,n+1;1,1;1)$, and $K_{n-2,n+2}$. We compute the spectral radius of the first two graphs by the above formula, and then use the facts $n^4 - 6n^2 + 8n - 3 + 4n^2 - 8n + 4 \geq n^4 - 8n^2 + 8n + 4$, for $n \geq 3$, and $\sqrt{n-4} = \sqrt{\frac{n^2-1+\sqrt{n^4-14n^2+49}}{2}}$ in order to compare all three graphs. This gives $G^{(3)} = G(n,n;1,1;1)$.

We next consider the maximal proper subgraphs of $G^{(i)}$ ($i = 1,2,3$) and the complete bipartite graph $K_{n-2,n+2}$ as candidates for $G^{(4)}$; we get $G^{(4)} = G(n-1,n+1;1,1;1)$.

By repeating the same procedure, that is, considering the maximal proper subgraphs of $G^{(1)}, G^{(2)}, \ldots, G^{(k)}$ and the complete bipartite graph $K_{p,q}$, we get $G^{(k+1)}$ for $4 \leq k \leq 9$. $\qquad\square$

Example 2.80   As an illustration we list the first 10 bipartite graphs with 40 vertices ordered by their spectral radii [73].

	Graph	λ_1
1.	$K_{20,20}$	20
2.	$K_{19,21}$	19.975
3.	$G(20,20;1,1;1)$	19.952
4.	$G(19,21;1,1;1)$	19.927
5.	$G(20,20;2,1;1)$	19.907
6.	$G(20,20;1,1;2)$	19.904
7.	$K_{18,22}$	19.900
8.	$G(19,21;2,1;1)$	19.882
9.	$G(19,21;1,2;1)$	19.881
10.	$G(19,21;1,1;2)$	19.879

$\square$

Exercises

2.1   Show that $\lambda_1 \geq \sqrt{\Delta}$ holds for any graph.

Hint: The star $K_{1,\Delta}$ is a subgraph (not necessarily induced) of any graph with maximal vertex degree Δ.

2.2   Show that the lower bound obtained by Hu [225],

$$\lambda_1 \geq \sqrt{\frac{\sum_{i=1}^{n} P_i^2}{\sum_{i=1}^{n} Q_i^2}}, \tag{2.86}$$

(where $P_i = \sum_{j \sim i} Q_j$, $Q_i = \sum_{j \sim i} t_j$, and t_j is the sum of degrees of the vertices adjacent to j) can be obtained by setting $r = 2, q = 7$ in the left inequality of (2.3).

Hint: Rewrite the above inequality as $\lambda_1 \geq \sqrt{\frac{\sum_{i=1}^{n}\left(\sum_{j \sim i}\left(\sum_{k \sim j} d_k m_k\right)\right)^2}{\sum_{i=1}^{n}\left(\sum_{j \sim i} d_j m_j\right)^2}}$ and then compute the corresponding walks in (2.3).

2.3   ([461, 463]) Prove that

$$\chi \leq \lambda_1 + 1 \tag{2.87}$$

holds for any graph G. If G is connected, equality holds if and only if G is the complete graph or an odd cycle.

2.4 ([462]) Prove that

$$\omega \geq \frac{\left(\sum_{i=1}^{n} x_i\right)^2}{\left(\sum_{i=1}^{n} x_i\right)^2 - \lambda_1} \tag{2.88}$$

holds for any graph, where x_i $(1 \leq i \leq n)$ are the coordinates of the principal eigenvector.

2.5 ([346]) Prove that equality in (2.88) is attained for complete ω-partite graphs.

2.6 Prove that the lower bound obtained by Shi [400],

$$\lambda_1 \geq \sqrt{\frac{\sum_{i=1}^{n} \left(\sum_{j \sim i} \sqrt{d_j}\right)^2}{2m}}, \tag{2.89}$$

holds for any connected graph.

2.7 Prove the lower bounds of (2.19).

2.8 Prove the lower bound for graphs without isolated vertices

$$\lambda_1 \geq \sqrt{2m - \Delta n + \Delta \delta + \Delta - \delta} \tag{2.90}$$

and show that equality holds if and only if G is regular.

Hint: Start with the left inequality of (2.19), show that

$$\min\left\{\sqrt{d_i m_i} \ : \ 1 \leq i \leq n\right\} = \min\left\{\sqrt{2m - d_i - \sum_{j \sim i} d_j} \ : \ 1 \leq i \leq n\right\},$$

and use $\sum_{j \sim i} d_j \geq (n - d_i - 1)\Delta$.

2.9 Prove the upper bound for the number of cliques given in (2.25).

2.10 Show that the Schwenk inequality (2.39) and the Stanley inequality (2.38) can be obtained from the inequality (2.35).

2.11 Prove Theorems 2.24 and 2.26.

2.12 ([86]) If a graph with n vertices, m edges, and diameter 3 contains t triangles, then show that

$$\lambda_1^3 - \lambda_1^2 \geq n - 1 - \frac{2m}{n} + \frac{6t}{n}. \tag{2.91}$$

Hint: Use the inequality (2.33).

2.13 Show that the spectral radius of the complete k-partite graph $K_{n_1,n_2,\ldots,n_k}$ satisfies

$$\lambda_1 \leq \frac{k-1}{k}n, \quad \text{where } n = \sum_{i=1}^{k} n_i.$$

2.14 Prove that any connected graph whose spectral radius lies in the interval $(2, \sqrt{2+\sqrt{5}})$ is a starlike tree with maximal vertex degree 3 or a caterpillar with at most two vertices of maximal vertex degree which is equal to 3.

2.15 ([102, 210]) Let G_u^n be the graph obtained from a connected graph G by attaching a path of length n at a vertex u of G, and suppose that $\lambda_1(G_u^n) \to \rho > 2$ as $n \to \infty$. Then show that ρ is the largest positive root of the equation

$$\frac{1}{2}\left(x + \sqrt{x^2 - 4}\right)P_G(x) - P_{G-u}(x) = 0.$$

2.16 ([416]) Use the previous exercise to prove

$$\lambda_1(T_n^4) < \zeta = \sqrt{\frac{14 + \sqrt[3]{188 + 12\sqrt{93}} + \sqrt[3]{188 - 12\sqrt{93}}}{6}}.$$

2.17 Prove Theorem 2.59.

2.18 Show that a connected bipartite graph of fixed order and size with maximal spectral radius must be a DNG.

Hint: Use Lemma 1.28 and follow the proof of Theorem 2.60.

2.19 ([120]) Prove that a graph of fixed order n and diameter D with maximal spectral radius is obtained from K_{n-D+2} by removing an edge uv and attaching a path $P_{\lfloor \frac{D}{2} \rfloor - 1}$ at u and $P_{\lceil \frac{D}{2} \rceil - 1}$ at v.

2.20 ([87]) Prove that among all graphs with $n \geq 18$ vertices and diameter $n - 5$, the graph $P_{2,2,n-4}^{3,n-6}$ (see the notation on page 69) has minimal spectral radius.

2.21 ([469]) Show that the maximal spectral radius for trees with k pendant vertices is obtained uniquely at the starlike tree $S_{i_1,i_2,...,i_k}$, where the values of $i_1, i_2, ..., i_k$ are almost equal.

2.22 Give the total ordering by the spectral radius of graphs with n vertices and diameter 3.

Notes

We list some results that are not exposed in the previous sections together with the corresponding literature.

Many results obtained before 1990 may be found in the survey paper [101], while the newest results can be seen in the recently published book [432]. The inequalities relating the matching number μ,

$$\lambda_1 \leq n - 1 - \left\lfloor \frac{n}{2} \right\rfloor + \mu \quad \text{and} \quad \lambda_1 \leq \mu \sqrt{n-1}, \qquad (2.92)$$

can be found in [431]. Both hold for connected graphs with at least four vertices. The equalities are attained respectively for a complete graph and a star.

The inequalities that include the vertex or edge connectivity can be found in [123], and those for graphs without paths or cycles of a specified length are derived in [345].

Planar graphs are considered in [137]. For example, it has been shown that for any such graph

$$\lambda_1 \leq \sqrt{8\Delta - 16} + 7.75. \qquad (2.93)$$

Planar graphs that can be drawn in a plane in such a way that all vertices belong to the unbounded face of the drawing with maximal spectral radius are considered in [390].

Fiedler and Nikiforov [163] studied the interplay between the bounds for λ_1 and the Hamiltonicity of the corresponding graphs. They proved that $\lambda_1(G) \geq n - 2$ implies the existence of a Hamiltonian path in G unless $G \cong K_{n-1} \cup K_1$. If strict inequality holds, then G contains a Hamiltonian cycle unless $G \cong P(n-1,1)$. Hamiltonian graphs are also considered in [389, 393, 460].

The technique of Maas, mentioned in the proof of Theorem 2.44 and based on intermediate eigenvalue problems of the second type, is also used in [388, 518].

Connected graphs with maximal spectral radius are also considered in the following papers: [286] (graphs with fixed degree sequence), [296] (graphs with fixed connectivity and minimal vertex degree), [39, 167] (Hamiltonian graphs with small number of cycles), [278] (trees, unicyclic and bicyclic graphs with fixed maximal vertex degree), [482] (unicyclic graphs), [469] (trees with fixed number of pendant vertices), [168, 281] (quasi-k-cyclic graphs[5] for $k \leq 3$ and quasi-trees). The last three papers consider the second maximal spectral radius as well.

Concerning connected graphs with minimal spectral radius, we mention the paper [32] (where the results exposed in Subsection 2.3.6 are extended by considering trees with minimal spectral radius and diameter at most 4) or [136] (considering graphs with small independence number).

A partial ordering of trees with perfect matchings or given diameter is given in [74, 187] (see also [412]).

---

[5] A quasi-k-cyclic graph contains a vertex whose removal gives a k-cyclic graph.

3

Least eigenvalue

In comparison with the spectral radius, the least eigenvalue has received less attention in the literature and most of the results deal with producing bounds. In Section 3.1 we consider the inequalities for the least eigenvalue. Section 3.2 deals with the graphs whose least eigenvalue is bounded from below by -2. In Section 3.3 we consider the graphs with minimal least eigenvalue.

3.1 Inequalities

Here we give a sequence of inequalities that involve the least eigenvalue of the adjacency matrix of a graph and consider its interaction with a number of graph invariants like the order, size, clique number, independence number or chromatic number.

For any graph on n vertices,

$$\lambda_n \geq -\sqrt{\left\lfloor \frac{n}{2} \right\rfloor \left\lceil \frac{n}{2} \right\rceil}, \tag{3.1}$$

where the complete bipartite graph $K_{\left\lfloor \frac{n}{2} \right\rfloor, \left\lceil \frac{n}{2} \right\rceil}$ is the only graph attaining this bound. This result is proved in various ways by Constantine [90], Hong [216], and Powers [371]. We also give an easy proof in Theorem 3.20.

An upper bound for non-complete graphs with at least four vertices is given by Yong [478],

$$\lambda_n < -\frac{1 + \sqrt{1 + 4\frac{n-3}{n-1}}}{2}. \tag{3.2}$$

This bound tends to $-\frac{1+\sqrt{5}}{2}$ when n tends to ∞, which is the largest limit point for the least eigenvalue [210].

Since any eigenvalue of a graph lies in $[-\lambda_1, \lambda_1]$, any upper bound for λ_1 obtained in the previous chapter can be considered as an opposite value of the lower bound for λ_n. Moreover, for bipartite graphs we have $\lambda_n = -\lambda_1$, and thus for these graphs all the results obtained in the previous chapter can be transferred here.

An eigenvector corresponding to the least eigenvalue of a bipartite graph is obtained from the Perron eigenvector by changing the sign of all coordinates that correspond to the vertices of one part (with the remaining coordinates unchanged). We prove the following result, which can be interpreted as a lower bound for λ_n as well.

Theorem 3.1 (cf. [219]) *For a connected graph G with $n \geq 3$ vertices there exists a connected bipartite spanning subgraph H such that*

$$\lambda_n(G) \geq \lambda_n(H),$$

with equality if and only if $G \cong H$.

Proof Let $\mathbf{x} = (x_1, x_2, \ldots, x_n)^T$ be a unit eigenvector corresponding to $\lambda_n(G)$ and H' be a graph obtained from G by deleting all edges $u_i u_j$ such that $x_i x_j \geq 0$. Clearly, H' is bipartite. Let further $A_{H'} = (a_{ij}(H'))$ be the adjacency matrix of H'. Then

$$
\begin{aligned}
\lambda_n(G) = \sum_{i=1}^{n} \sum_{j=1}^{n} a_{ij} x_i x_j &\geq \sum_{x_i x_j < 0} a_{ij} x_i x_j \\
&= \sum_{i=1}^{n} \sum_{j=1}^{n} a_{ij}(H') x_i x_j \geq \min_{\mathbf{y} \in \mathbb{R}^n, \|\mathbf{y}\|=1} \left(\sum_{i=1}^{n} \sum_{j=1}^{n} a_{ij}(H') y_i y_j \right) \\
&= \lambda_n(H').
\end{aligned}
$$

If H' is connected then $H \cong H'$, otherwise there is a component H of H' such that $\lambda_n(H) = \lambda_n(H')$. $\square$

3.1.1 Bounds in terms of order and size

There is a simple consequence of the previous theorem.

Corollary 3.2 *For a graph G,*

$$\lambda_n \geq - \max_{V_1 \sqcup V_2 = V} \sqrt{m(V_1, V_2)}, \tag{3.3}$$

where $m(V_1, V_2)$ is the number of edges between two disjoint sets V_1 and V_2 making up a partition of $V(G)$.

Proof By the previous theorem $\lambda_n(G) \geq \lambda_n(H) = -\lambda_1(H)$, for any bipartite subgraph H of G. By Theorem 2.38, we have $-\lambda_1(H) \geq -\sqrt{m(H)}$, and the proof follows. □

We now give the two upper bounds for λ_n. The first is a consequence of a more general result concerning the eigenvalues of Hermitian matrices.

Theorem 3.3 (Bollobás and Nikiforov [49]) *For a non-trivial graph G with the partition $V(G) = V_1 \sqcup V_2$,*

$$\lambda_n \leq \frac{2m(V_1)}{|V_1|} + \frac{2m(V_2)}{|V_2|} - \frac{2m}{n}. \tag{3.4}$$

Remark 3.4 For any non-trivial bipartite graph, Corollary 3.2 and Theorem 3.3 give

$$-\sqrt{m} \leq \lambda_n \leq -\frac{2m}{n}, \tag{3.5}$$

which also follows from Theorem 2.38 and the corresponding inequality of (2.2). Observe that the bipartiteness is required only for the right inequality to hold. □

For the second result we need the following notation. By $t(G)$ we denote the number of triangles in G, and by $t''(G)$ we denote the number of its induced subgraphs of order 3 and size 1. For every vertex i $(1 \leq i \leq n)$, let $t(i) = m(N(i))$ and $t''(i) = m(V(G) \backslash N(i))$. Observe the following equalities:

$$3t(G) = \sum_{i=1}^{n} t(i), \ \ t''(G) = \sum_{i=1}^{n} t''(i), \ \ 3t(G) = \sum_{i=1}^{n} d_i^2 - nm + t''(G). \tag{3.6}$$

Theorem 3.5 (Nikiforov [343]) *For a graph G with no isolated vertices,*

$$\lambda_n \leq \frac{2n}{(n^2 - 2m)} \sum_{i=1}^{n} \frac{t(i)}{d_i} - \frac{4m^2}{n(n^2 - 2m)}. \tag{3.7}$$

Proof For every i and the bipartition $V_1 = N(i)$ and $V_2 = V(G) \backslash N(i)$, the inequality (3.4) gives

$$\lambda_n \leq \frac{2m(V_1)}{d_i} + \frac{2m(V_2)}{n - d_i} - \frac{2m}{n} = \frac{2t(i)}{d_i} + \frac{2t''(i)}{n - d_i} - \frac{2m}{n}$$

and so

$$\lambda_n(n - d_i) \leq \frac{2t(i)}{d_i}(n - d_i) + 2t''(i) - \frac{2m}{n}(n - d_i).$$

Summing this inequality for all i and using the third equality of (3.6), we get the assertion. $\qquad\square$

3.1.2 Inequalities in terms of clique number, independence number or chromatic number

The next result is proved using Theorem 3.5.

Theorem 3.6 (Nikiforov [343]) *For a graph G with clique number ω,*

$$\lambda_n < -\frac{2^{\omega+1} m^{\omega}}{\omega n^{2\omega-1}}. \tag{3.8}$$

Remark 3.7 Considering the sharpness of the inequality obtained, Nikiforov has shown that it is sharp up to a constant factor at least for $\omega = 2$ and a certain range of m. The previous result is also used in proving the following (see [339]). For a graph G with independence number α,

$$\alpha > \left(\frac{n}{\overline{d}+1} - 1\right)\left(\ln\frac{\overline{d}+1}{\overline{\lambda}} - \ln\ln(\overline{d}+1)\right), \tag{3.9}$$

where $\overline{d} = \frac{2m}{n} \geq 2$ and $\overline{\lambda} = |\lambda_n(\overline{G})|$.

A simpler version of (3.8) is given in [339]:

$$\omega \geq 1 + \frac{\overline{d}n}{(n - \overline{d})(\overline{d} - \lambda_n)}, \tag{3.10}$$

where $\bar{d}$ is the same as above. Equality holds if and only if G is a regular complete ω-partite graph. If G is triangle-free, the following holds:

$$\lambda_n \leq -\frac{\bar{d}^2}{n-\bar{d}}, \tag{3.11}$$

which is a slightly better result than (3.8). □

Example 3.8 By the Turán theorem [98, Theorem 7.25], the triangle-free graph with maximal number of edges is a complete bipartite graph in which the numbers of vertices in each part are as equal as possible. If we consider a complete bipartite graph with $\frac{n}{2}$ vertices in each part, then, by setting $\omega = 2$ in (3.8), we get $\lambda_n < -\frac{n}{4}$. For the same graph, the inequality (3.11) gives $\lambda_n \leq -\frac{n}{2}$ (in fact, equality is attained). □

Remark 3.9 A bound for the independence number is mentioned in Remark 3.7. Prior to this result we note the result of Godsil and Newman [169] (which also follows from (3.4)). For any independent set S of cardinality s, we have

$$s \leq n\frac{-\lambda_n}{k_S - \lambda_n} \tag{3.12}$$

where $k_S = 2\bar{d}_S - \frac{2m}{n}$ and $\bar{d}_S$ is the average vertex degree of the vertices belonging to S.

This inequality generalizes the result of Delsarte and Hoffman (see again [169]):

$$s \leq n\frac{-\lambda_n}{r - \lambda_n} \tag{3.13}$$

for any r-regular graph.

In particular, we can insert α instead of s to obtain an upper bound for the independence number. □

Example 3.10 Considering the independence number α instead of s in (3.12) and (3.13), it can easily be seen that for the Petersen graph both equalities are attained (its independence number is 4). □

Using the Rayleigh principle, we get a lower bound in terms of order and chromatic number.

Theorem 3.11   (Fan et al. [155]) *For a graph G with chromatic number χ ($3 \le \chi \le \frac{n}{2}$),*

$$\lambda_n \ge -\frac{n - \chi + 2 + \sqrt{(n - \chi - 2)^2 + 4\chi(n - \chi)}}{4}, \qquad (3.14)$$

with equality if and only if $G \cong \left(K_{\frac{\chi}{2}} \cup \frac{\chi}{2}K_1\right) \nabla \left(K_{\frac{n-\chi}{2}} \cup \frac{n-\chi}{2}K_1\right)$, where n, χ are both even.

3.2 Graphs whose least eigenvalue is at least -2

The determination of graphs whose least eigenvalue is bounded from below by -2 is an old problem in spectral graph theory attributed to Hoffman in the early 1960s. Before we discuss these graphs, we prove a partial result which can be found in various literature, see for example [102, p. 9] or [107, Remark 1.1.7].

Theorem 3.12   *The least eigenvalue of any generalized line graph is greater than or equal to -2.*

Proof Define the vertex–edge incidence matrix R of a multigraph $\widehat{H}$ described in the definition of a generalized line graph G (page 6) as a usual incidence matrix with one exception: if e and f are the edges between u and v in a petal at u then exactly one of the entries (v, e), (v, f) is 1 and the other is -1. Then the adjacency matrix of G is $R^T R - 2I$, which yields $\lambda_n(G) \ge -2$. $\qquad \square$

A graph is said to be *exceptional* if it is connected, its least eigenvalue is not less than -2, and it is not a generalized line graph. Therefore, in order to determine all graphs with the property $\lambda_n \ge -2$ it remains to determine all exceptional graphs. This problem was open for many years, and, according to [107, p. 24], it was finally solved independently by a group of mathematicians in 1999. Three important results concerning exceptional graphs preceded this outcome. (1) In 1976, root systems were used to show that an exceptional graph has at most 36 vertices [67]. (2) In 1979, the exceptional graphs with $\lambda_n > -2$ were determined. There are exactly 573 such graphs: 20 with 6 vertices, 110 with 7 vertices, and 443 with 8 vertices [135]. (3) In 1998, the star complement technique was introduced into the study of graphs with least eigenvalue -2 [105].

We have the following theorem.

Theorem 3.13   (cf. [102, Theorem 5.4.1] or [107, Theorem 5.3.1]) *Let G be a graph with least eigenvalue* −2, *then G is exceptional if and only if it has an exceptional star complement for* −2.

In the previous theorem, the candidates for an exceptional star complement are precisely the 573 exceptional graphs with least eigenvalue greater than −2. If G is a maximal exceptional graph then it is a maximal graph with some prescribed exceptional star complement. In addition, it turns out that if G is a maximal graph with a prescribed star complement then it is a maximal exceptional graph only if the corresponding star complement has eight vertices, and so there are 443 possibilities. For each of these graphs, Lepović (cf. [102, p. 155]) used a computer implementation of the star complement technique to determine the maximal exceptional graphs which arise. There are 473 such graphs, with the following distribution of the number of vertices [102]:

number of vertices	22	28	29	30	31	32	33	34	36
number of graphs	1	1	432	25	7	3	1	2	1

According to Theorem 3.12 and the above computational results, we may state the following result.

Theorem 3.14   *If G is a connected graph with least eigenvalue* −2 *then G is a generalized line graph or it is an induced subgraph of at least one of* 473 *maximal exceptional graphs obtained.*

The whole book [107] is devoted to graphs with the property $\lambda_n \geq -2$. A number of additional results can be found there, including the historical review of this investigation (pp. 22–24), the 31 forbidden subgraphs for generalized line graphs (p. 35), the complete list of all maximal exceptional graphs (pp. 198–212) or all regular graphs with $\lambda_n \geq -2$. Precisely, the following statement holds.

Theorem 3.15   (cf. [107, Theorem 4.1.1]) *If G is a connected regular graph with least eigenvalue* −2, *then one of the following holds:*

  (i) G is a line graph;
  (ii) G is a cocktail party graph;
  (iii) G is one of 187 *exceptional regular graphs.*

The list of all exceptional regular graphs is given in the same book (pp. 218–227).

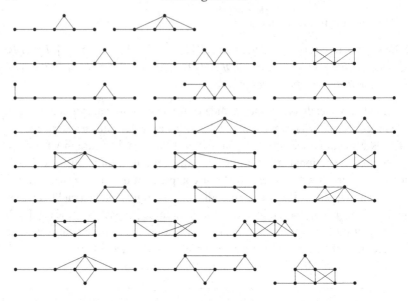

Figure 3.1　Controllable line graphs on 6, 8, and 9 vertices (cf. [109]).

In Remark 2.57 we considered controllable graphs whose spectral radius lies in the specified interval. Here we reproduce some results obtained in a joint work with Cvetković et al. [109]. In particular, we consider controllable graphs whose least eigenvalue is at least −2. We first treat generalized line graphs.

Theorem 3.16　([109]) *Controllable generalized line graphs are line graphs of either trees or odd unicyclic graphs.*

Proof　By definition, any proper generalized line graph contains at least one pair of pendant vertices with a common neighbour, implying non-controllability (since it is not asymmetric). By [103, Theorem 2.6.4], −2 is never a main eigenvalue of a line graph. Finally, it is known ([105, Theorem 2.3.20], see also Exercise 3.6) that connected line graphs with least eigenvalue greater than −2 are line graphs of either trees or odd unicyclic graphs. This completes the proof.　　　　□

In relation to the previous theorem, we note that the line graph of a tree or an odd unicyclic graph is not necessarily controllable. There are exactly 24 such graphs with 9 or fewer vertices. One of them is K_1, while the remaining graphs are depicted in Fig. 3.1; there are two graphs on 6 vertices, none on 7 vertices, three graphs on 8 vertices, and 18 graphs on 9 vertices.

Now we turn to exceptional graphs. Clearly, the exceptional controllable graph whose least eigenvalue is greater than -2 is one of the mentioned 573 exceptional graphs with 6, 7 or 8 vertices, and therefore it can be determined by considering the corresponding lists of these graphs (cf. [105]). In this way we establish the following result.

Theorem 3.17   ([109]) *There are exactly* 119 *exceptional controllable graphs whose least eigenvalue is greater than* -2*: two graphs have* 6 *vertices,* 17 *graphs have* 7 *vertices, and* 100 *graphs have* 8 *vertices.*

Exact identifications of the graphs mentioned in the previous theorem are given in [109]. We next consider the exceptional controllable graphs whose least eigenvalue is equal to -2.

Theorem 3.18   ([109]) *There are exactly* 587 *exceptional controllable graphs whose least eigenvalue is equal to* -2*: three graphs have* 7 *vertices,* 45 *graphs have* 8 *vertices, and* 539 *graphs have* 9 *vertices.*

Proof   The graphs in question are 1-vertex extensions of the exceptional graphs with least eigenvalue greater than -2, for otherwise -2 is a multiple eigenvalue. Considering these graphs we obtain the result by computer. $\square$

Amalgamating the results above, we obtain the following theorem.

Theorem 3.19   ([109]) *There are exactly* 706 *exceptional controllable graphs. Any other controllable graph with least eigenvalue at least* -2 *is the line graph of a tree or an odd-unicyclic graph.*

3.3   Graphs with minimal least eigenvalue

The graph of fixed order n with maximal least eigenvalue is the totally disconnected graph nK_1, while in the set of connected graphs this is the complete graph K_n. The graph with minimal least eigenvalue is given in the following theorem.

Theorem 3.20   *The graph* $K_{\lfloor \frac{n}{2} \rfloor, \lceil \frac{n}{2} \rceil}$ *is the unique graph with minimal least eigenvalue within the set of graphs with n vertices.*

Proof   Let G be a graph that minimizes the least eigenvalue, and let $\mathbf{x}$ be a unit eigenvector corresponding to $\lambda_n(G)$. Denote

$$V^+(\mathbf{x}) = \{u \in V(G) : x_u > 0\},$$
$$V^-(\mathbf{x}) = \{u \in V(G) : x_u < 0\}, \qquad (3.15)$$
$$V^0(\mathbf{x}) = \{u \in V(G) : x_u = 0\}.$$

Clearly, $V^+ \neq \emptyset \neq V^-$. The graph G has no edges joining vertices within V^+ and V^-, since otherwise by deleting these edges we would get a graph $\widetilde{G}$ with $\mathbf{x}^T A(\widetilde{G})\mathbf{x} < \mathbf{x}^T A(G)\mathbf{x}$, and so $\lambda_n(\widetilde{G}) < \lambda_n(G)$. Similarly, G has edges joining every vertex of V^+ with every vertex of V^-. Thus, G contains a complete bipartite subgraph with vertex partition (V^+, V^-).

If $V^0 = \emptyset$ then G is itself a complete bipartite graph, and surely $G \cong K_{\lfloor \frac{n}{2} \rfloor, \lceil \frac{n}{2} \rceil}$. Assume that $V^0 \neq \emptyset$. Deleting all edges between V^0 and V^+ and adding all possible edges between V^0 and V^-, we get a connected bipartite graph $\widetilde{G}$ satisfying

$$\lambda_n(\widetilde{G}) \leq \mathbf{x}^T A(\widetilde{G})\mathbf{x} = \mathbf{x}^T A(G)\mathbf{x} = \lambda_n(G).$$

Hence, $\mathbf{x}$ is also an eigenvector for $\lambda_n(\widetilde{G})$, which is impossible as any eigenvector corresponding to the least eigenvalue of a connected bipartite graph contains no zero entries. $\qquad \square$

3.3.1 Least eigenvalue under small graph perturbations

In what follows we consider the least eigenvalue of a graph obtained by some vertex or edge perturbations. The first result is analogous to Lemma 1.28, and the proof is similar. It considers two relocation types mentioned above that lemma.

Lemma 3.21 (Bell et al. [37]) *Let* $\mathbf{x} = (x_1, x_2, \ldots, x_n)^T$ *be an eigenvector corresponding to the least eigenvalue of* G *and* G' *be the graph obtained from* G *by relocation* $\mathscr{R}$, *then*

> (i) $\lambda_n(G') < \lambda_n(G)$ *if* $x_u < 0$ *and* $x_v \leq x_{u'}$ *or* $x_u = 0$ *and* $x_v \neq x_{u'}$ *or* $x_u > 0$ *and* $x_v \geq x_{u'}$,
> (ii) $\lambda_n(G') \leq \lambda_n(G)$ *if* $x_u = 0$ *and* $x_v = x_{u'}$.

If G' *is obtained by relocation* $\mathscr{S}$, *then*

> (i') $\lambda_n(G') < \lambda_n(G)$ *if* $x_{u'}x_{v'} < x_u x_v$,
> (ii') $\lambda_n(G') \leq \lambda_n(G)$ *if* $x_{u'}x_{v'} = x_u x_v$, *and in this situation* $\lambda_n(G') = \lambda_n(G)$ *only if* $x_u = x_v = x_{u'} = x_{v'}$.

We have the following result, which may be proved using the previous lemma.

Theorem 3.22 (Fan et al. [152]) *Let G_1 and G_2 be two non-trivial connected graphs and let $v_1, v_2 \in V(G_1)$, $u \in V(G_2)$. Let G (resp. H) be obtained from G_1, G_2 by identifying the vertex v_2 (resp. v_1) with u. If there exists an eigenvector $\mathbf{x}$ of G corresponding to its least eigenvalue such that $|x_{v_1}| \geq |x_{v_2}|$, then*

$$\lambda_n(H) \leq \lambda_n(G),$$

where $n = |V(G)| = |V(H)|$. Equality holds if and only if $\mathbf{x}$ is an eigenvector of H corresponding to its least eigenvalue, $x_{v_1} = x_{v_2}$, and $\sum_{w \in V(G_2), w \sim u} x_w = 0$.

Using the previous theorem we may consider the least eigenvalue of the graphs obtained by two specific graph operations.

Corollary 3.23 ([152]) *Let G_0, G_1, and G_2 be non-trivial connected graphs and let $v_1, v_2 \in V(G_0)$, $u_1 \in V(G_1)$, $u_2 \in V(G_2)$. Let G be obtained by identifying the vertex v_1 with the vertex u_1 and identifying the vertex v_2 with the vertex u_2. Let H be obtained by identifying the vertex v_1 first with u_1 and then with u_2. If there exists an eigenvector $\mathbf{x}$ of G corresponding to its least eigenvalue such that $|x_{v_1}| \geq |x_{v_2}|$, then*

$$\lambda_n(H) \leq \lambda_n(G),$$

where $n = |V(G)| = |V(H)|$. Equality holds if and only if $\mathbf{x}$ is an eigenvector of H corresponding to its least eigenvalue, $x_{v_1} = x_{v_2}$, and $\sum_{w \in V(G_2), w \sim u} x_w = 0$.

Proof Let F be the graph obtained from G_0, G_1 by identifying the vertices v_1 and u_1. The result follows from the previous theorem by viewing F, G_2 as G_1, G_2. □

We are now in a position to give a proof of case (ii) in Lemma 1.29. We repeat its formulation here.

Corollary 3.24 ([152]) *Let G be a non-trivial connected graph and let $G(k,l)$ $(k \geq l \geq 1)$ be a graph obtained from G by attaching two hanging paths, P_k and P_l, at its vertex u. Then*

$$\lambda_n(G(k,l)) \leq \lambda_n(G(k+1,l-1)),$$

where n plays the same role as above.

Proof Let v be the kth vertex of P_{k+1} (counted from u). The inequality follows from the previous corollary by viewing P_k, G, and P_l as G_0, G_1, and G_2 (for $|x_u| \geq |x_v|$) or G_0, G_2, and G_1 (for $|x_u| < |x_v|$), respectively. □

3.3.2 Graphs of fixed order and size

These graphs with minimal least eigenvalue are considered by Bell et al. [37]. The main result reads as follows.

Theorem 3.25 ([37]) *Let G be a connected graph whose least eigenvalue is minimal within the set of connected graphs of order n and size m $\left(0 < m < \binom{n}{2}\right)$, then G is either*

> *(i) a bipartite graph or*
> *(ii) a join of two nested split graphs (not both totally disconnected).*

In order to prove the theorem we need the following results. For an eigenvector $\mathbf{x} = (x_1, x_2, \ldots, x_n)^T$ corresponding to the least eigenvalue λ_n, we use the notation given in (3.15).

Lemma 3.26 ([37]) *If $V^0(x) \neq \emptyset$, then $d_u = n - 1$ for any vertex $u \in V^0(x)$.*

We omit the technical proof based on Lemma 3.21 and refer the reader to the corresponding reference. We have the following corollary.

Corollary 3.27 ([37]) *Let G be a graph whose least eigenvalue λ_n is minimal among the connected graphs of order n and size $m < \binom{n}{2}$. Then λ_n is a simple eigenvalue.*

Proof Assume that λ_n has multiplicity at least two. Then for any vertex i there exists an eigenvector $\mathbf{x}$ whose ith coordinate is equal to zero. Since G is not complete, we may choose i to be a vertex such that $d_i < n - 1$, which contradicts the previous lemma. □

If $V^0 \neq \emptyset$ then $G = K \nabla H$, where K is a complete graph whose vertices correspond to zero coordinates of $\mathbf{x}$, and H is a subgraph of G induced by the vertices in $V^- \cup V^+$. Denote by H^- and H^+ the subgraphs of H induced by the vertices in V^- and V^+, respectively. We prove a lemma.

Lemma 3.28 ([37]) *Both H^- and H^+ are NSGs.*

Proof Let $V^+ = \{1, 2, \ldots, k\}$, where $x_1 \leq x_2 \leq \cdots \leq x_k$. We prove that if jq is an edge of G then ip is also an edge of G whenever $1 \leq i \leq j \leq k$ and $1 \leq p \leq q \leq k$. Assume to the contrary. A deletion of the edge jq and an addition of the edge ip gives the graph G'. By Lemma 3.21, we have

$$0 \leq \lambda_n(G') - \lambda_n(G) = 2(x_i - x_j)x_p + 2(x_p - x_q)x_j \leq 0,$$

and thus $x_i = x_j, x_p = x_q$ and $\mathbf{x}$ also corresponds to $\lambda_n(G') = \lambda_n(G)$, which contradicts Corollary 3.27. Hence, H^+ is an NSG, and in the same way we derive the same conclusion for H^-. □

We arrive at the following conclusion.

Lemma 3.29 ([37]) *If at least one of the graphs H^- or H^+ is non-trivial then $H = H^- \nabla H^+$; otherwise, H is bipartite.*

Proof If any of the sets V^- or V^+ induces an edge ij then $H = H^- \nabla H^+$ since otherwise, by Lemma 3.21, a deletion of the edge ij gives a graph with smaller least eigenvalue. □

We are now ready to prove the initial theorem.

Proof of Theorem 3.25 We first prove that if $V^0 \neq \emptyset$ then $H = H^- \nabla H^+$. Choose the vertices $a \in V^0$, $b \in V^- \cup V^+$, $c \in V^-$, and $d \in V^+$ (such that c and d are non-adjacent). By Lemma 3.26, a is adjacent to b, and ab is not a cut edge. Moreover, by Lemma 3.21, $x_c x_d < x_a x_b$. By the same lemma, if we replace the edge ab with cd we get the connected graph G' with $\lambda_n(G') < \lambda_n(G)$, which contradicts the minimality of $\lambda_n(G)$, and so $H = H^- \nabla H^+$.

Moreover, by Lemma 3.26, $V^0 \neq \emptyset$ implies that G has the form $F_1 \nabla F_2$ where both F_1 and F_2 are NSGs with vertex sets $V^- \cup X$ and $V^+ \cup Y$ ($X \cup Y$ being any bipartition of V^0). If $V^0 = \emptyset$ then G coincides with H and the result follows from Lemma 3.29. □

Considering the structure of the two types of minimizing graph, it has been shown for case (ii) of Theorem 3.25 (see [37]) that if $\mathbf{x}$ contains no zero coordinates, then the join is obtained from two NSGs with equal maximal spectral radius. If $\binom{n}{2} - m$ is even then G is a complement of $N \cup 2G(h)$, where N consists of the appropriate number of isolated vertices and $G(h)$ is the unique graph with maximal spectral radius among graphs with h edges and no isolated vertices obtained by Rowlinson (see page 66).

Recently, Jovović et al. [235] identified the unique graph with minimal least eigenvalues within the set of connected graphs of fixed order n and size m for $m = \lceil \frac{n}{2} \rceil \lfloor \frac{n}{2} \rfloor + k$, where k is a fixed constant in the segment $[1, \lceil \frac{n}{2} \rceil]$. This is $\left(K_{1,k} \cup (\lceil \frac{n}{2} \rceil - k - 1)K_1 \right) \nabla \lceil \frac{n}{2} \rceil K_1$.

Case (i) of Theorem 3.25 is considered in [38], where it is proved that G must be a DNG. Moreover, if G is the same as in Theorem 3.25, then

(i) if $n-1 \le m \le \lfloor \frac{n}{2} \rfloor \lceil \frac{n}{2} \rceil$ and $m \ne t(n-t)+1$ for all $t \in \{1,2,\ldots, \lfloor \frac{n}{2} \rfloor - 1\}$ then G is a DNG,

(ii) if $m \le \lfloor \frac{n}{2} \rfloor \lceil \frac{n}{2} \rceil$ and $m = t(n-t)+1$ for some $t \in \{1,2,\ldots, \lfloor \frac{n}{2} \rfloor - 1\}$ then G is either a DNG or $K_{t,n-t} - e$, where e is an edge joining two vertices of degree $\min\{t, n-t\}$ in $K_{t,n-t}$,

(iii) if $\lfloor \frac{n}{2} \rfloor \lceil \frac{n}{2} \rceil < m < \binom{n}{2}$ then G is a join of two NSGs.

In relation to the previous results, we give another one concerning bipartite graphs with minimal least eigenvalue [363]. If G is such a graph of order n and size $m = n+k$ (with $0 \le k \le 4$ and $k+5 \le n$) then G is a DNG with parameters $(m_1, m_2, \ldots, m_h; n_1, n_2, \ldots, n_h)$, where

- $h > 1$,
- exactly one of the parameters m_1 and n_1 is equal to 1,
- if $h = 2$ then $m_1 = m_2 = 1, n_1 = k+2$, and $n_2 = n-k-4$,
- $h \ne 3$.

This gives the structure of the k-cyclic bipartite graphs that minimize the least eigenvalue. It is conjectured in [363] that for fixed k and n sufficiently large, the case $h > 3$ is not possible.

3.3.3 Graphs with prescribed properties

We give a brief review of graphs with prescribed properties that minimize the least eigenvalue. All of them are obtained using previously mentioned statements (from Lemma 3.21 to Corollary 3.24) considering the behaviour of the least eigenvalue under some graph perturbations.

Graphs with a fixed number of cut vertices, cut edges or given connectivity which minimize the least eigenvalue are very close to complete bipartite graphs.

Let $G(a, a', b, b')$ denote the graph obtained from $K_{s,t} \cong (sK_1) \nabla (tK_1)$ by attaching a pendant vertices at a vertices of sK_1 and b pendant vertices at b vertices of tK_1, where $a' = s - a \ge 0$ and $b' = t - b \ge 0$. By Wang and Fan [454], if G has minimal least eigenvalue within the set of connected graphs with $n \ge 8$ vertices and k $(1 \le k \le \frac{n}{2})$ cut vertices then

(i) if $n - k$ is even we have $G \cong G\left(\lfloor \frac{k}{2} \rfloor, \lceil \frac{n-2k}{2} \rceil, \lceil \frac{k}{2} \rceil, \lfloor \frac{n-2k}{2} \rfloor\right)$,

(ii) if $n - k$ is odd and $k \le \lfloor \frac{n-k}{2} \rfloor$ we have $G \cong G\left(0, \lceil \frac{n-k}{2} \rceil, k, \lfloor \frac{n-k}{2} \rfloor - k\right)$,

(iii) if $n - k$ is odd and $k \ge \lceil \frac{n-k}{2} \rceil$ we have $G \cong G\left(k - \lfloor \frac{n-k}{2} \rfloor, n - 2k, \lfloor \frac{n-k}{2} \rfloor, 0\right)$.

This result improves the result of Wang, et al. [457] concerning graphs with exactly one cut vertex.

By Wang and Fan [455], for $n \geq 13$ and $1 \leq k \leq \frac{n}{2}$, the graph with minimal least eigenvalue within the set of connected graphs with n vertices and k cut edges is obtained from $K_{\lfloor \frac{n-k}{2} \rfloor, \lceil \frac{n-k}{2} \rceil} \cong \left(\lfloor \frac{n-k}{2} \rfloor K_1 \right) \nabla \left(\lceil \frac{n-k}{2} \rceil K_1 \right)$ by attaching k pendant vertices at a single vertex of $\lfloor \frac{n-k}{2} \rfloor K_1$. Similarly, by Ye et al. [476], for $1 \leq k \leq \frac{n}{2}$, the graph obtained from the same graph $K_{\lfloor \frac{n-k}{2} \rfloor, \lceil \frac{n-k}{2} \rceil} \cong \left(\lfloor \frac{n-k}{2} \rfloor K_1 \right) \nabla \left(\lceil \frac{n-k}{2} \rceil K_1 \right)$ but by joining a single vertex with k vertices of $\lfloor \frac{n-k}{2} \rfloor K_1$ is the unique graph with minimal least eigenvalue within the set of graphs with n vertices and vertex connectivity k, and also within the set of graphs with n vertices and edge connectivity k.

Following Tan and Fan [440], we may consider graphs with given matching number μ, independence number α, vertex cover β or edge cover β'.

Theorem 3.30 [440] *Given a graph G with n vertices, let $\xi \in \{\alpha, \mu, \beta, \beta'\}$ and $\alpha \geq \frac{n}{2} \geq \beta$ then*

$$\lambda_n \geq -\sqrt{\xi(n-\xi)}$$

with equality if and only if $G \cong K_{\xi, n-\xi}$. Or, alternatively,

$$\alpha, \beta' \leq \frac{n + \sqrt{n^2 - 2\lambda_n^2}}{2}, \tag{3.16}$$

$$\mu, \beta \geq \frac{n - \sqrt{n^2 - 2\lambda_n^2}}{2}. \tag{3.17}$$

In other words, $K_{\xi, n-\xi}$ is the graph with minimal least eigenvalue within the set of graphs with n vertices and fixed value of ξ (where ξ is described in the last theorem).

To consider cacti we need the graphs given in Fig. 3.2. There, $B(k-s, s, n-(2k+s+1))$, for $0 \leq s \leq t$, denotes a graph which contains $k-s$ cycles of length 3, s cycles of length 4 (all sharing the same vertex u), and $n - (2k+s+1)$ pendant vertices attached at u. Here, t stands for the maximal number of cycles of length 4 (it holds that $t = \min\{k, n-(2k+1)\}$). Using Theorem 3.22 and the corresponding corollaries, Petrović et al. [364] obtained the following results.

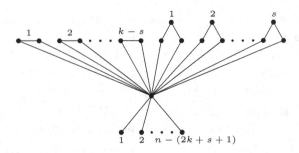

Figure 3.2 Graphs $B(k-s,s,n-(2k+s+1))$.

Theorem 3.31   ([364]) *Let G have minimal least eigenvalue within the set of connected cacti with $n \geq 3$ vertices. Then*

> *(i) if $n = 3$ or $n \geq 12$ we have $G \cong B(1,0,n-3)$,*
> *(ii) if $4 \leq n \leq 6$ or $8 \leq n \leq 11$ we have $G \cong B(0,1,n-4)$,*
> *(iii) if $n = 7$ we have $G = B(0,1,3)$ or $G \cong B(0,2,0)$.*

The last theorem improves the result of Fan et al. [152] concerning unicyclic graphs with minimal least eigenvalue.

Theorem 3.32   ([364]) *Let G have minimal least eigenvalue within the set of connected cacti with $n \geq 3$ vertices and k cycles ($k \geq 1, n \geq 2k+1$). Denote $t = \min\{k, n-(2k+1)\}$ and let $n_0(k) = 12$ for $1 \leq k \leq 2$, $n_0(k) = 13$ for $3 \leq k \leq 4$, and $n_0(k) = 14$ for $5 \leq k \leq 6$. Then*

> *(i) if $k \leq 6$ and $n < n_0(k)$ we have $G \cong B(k-t,t,n-(2k+t+1))$,*
> *(ii) if $k \leq 6$ and $n \geq n_0(k)$ or $k > 6$ we have $G \cong B(k,0,n-(2k+1))$.*

Exercises

3.1   ([211]) Show that if H is an induced subgraph of G, $\lambda_n(H) = \lambda_n(G)$, $\mathbf{x_H}$ is an eigenvector of H with corresponding eigenvalue $\lambda_n(H)$, and $\mathbf{x}$ is formed from $\mathbf{x_H}$ by adding coordinates equal to zero, then $\mathbf{x}$ is an eigenvector of G with corresponding eigenvalue equal to $\lambda_n(G)$.

3.2   Show that the upper bound (3.11) which concerns triangle-free graphs will always give a better result than the general upper bound (3.8).

3.3   Prove Theorem 3.11.

3.4 For connected graphs, show that the lower bound for λ_n, obtained by Alon and Sudakov [11]

$$\lambda_n \geq -\Delta + \frac{1}{n(D+1)}, \tag{3.18}$$

is a consequence of both upper bounds for λ_1 (2.27) and (2.28).

Hint: Show that $\frac{n\Delta - 2m}{n(D(n\Delta - 2m) + 1)}$ is monotone increasing in $n\Delta - 2m$, and thus

$$-\lambda_n \leq \lambda_1 < \frac{n\Delta - 2m}{n(D(n\Delta - 2m) + 1)} \leq \Delta - \frac{1}{n(D+1)}.$$

3.5 ([134]) Show that if G is a graph of diameter D and Line(G) its line graph on n vertices, then $-2 \leq \lambda_n(\text{Line}(G)) \leq -2\cos\frac{\pi}{D+1}$.

3.6 ([105, Theorem 2.3.20]) Prove that a connected graph with least eigenvalue greater than -2 is

    (i) Line($T; 1, 0, \ldots, 0$), where T is a tree,
    (ii) Line(H), where H is a tree or an odd unicyclic graph or
    (iii) an exceptional graph with $n \in \{6, 7, 8\}$ vertices.

3.7 Show that if G is a regular graph of degree at least 13 and if the least eigenvalue of G is -2, then G does not contain $K_{1,3}$ as an induced subgraph.

3.8 Prove Theorem 3.22.

3.9 Show that a connected bipartite graph of fixed order and size with minimal least eigenvalue must be a DNG.

3.10 Prove Theorems 3.31 and 3.32.

Notes

The least eigenvalue of graphs embedded in (non)-orientable surfaces is considered in [219]. The following lower bound for planar graphs has arisen as a consequence of a general result concerning the Euler characteristic of a graph:

$$\lambda_n \geq -\sqrt{2n - 4}, \tag{3.19}$$

with equality only for $K_{2,n-2}$.

Some lower and upper bounds for λ_n in terms of the coordinates of the corresponding eigenvector are given in [471].

By the results of Section 3.2, all graphs whose least eigenvalue is bounded from below by -2 are determined. The natural next step is to decrease the corresponding bound. Woo and Neumaier [466] introduced the concept of Hoffman graphs and defined the generalization of line and generalized line graphs based on the family of graphs introduced. By the same authors, if the least eigenvalue of a graph G exceeds a fixed number larger than the smallest root of the polynomial $x^3 + 2x^2 - 2x - 2$ (close to -2.4812) and if every vertex of G has sufficiently large degree then the least eigenvalue of G is at least $-1 - \sqrt{2}$. In addition, the structure of G is completely characterized through the generalization of line graphs. Additional results on this topic can be found in [442].

All the results concerning graphs with minimal least eigenvalue exposed in the previous section have been obtained recently. Similar results concerning unicyclic graphs with fixed number of pendant vertices or unicyclic graphs with fixed diameter can be found in [289, 495]. For bicyclic graphs, see [294, 365]. The least eigenvalue of cacti with pendant vertices is considered in [472]. The unique graph with minimal least eigenvalue among all graphs of fixed order whose complements have vertex or edge connectivity 2 is characterized in [488]. Additional results on the minimal least eigenvalue of complements of graphs are given in [156, 500].

The ordering of graphs with respect to their least eigenvalue has been investigated less but, for example, the first five unicyclic graphs of order n in terms of their smaller least eigenvalues are identified in [474].

4

Second largest eigenvalue

The second largest eigenvalue gives information about the structure (in particular, connectivity, expansion or randomness properties) of graphs. Inequalities that involve this invariant are given in Section 4.1. Graphs with bounded second largest eigenvalue are considered in Section 4.2. Some additional data is given in Section 4.3 (Appendix).

4.1 Inequalities

Hong [216] and Powers [371] independently gave the following magnitude for the second largest eigenvalue of an arbitrary graph with $n \geq 2$ vertices:

$$-1 \leq \lambda_2 \leq \begin{cases} \frac{n-3}{2}, & \text{if } n \text{ is odd,} \\ \frac{n-2}{2}, & \text{if } n \text{ is even.} \end{cases} \tag{4.1}$$

The left equality occurs for the complete graph K_n. For n even, the upper bound occurs only for $2K_{\frac{n}{2}}$. Later, Zhai et al. [494] proved that, for n odd, the upper bound is attained only for some of the graphs obtained from two disjoint copies of $K_{\frac{n-1}{2}}$ by adding some edges joining their vertices with an additional isolated vertex. Notice that the above inequalities hold for all (not necessarily connected) graphs. The same authors proved that the second largest eigenvalue of any connected graph with even number of vertices does not exceed the second largest eigenvalue of the double kite $DK\left(\frac{n}{2}, 0\right)$.

For connected graphs Hong [216] proposed

$$\lambda_2 \leq \frac{\sqrt{n^2 - 4}}{2} - 1. \tag{4.2}$$

Concerning bipartite graphs, Powers [371] gave the following upper bound:

$$\lambda_2 \leq \begin{cases} k, & \text{if } n = 4k \text{ or } n = 4k+1, \\ \sqrt{k(k+1)}, & \text{if } n = 4k+2 \text{ or } n = 4k+3, \end{cases} \tag{4.3}$$

where $k = \lfloor \frac{n}{4} \rfloor$. Similarly to the above, it is proved in [494] that for $n = 4k+1$ (resp. $n = 4k+3$) equality holds only for some of the graphs obtained from two disjoint copies of $K_{k,k}$ (resp. $K_{\lfloor \frac{2k+1}{2} \rfloor, \lceil \frac{2k+1}{2} \rceil}$) by adding some edges joining their vertices with an additional isolated vertex. The second largest eigenvalue of a connected graph with $n = 4k$ (resp. $n = 4k+2$) vertices does not exceed the second largest eigenvalue of the graph obtained from two disjoint copies of $K_{k,k}$ (resp. $K_{k,k+1}$) by inserting an edge between their larger parts.

In what follows, we consider regular graphs and trees.

4.1.1 Regular graphs

By Theorem 1.10, for any regular graph G, $\lambda_2(G) + \lambda_n(\overline{G}) = -1$ and therefore, in the case of regular graphs, all the results obtained in the previous chapter can easily be transferred here. For example, if α denotes the independence number of an r-regular graph then, using the upper bound (3.8), we get

$$\lambda_2 > -1 + \frac{2(n-1-r)^{\alpha}}{\alpha n^{\alpha-1}}. \tag{4.4}$$

It can be observed that for small values of λ_2 regular graphs often have a more 'round' shape (smaller diameter and higher connectivity), while for large values of λ_2 they have a more 'path-like' shape (larger diameter and lower connectivity). This property is explained in the following theorem.

Theorem 4.1 (Cvetković [95] or [102, Theorem 3.5.1]) *Let G be an r-regular graph with n vertices. Let v be any vertex of G and let $\overline{d}$ be the average vertex degree of the subgraph induced by the vertices not adjacent to v. Then*

$$\overline{d} \leq r \frac{\lambda_2^2 + \lambda_2(n-r)}{\lambda_2(n-1) + r}. \tag{4.5}$$

The proof is based on the interlacing theorem and it can be found in any of the corresponding references. Rather than repeating it, we offer a similar proof of the subsequent Theorem 4.6. Note that the upper bound for $\overline{d}$ decreases as λ_2 decreases. In addition, a decrease in $\overline{d}$ increases the number of edges between the subgraph induced by the vertices adjacent to v and the subgraph induced by the vertices not adjacent to v. Simultaneously, we have fewer edges between these subgraphs, giving a more 'round' shape of graph.

Regular triangle-free graphs

We consider the interplay between the second largest eigenvalue of regular triangle-free graphs and their structural properties. Koledin and Stanić [248] obtained the following results.

Theorem 4.2 ([248]) *Given an r-regular triangle-free graph G with n vertices, then*

$$n \le \frac{r^2(\lambda_2+2) - r\lambda_2(\lambda_2+1) - \lambda_2^2}{r - \lambda_2^2} \tag{4.6}$$

whenever the right-hand side is positive.

Proof We partition the vertices of G into three parts: (i) an arbitrary vertex v, (ii) the vertices adjacent to v, and (iii) the remaining vertices, and consider the corresponding blocking $A = (A_{ij})$, $1 \le i, j \le 3$, of its adjacency matrix. If $\overline{d}$ denotes the average vertex degree of the subgraph induced by the vertices not adjacent to v then, since G is triangle-free, the matrix whose entries are the average row sums in A_{ij} has the form

$$B = \begin{pmatrix} 0 & r & 0 \\ 1 & 0 & r-1 \\ 0 & r-\overline{d} & \overline{d} \end{pmatrix}.$$

By Theorem 1.25, the eigenvalues of B interlace those of A. Since the characteristic polynomial of B is $(x-r)(x^2 - (\overline{d} - r)x - \overline{d})$, we get that $x^2 - (\overline{d} - r)x - \overline{d}$ must be positive in λ_2, that is, we have

$$\lambda_2^2 - (\overline{d} - r)\lambda_2 - \overline{d} \ge 0.$$

Equivalently,

$$\overline{d} \le \frac{\lambda_2(\lambda_2 + r)}{\lambda_2 + 1}.$$

Since there are exactly r^2 edges between the set of vertices adjacent to v and the set of remaining vertices of G, we get

$$\bar{d} = \frac{r(n-2r)}{n-r-1}.$$

(Since G is triangle-free, we have $n - 2r \geq 0$.) Substituting this into the previous inequality, we get

$$r(n-2r)(\lambda_2 + 1) \leq (n-r-1)\lambda_2(\lambda_2 + r).$$

After simplifying, we obtain the result.   □

In the next theorem we use the following technical lemma.

Lemma 4.3 ([248]) *Let G be a connected r-regular non-bipartite triangle-free graph satisfying $\lambda_2 \leq c$. If $r > 2c^2 + c + 1$ then G contains C_5 as an induced subgraph.*

We also use the computational result for the second largest eigenvalue of the double comet $DC(k,2)$, $\lambda_2(DC(k,2)) = \frac{-1+\sqrt{4k+1}}{2}$.

Theorem 4.4 ([248]) *Let G be a connected r-regular non-bipartite triangle-free graph satisfying $\lambda_2 \leq c$ (where $c > 0$). Then $r \leq c^4 + 2c^3 + 4c^2 + 2c + 3$.*

Proof Note first that $\lambda_2 \leq c$ implies that G cannot contain $DC\left(\lfloor c^2+c \rfloor + 1, 2\right)$ as an induced subgraph.

Assume that $r > c^4 + 2c^3 + 4c^2 + 2c + 3$, and also let us assume that there are two vertices of G at distance 2, u and v, that have at most $\lfloor c^2+c \rfloor$ common neighbours.

There are at least $\lceil c^4 + 2c^3 + 3c^2 + c + 3 \rceil$ neighbours of u not adjacent to v, and so there are more than $\lfloor c^2 + c \rfloor + 1$ groups of $\lceil c^2 \rceil$ neighbours of u not adjacent to v. Vertices in each of these groups must be adjacent to more than $\lceil c^4 + 2c^3 + 2c^2 + c + 1 \rceil$ neighbours of v (otherwise, G contains P_3 with $\lceil c^2 \rceil$ vertices attached at one of its ends and $\lceil c^2 \rceil + 1$ at another, causing $\lambda_2(G) > c$).

So, in every group there is at least one vertex adjacent to at least $\lfloor c^2 + c \rfloor + 1$ neighbours of v. Fix exactly $\lfloor c^2 + c \rfloor + 1$ such vertices, and let w denote one common neighbour of u and v. Any of the fixed vertices can have at most $\lfloor c^2 + c \rfloor$ common neighbours with w (otherwise, $DC\left(\lfloor c^2+c \rfloor + 1, 2\right)$ is contained in G), but then there are more than $\lfloor c^2 + c \rfloor + 1$ neighbours of w which are not adjacent to any of those $\lfloor c^2 + c \rfloor + 1$ vertices. Thus, G contains $DC\left(\lfloor c^2+c \rfloor + 1, 2\right)$ as an induced subgraph, and consequently if $r >$

$c^4 + 2c^3 + 4c^2 + 2c + 3$ then the second assumption cannot hold. In other words, every two vertices at distance 2 must have at least $\lfloor c^2 + c \rfloor + 1$ common neighbours.

By Lemma 4.3, G contains C_5 as an induced subgraph. Considering two adjacent vertices of C_5 and their neighbours, we get that G must contain the double comet $DC(\lfloor c^2 + c \rfloor + 1, 2)$, which gives the assertion. $\qquad\square$

We have two simple consequences of Lemma 4.3 and Theorem 4.4 that provide a spectral condition for bipartiteness within the set of connected regular triangle-free (triangle-free and C_5-free) graphs. In both cases the second largest eigenvalue is used to check whether a given connected regular triangle-free or regular triangle-free and C_5-free graph is bipartite.

Corollary 4.5 ([248]) *Let G be a connected r-regular triangle-free graph satisfying $r > \lambda_2^4 + 2\lambda_2^3 + 4\lambda_2^2 + 2\lambda_2 + 3$. Then G is bipartite.*

Corollary 4.6 ([248]) *Let G be a connected r-regular $\{C_3, C_5\}$-free graph satisfying $r > 2\lambda_2^2 + \lambda_2 + 1$. Then G is bipartite.*

If G is a bipartite graph with parts X and Y then its *bipartite complement* is a bipartite graph $\overline{\overline{G}}$ with the same parts having the edge between X and Y exactly where G does not. Observe that any connected bipartite graph has a unique bipartite complement.

If G is a connected r-regular bipartite graph with $2n$ vertices then $\overline{\overline{G}}$ is an $(n - r)$-regular bipartite graph. It can easily be verified that the characteristic polynomials of G and $\overline{\overline{G}}$ satisfy

$$\frac{P_G(x)}{x^2 - r^2} = \frac{P_{\overline{\overline{G}}}(x)}{x^2 - (n-r)^2} \tag{4.7}$$

(cf. [444]). So, apart from the eigenvalues $\pm r$ of G and $\pm(n - r)$ of $\overline{\overline{G}}$, the spectra of G and $\overline{\overline{G}}$ are the same. If G is a disconnected regular bipartite graph then its bipartite complement is not uniquely determined, but even then the above formula remains unchanged.

We now consider regular bipartite graphs. Here is an upper bound for λ_2.

Theorem 4.7 ([248]) *Let G be a connected r-regular bipartite graph with n vertices. Then*

$$\lambda_2 \leq \frac{n}{2} - r. \tag{4.8}$$

Proof The bipartite complement of G is an $(\frac{n}{2} - r)$-regular bipartite graph, and thus its second largest eigenvalue can be at most $\frac{n}{2} - r$. Since the second largest eigenvalues of a regular bipartite graph and its bipartite complement are the same (see (4.7)), the proof follows. $\qquad\square$

It is known from [138] that if G is a regular graph of order n and degree r, then the sum of its two largest eigenvalues r and λ_2 is at most $n-2$. Moreover, $r + \lambda_2 = n - 2$ if and only if the complement of G has a connected component which is bipartite. A direct consequence of the previous theorem is a similar characterization if G is r-regular and bipartite, with equality if and only if $\overline{\overline{G}}$ is disconnected.

The next result bounds the order of an r-regular bipartite graph whose diameter is equal to 3.

Theorem 4.8 ([248]) *Let G be a connected r-regular bipartite graph with n vertices and diameter 3. Then*

$$n \le 2\frac{r^2 - \lambda_2^2}{r - \lambda_2^2}, \tag{4.9}$$

whenever the right-hand side is positive.

Proof Consider the distance partition of the set of vertices of G. It is an equitable partition with quotient matrix

$$B = \begin{pmatrix} 0 & r & 0 & 0 \\ 1 & 0 & r-1 & 0 \\ 0 & \frac{2r(r-1)}{n-2} & 0 & \frac{r(n-2r)}{n-2} \\ 0 & 0 & r & 0 \end{pmatrix}.$$

By Theorem 1.25, $\lambda_2(G) \ge \lambda_2(B) = \sqrt{\frac{r(n-2r)}{n-2}}$, which, after simplifying, leads to the result. $\qquad\square$

Remark 4.9 Instead of numerical examples, we list some graphs for which the equalities in the previous theorems are attained.

Theorem 4.2 – This bound is attained for the Petersen graph or the complement of the Clebsch graph.[1]

Theorem 4.7 – Equality holds for any regular complete bipartite graph with at least 4 vertices, or $\overline{(r+1)K_2}$ ($r \ge 1$) (regular complete bipartite graphs with

---

[1] The Clebsch graph is a strongly regular graph with parameters $(16, 5, 0, 2)$, see [98, p. 184] or [102, p. 10].

a perfect matching removed). It is also attained for $\overline{2G}$, where G is any regular bipartite graph.

Theorem 4.8 – Equality holds for the Heawood graph[2] and its bipartite complement. It is also attained for $\overline{(r+1)K_2}$ $(r \geq 1)$. □

Expanders

Let G be a connected regular graph with n vertices. For $S, T \subseteq V(G)$, denote the set of all edges between S and T by $m(S,T) = \{uv \in E(G) : u \in S, v \in T\}$, where the edges whose ends are both in $S \cap T$ are counted twice. By setting $T = \overline{S}$ we exclude this possibility. Moreover, in this case the set $m(S, \overline{S})$ is often denoted by ∂S and called the *edge boundary* of S.

The *expansion parameter* of G is defined as

$$h(G) = \min_{1 \leq |S| \leq \frac{n}{2}} \frac{|\partial S|}{|S|}. \tag{4.10}$$

A family of *expanders* $\{G_i\}$, for $i \in \mathbb{N}$, is a collection of graphs with the following properties:

- The connected graph G_i is r-regular of order n_i; $\{n_i\}$ is a monotone increasing series that does not increase too fast (e.g., $n_{i+1} \leq n_i^2$).
- For all $i \in \mathbb{N}$, $h(G_i) \geq \varepsilon > 0$.

The following result is the main connection between the expansion parameter of a regular graph and its second largest eigenvalue.

Theorem 4.10 (Tanner et al., cf. [9, 222]) *For a connected r-regular graph,*

$$\frac{r - \lambda_2}{2} \leq h(G) \leq \sqrt{2r(r - \lambda_2)}. \tag{4.11}$$

The theorem actually proves that the spectral gap $\eta = r - \lambda_2$ can give a good estimate of the expansion of a graph, since a smaller spectral gap makes a smaller upper bound for $h(G)$. Moreover, regular graphs $\{G_i\}$ are expanders if for all $i \in \mathbb{N}$, $\eta(G_i) \geq \varepsilon > 0$.

---

[2] The Heawood graph is a connected cubic graph on 14 vertices (the graph G_4 of Table A.3 in Section 4.3). See also the figure in [98, p. 225].

Remark 4.11   In some literature, the expansion parameter given in (4.10) is called the *Cheeger constant*. At the same time, the inequalities (4.11) are called the *Cheeger inequalities*.

In what follows we consider the eigenvalue with second largest modulus of regular graphs, that is, $\Lambda = \max_i\{|\lambda_i| \ : \ \lambda_i \neq \pm r\}$. Clearly, $\Lambda = |\lambda_2|$, or $\Lambda = |\lambda_n|$. We prove a variant of the expander mixing lemma.

Lemma 4.12   (Alon and Chung [8], cf. [58, Proposition 4.3.2] or [102, Lemma 3.5.2]) *For a connected r-regular graph G, let* $\Lambda = \max_i\{ |\lambda_i| \ : \ \lambda_i \neq \pm r\}$ *and* $S, T \subseteq V(G)$, *with* $|S| = s, |T| = t$. *Then*

$$\left| m(S,T) - \frac{rst}{n} \right| \leq \Lambda\sqrt{st\left(1 - \frac{s}{n}\right)\left(1 - \frac{t}{n}\right)} \leq \Lambda\sqrt{st}. \qquad (4.12)$$

Proof   Following [58], we consider the characteristic vectors $\mathbf{y_S}$ and $\mathbf{y_T}$ of the sets S and T. Both of these can be expressed as a linear combination of a set of orthonormal eigenvectors of the adjacency matrix A: $\mathbf{y_S} = \sum_{i=1}^{n} \alpha_i \mathbf{x_i}$ and $\mathbf{y_T} = \sum_{i=1}^{n} \beta_i \mathbf{x_i}$. Then

$$m(S,T) = \mathbf{y_S}^T A \mathbf{y_T} = \sum_{i=1}^{n} \alpha_i \beta_i \lambda_i.$$

Since $\alpha_1 = \frac{s}{\sqrt{n}}$, $\beta_1 = \frac{t}{\sqrt{n}}$, and $\lambda_1 = r$, we can exclude these parameters from the above sum. In this way we get

$$\left| m(S,T) - \frac{rst}{n} \right| = \left| \sum_{i=2}^{n} \alpha_i \beta_i \lambda_i \right| \leq \Lambda \sum_{i=2}^{n} |\alpha_i \beta_i|,$$

with $\sum_{i=2}^{n} \alpha_i^2 \leq \mathbf{y_S}^T \mathbf{y_S} - \frac{s^2}{n} = \frac{s(n-s)}{n}$ and $\sum_{i=2}^{n} \beta_i^2 \leq \frac{t(n-t)}{n}$, which gives the left inequality of (4.12). The right inequality follows directly, and the proof is complete. $\square$

This lemma can be viewed as relating Λ to the question of how 'random' the graph is. The left-hand side of (4.12) compares the expected number of edges between S and T in a random graph and the actual number of edges between these sets. This difference is small when Λ is small. Recently, Nikiforov [334] gave a valuable discussion on the expander mixing lemma including its extension to any matrices.

If $N(S)$ denotes the set of vertices of a graph G that have at least one neighbour in S, then a different way to define the expansion of regular graphs is to consider the quotient $\frac{|N(S)|}{|S|}$ instead of $\frac{|\partial S|}{|S|}$ in (4.10). In this way we get the *vertex expansion parameter* $\min_{1 \leq |S| \leq \frac{n}{2}} \frac{|N(S)|}{|S|}$, along with the *vertex expanders* defined in the same way as above. In relation to this definition we mention the following result.

Theorem 4.13 (Tanner [441] or [102, Theorem 3.5.3]) *For a connected r-regular graph G, let* $\Lambda = \max_i\{|\lambda_i| : \lambda_i \neq \pm r\}$ *and* $S \subseteq V(G)$. *Then*

$$\frac{|N(S)|}{|S|} \geq \frac{r^2}{\Lambda^2 + \frac{(r^2-\Lambda^2)|S|}{n}}. \tag{4.13}$$

Alon [6] conjectured and Friedland [164] proved that large random r-regular graphs have second largest eigenvalue smaller than $2\sqrt{r-1} + \varepsilon$. Friedland observed that numerical experiments seem to indicate that random r-regular graphs in fact satisfy $\lambda_2 < 2\sqrt{r-1}$.

A connected r-regular graph is called a *Ramanujan graph* (this terminology was introduced in [302]) if $\Lambda \leq 2\sqrt{r-1}$. It is not difficult to find such graphs. For example, any non-trivial complete graph is a Ramanujan graph. Some less trivial examples are indicated in [248]. A more difficult task is the construction of infinite sequences of Ramanujan graphs with given degree r and order n tending to infinity. According to [102, p. 68], the first such sequence was constructed by Lubotzky et al. [302] in 1988.

The relations between expanders and second smallest L-eigenvalues are considered in Subsection 6.6.3.

4.1.2 Trees

We first point out the two general results mentioned later on page 148, which coincide mutually for λ_2. It is also noteworthy to say that the upper bound

$$\lambda_2(T) \leq \sqrt{\frac{n-3}{2}} \tag{4.14}$$

obtained by Neumaier [329] holds only for n odd (as shown in [217] with a counterexample concerning an even tree).

We prove the following theorem.

Theorem 4.14 (Hu [226]) *Let T be a tree with $n \geq 2$ vertices. Then*

$$\lambda_2(T) \geq \begin{cases} \sqrt{\dfrac{d_1+d_2-1-\sqrt{(d_1+d_2-1)^2-4(d_1-1)(d_2-1)}}{2}}, & \text{if } d(T) = 1, \\[3mm] \sqrt{\dfrac{d_1+d_2-\sqrt{(d_1-d_2)^2+4}}{2}}, & \text{if } d(T) = 2, \\[3mm] \sqrt{d_1-1}, & \text{if } d(T) = 3 \text{ and } d_1 = d_2, \\[3mm] \sqrt{d_2}, & \text{otherwise,} \end{cases}$$

$$(4.15)$$

where d_1, d_2 are the largest and second largest degree in T, respectively. $d(T)$ is the distance between the largest and second largest degree vertices. Equality holds if T is a tree T_1, T_2 or T is T_4 and $d_1 = d_2$, where T_i is a tree obtained from the stars K_{1,d_1-1} and K_{1,d_2-1} by inserting a path of length i between their centres.

Proof For $d(T) = 1$ (resp. $d(T) = 2$), T contains the tree T_1 (resp. T_2) as an induced subgraph. We compute the second largest eigenvalue of both trees T_1 and T_2 and obtain the first two values on the right-hand side of (4.15), and then the result follows from the interlacing theorem.

If $d(T) = 3$ and $d_1 = d_2$, we have $\lambda_2(T) \geq \lambda_2(T_3) > \sqrt{d_1-1}$.

If $d(T) = 3$ and $d_1 > d_2$, or $d(T) \geq 4$, using the interlacing theorem, we get $\lambda_2(T) \geq \lambda_1(K_{1,d_2}) = \sqrt{d_2}$.

Equalities are verified by direct computation. □

Example 4.15 Consider the caterpillars with two vertices of degree three, illustrated in Fig. 4.1. By the notation introduced on page 5, any such caterpillar can be denoted by $T_n^{i,j}$. Here is the numerical computation that illustrates the result of the previous theorem:

Tree	(4.15)	λ_2	Tree	(4.15)	λ_2	Tree	(4.15)	λ_2
$T_6^{2,3}$	1	1	$T_7^{2,4}$	1.414	1.414	$T_8^{2,5}$	1.414	1.414
$T_7^{2,3}$	1	1.207	$T_8^{2,4}$	1.414	1.520	$T_9^{2,5}$	1.414	1.691
$T_{14}^{6,7}$	1	1.784	$T_{15}^{6,8}$	1.414	1.848	$T_{16}^{6,9}$	1.414	1.906

□

Finally, we note without proof a result of An [17] concerning trees on $2n$ vertices and with perfect matchings:

Figure 4.1 Caterpillars for Example 4.15.

$$\lambda_2(T) \leq \begin{cases} \frac{\sqrt{k-1}+\sqrt{k+3}}{2}, & \text{if } n = 2k, \\ \frac{\sqrt{k}+\sqrt{k+4}}{2}, & \text{if } n = 2k+1. \end{cases} \qquad (4.16)$$

4.2 Graphs with small second largest eigenvalue

There is a wide literature concerning graphs whose second largest eigenvalue is comparatively small. The results obtained before 2001 have been given an adequate attention in the survey paper [113] and the book [369]. Here we give a brief review of these results; for more details, including some omitted proofs, we refer the reader to the corresponding references. In particular, all results concerning graphs whose second largest eigenvalue does not exceed $\frac{1}{3}, \sqrt{2}-1$ or $\frac{\sqrt{5}-1}{2}$, and some results concerning $\lambda_2 \leq 1$, can be found in at least one of these references. We pay more attention to the results obtained after 2001, which treat bounds equal to 1, $\sqrt{2}, \sqrt{3}$ or 2. Regular graphs with bounded second largest eigenvalue are considered in a separate subsection (Subsection 4.2.6).

For any real number k, the property $\lambda_2 \leq k$ is a hereditary property meaning if, for any graph G, $\lambda_2(G) \leq k$ then, for any induced subgraph H of G, $\lambda_2(H) \leq k$ (this fact is a direct consequence of the interlacing theorem). If it occurs that $\lambda_2(G) \leq k$, but at the same time no supergraph of G satisfies the same inequality, then G is called the *maximal* graph for $\lambda_2 \leq k$. Similarly, if $\lambda_2(G) > k$ and at the same time no induced subgraph of G satisfies the same inequality, then G is called the *minimal forbidden graph* for $\lambda_2 \leq k$.

4.2.1 Graphs with $\lambda_2 \leq \frac{1}{3}$ or $\lambda_2 \leq \sqrt{2}-1$

According to [413], a non-trivial connected graph has exactly one positive eigenvalue if and only if it is a complete multipartite graph. In other words, if a connected graph is not complete multipartite then its second largest eigenvalue

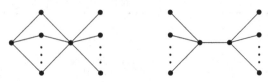

is positive. In order to consider the graphs in the title, we need the following result.

Theorem 4.16   (Howes [224], cf. [113] or [369, Theorem 3.1]) *Let $\mathcal{G}$ be an infinite family of graphs. The following statements are equivalent:*

(i) *There is a real number k such that $\lambda_2(G) \leq k$ for every $G \in \mathcal{G}$.*
(ii) *There is a positive integer s such that for each $G \in \mathcal{G}$ the graphs $(K_1 \cup K_s)\nabla K_s$, $(sK_1 \cup K_{1,s})\nabla K_1$, $(sK_1 \cup K_{s-1})\nabla K_1$, $K_s \cup K_{1,s}$, $2K_{1,s}$, $2K_s$, and the graphs illustrated in Fig. 4.2 (each obtained from two copies of $K_{1,s}$ by adding extra edges) are not subgraphs of G.*

The next two theorems are proved using the interlacing theorem and the previous result.

Theorem 4.17   (Cao and Hong [70], cf. [369, Theorem 3.2]) *If G is a connected graph with $0 < \lambda_2(G) \leq \frac{1}{3}$ then G has at most three vertices or, for $n \geq 4$, $G \cong ((n-3)K_1)\nabla(K_1 \cup K_2)$.*

Theorem 4.18   (Petrović [359], cf. [369, Theorem 3.3]) *G is a connected graph with $0 < \lambda_2(G) \leq \sqrt{2} - 1$ if and only if one of the following holds.*

(i) *$G \cong \overline{s_0 P_3}\nabla K_{s_1,s_2,\dots,s_m}$.*
(ii) *$G \cong (K_1 \cup K_{r,s})\nabla \overline{K_t}$ and the parameters $r,s,$ and t satisfy one of the following conditions:*

- $r > 1$, $s \geq r$, $t = 1$,
- $r = 1$, $s \geq 1$, $t \geq 2$,
- $r = 2$, $s \geq 2$, $t = 2$,
- $r = 2$, $2 \leq s \leq 3$, $t \geq 3$,
- $r = 2$, $s = 4$, $3 \leq t \leq 7$,

- $r = 2$, $s = 5$, $3 \leq t \leq 4$,
- $r = 2$, $6 \leq s \leq 8$, $t = 3$,
- $r = 3$, $s = 3$, $2 \leq t \leq 4$,
- $r = 3$, $4 \leq s \leq 7$, $t = 2$,
- $r = 4$, $s = 4$, $t = 2$.

(iii) *$G \cong (K_1 \cup K_{1,r})\nabla K_{s,t}$ and the parameters $r,s,$ and t satisfy one of the following conditions:*

- $r = 1$, $s \geq 1$, $t \geq s$,
- $r = 2$, $1 \leq s \leq 2$, $t \geq s$,
- $r = 2$, $s = 3$, $3 \leq t \leq 7$,
- $r = 2$, $s = 4$, $t = 4$,
- $r = 3$, $s = 1$, $t = 1$.

4.2.2 The golden section bound

Denote $\sigma = \frac{\sqrt{5}-1}{2}$ (the golden section). Computational results show that σ appears in the spectra of many graphs, for example $\lambda_2(P_4) = \sigma$. We give some results concerning those graphs with $\lambda_2 \leq \sigma$. As we will see, the graphs satisfying this property are not completely determined.

In order to consider the graphs with $\lambda_2 < \sigma$, we need the family of graphs denoted by Φ, which is the smallest family of graphs that contains K_1 and is closed under adding isolated vertices and taking joins of graphs.

It can be checked that any graph G without isolated vertices whose complement is a connected graph must contain P_4 or $2K_2$ as an induced subgraph (if necessary, see [464]). Therefore, if $\lambda_2(G) < \sigma$ and $\overline{G}$ is connected then G must have at least one isolated vertex. Otherwise, if $\overline{G}$ is disconnected then, by Theorem 4.16, G is a join of at least two graphs. In other words, any graph whose second largest eigenvalue is less than σ belongs to Φ (but not vice versa).

The family Φ was introduced in [406], where weighted rooted trees were used to represent graphs in Φ. Namely, to any graph G from Φ we associate a weighted rooted tree T_G in the following way: if $H = (H_1 \nabla H_2 \nabla \cdots \nabla H_m) \cup nK_1$ is any subrepresentation of G then a subtree T_H with a root v (of weight n) represents H whereas for each i ($1 \leq i \leq m$) there is a vertex v_i (a direct successor of v in T_H) representing a root of H_i. It turns out that the set of graphs with $\lambda_2 < \sigma$ falls into a finite number of structured types of graph in Φ. These types are illustrated in Fig. 4.3 by the corresponding representing trees with emphasized roots.

It has been proved in [406] that the set of minimal forbidden subgraphs for $\lambda_2 < \sigma$ is finite. They all belong to Φ except for P_4 and $2K_2$.

Concerning the graphs with $\lambda_2 \leq \sigma$, we give the following two results.

Theorem 4.19 (Cvetković and Simić [112], cf. [113] or [369, Theorem 3.4]) *A graph with $\lambda_2 \leq \sigma$ has at most one non-trivial component G for which one of the following holds:*

(i) *G is a complete multipartite graph,*

(ii) *G is an induced subgraph of C_5,*

(iii) *G contains a triangle.*

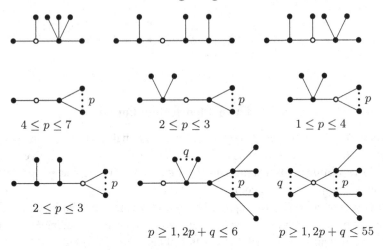

$$4 \leq p \leq 7 \qquad 2 \leq p \leq 3 \qquad 1 \leq p \leq 4$$

$$2 \leq p \leq 3$$

$$p \geq 1, 2p + q \leq 6 \qquad p \geq 1, 2p + q \leq 55$$

Figure 4.3 Rooted trees associated with graphs satisfying $\lambda_2 < \sigma$.

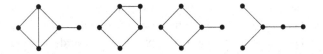

Figure 4.4 Minimal forbidden subgraphs for $\lambda_2 \leq \sigma$.

For the graphs with $\lambda_2 \leq \sigma$ that contain a triangle, see [112] or [369, Theorem 3.5]. It has also been proved that the set of minimal forbidden subgraphs for $\lambda_2 \leq \sigma$ is finite. The next theorem provides more details.

Theorem 4.20 ([112], cf. [113] or [369, Theorem 3.6]) *If G is a minimal forbidden subgraph for $\lambda_2 \leq \sigma$ then either*

(i) *G is $2K_2$ or one of the graphs depicted in Fig. 4.4, or*
(ii) *G belongs to Φ.*

4.2.3 Graphs whose second largest eigenvalue does not exceed 1

Hoffman posed the problem of characterizing the graphs with property $\lambda_2 \leq 1$, and Cvetković partially solved this problem in the following theorems.

Theorem 4.21 ([94], cf. [369, Theorem 3.8]) *If $\lambda_2(G) \leq 1$ then $\overline{G}$ satisfies one of the following conditions:*

(i) *the least eigenvalue of $\overline{G}$ is at least -2,*

(ii) exactly one eigenvalue of $\overline{G}$ is less than -2.

In addition, if G belongs to (i), then $\lambda_2(G) \leq 1$.

Proof Setting $i = n, j = 2$ in the first of the Courant–Weyl inequalities, we get $\lambda_2(G) + \lambda_{n-1}(\overline{G}) \geq -1$, and thus $\lambda_2(G) \leq 1$ implies $\lambda_{n-1}(\overline{G}) \geq -2$, which proves the first part.

If G belongs to (i), setting $i = 2, j = n$ in the second of the Courant–Weyl inequalities, we get $\lambda_2(G) + \lambda_n(\overline{G}) \leq -1$, and thus

$$\lambda_n(\overline{G}) \geq -2 \qquad (4.17)$$

implies $\lambda_2(G) \leq 1$. $\qquad \square$

Corollary 4.22 ([94], cf. [369, Theorem 3.9]) *Complements of graphs whose least eigenvalue is greater than -2 have second largest eigenvalue less than 1.*

Proof The proof is a direct implication of (4.17). $\qquad \square$

Recall that all the connected graphs with least eigenvalue greater than -2 have been described (see Exercise 3.6 in the previous chapter). The exact characterization of graphs with the second largest eigenvalue at most 1 still remains an open question. In what follows we determine all connected graphs with the property $\lambda_2 \leq 1$ within some classes of graphs.

Bipartite graphs and generalized line graphs

Bipartite graphs with the property $\lambda_2 \leq 1$ were characterized by Petrović in 1991. He used the concept of forbidden subgraphs and eigenvalue interlacing. To present the result we need the following notation. A rectangle with an arbitrary number of vertices inside it denotes the graph consisting of only isolated vertices (see Fig. 4.5). Two such rectangles joined with exactly one full line denote a complete bipartite graph, while being joined with a sequence of k $(k \geq 1)$ dashed parallel lines, they denote a complete bipartite graph with $2k$ vertices and exactly k mutually independent edges excluded. If two rectangles are joined with a sequence of k $(k \geq 1)$ full parallel lines, it signifies a bipartite graph with $2k$ vertices in which two vertices are adjacent if and only if they are joined by such a line.

We have the following result.

Theorem 4.23 ([358], cf. [369, Theorem 3.13]) *A connected bipartite graph has the property $\lambda_2 \leq 1$ if and only if it is an induced subgraph of some of the graphs illustrated in Fig. 4.5.*

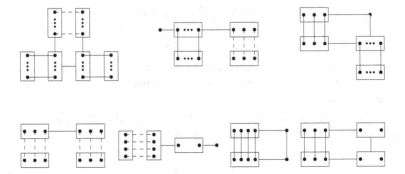

Figure 4.5 Graphs for Theorem 4.23. Adapted from [358] with permission from Elsevier.

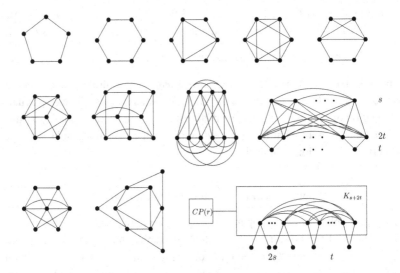

Figure 4.6 Graphs for Theorem 4.24.

Generalized line graphs are considered in the next theorem.

Theorem 4.24   (Petrović and Milekić [368], cf. [369, Theorem 3.21]) *A connected generalized line graph has the property* $\lambda_2 \leq 1$ *if and only if it is an induced subgraph of some of the graphs illustrated in Fig.* 4.6.

Note that the last graph in the second line of Fig. 4.6 contains the complete graph K_{s+2t} $(s, t \geq 1)$ as an induced subgraph. For the last graph in the third line, the line between two rectangles denotes the join of the cocktail party $CP(r)$ and the complete graph K_{s+2t}.

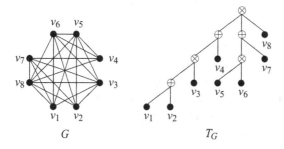

Figure 4.7 An arbitrary co-graph G and its representation T_G. Reprinted from [420] with permission from Elsevier.

It is also noteworthy to say that all graphs given in the first two lines of Fig. 4.6 are just line graphs. Therefore, any line graph with $\lambda_2 \leq 1$ is an induced subgraph of some of these graphs.

Nested split graphs

From this point on we consider results that cannot be found in [113, 369]. The NSGs with $\lambda_2 < 1$ are determined by Stanić [420], and those with $\lambda_2 = 1$ are determined by Milatović and Stanić [321].

We need a representation of NSGs which differs from the representation introduced on page 7. Recall that an NSG is a $\{P_4, C_4, 2K_2\}$-free graph. In addition, any P_4-free graph is called a *co-graph*. By [47, 420], co-graphs are represented by so-called *co-trees* in the following way. Let $\oplus$ and $\otimes$ stand for the (disjoint) union and join of two graphs. The co-tree T_G is a rooted tree in which any interior vertex w is either of $\oplus$-type (corresponds to union) or $\otimes$-type (corresponds to join). A terminal vertex represents itself in G. Any interior vertex, say w, represents a subgraph of G induced by the terminal successors of w, and is denoted by G_w. The direct non-terminal successor of any interior vertex w has a type which differs from the type of w. In addition, each non-terminal vertex has at least two direct successors. It is worth mentioning that the described representation is unique. An example consisting of a co-graph G and its representation T_G is given in Fig. 4.7.

Apparently, each NSG is a co-graph. Therefore, we present a representation of NSGs which is a special case of the representation of co-graphs. First, we have the following lemma.

Lemma 4.25 *Let G be an arbitrary NSG and let T_G be its (unique) repre-*

sentation. Then each non-terminal vertex of T_G has at most one non-terminal direct successor.

Proof Let w be a non-terminal vertex of T_G of $\oplus$-type. Assume to the contrary, and let w_1 and w_2 be non-terminal direct successors of w. Then both w_1 and w_2 are of $\otimes$-type and each of them has at least two adjacent terminal successors. But these two pairs of terminal vertices make a $2K_2$, which is an induced subgraph of G. A contradiction. The second possibility is considered in a similar way. □

By the previous lemma, a representation T_G of an arbitrary NSG G has a simple form so we can avoid drawing trees in its presentation. Namely, it is sufficient to say whether G is connected or not and to list the numbers of terminal successors of each non-terminal vertex of T_G (in natural order). Therefore, we use $C(a_1, a_2, \ldots, a_n)$ to denote an NSG such that the tree $T_{C(a_1, a_2, \ldots, a_n)}$ has exactly n non-terminal vertices, while its root is of $\otimes$-type and has exactly a_1 direct terminal successors; each non-terminal successor of the root has exactly a_2 direct terminal successors, etc. A disconnected NSG is denoted by $D(a_1, a_2, \ldots, a_n)$. Recall that each non-terminal vertex has at least two direct successors. Thus, we assume that $a_1, a_2, \ldots, a_{n-1}$ are positive integers, while $a_n \geq 2$. If for the n-tuple $(a_1, a_2, \ldots, a_n)$, $a_i = a_{i+1} = \cdots = a_{i+k}$, $1 \leq i$, $i+k \leq n$, we write $(a_1, a_2, \ldots, a_i^{k+1}, a_{i+k+1}, \ldots, a_n)$.

Since each NSG has at most one non-trivial component, its second largest eigenvalue is equal to the second largest eigenvalue of that component. Thus, it is sufficient to consider connected NSGs.

Lemma 4.26 ([420]) *Let $C(a_1, a_2, \ldots, a_n)$ be a connected NSG with the property $\lambda_2 \leq 1$. Then $n \leq 10$ holds.*

Proof If $n > 10$ then an NSG $C(a_1, a_2, \ldots, a_n)$ contains the NSG $C(1^{10}, 2)$ as an induced subgraph. Since $\lambda_2(C(1^{10}, 2)) > 1$, the proof follows from the interlacing theorem. □

Let $G = C(a_1, a_2, \ldots, a_n)$ be an arbitrary connected NSG, and let V_i denote the set of vertices corresponding to a_i $(1 \leq i \leq n)$. Hence, $|V_i| = a_i$ $(1 \leq i \leq n)$. It is easy to check that the partition of the vertex set of G into non-empty subsets $V_1, V_2, \ldots, V_n$ determines a divisor H of G.

The following theorem is crucial for our consideration.

Theorem 4.27 ([420]) *Let λ be an eigenvalue of a connected NSG $G = C(a_1, a_2, \ldots, a_n)$ distinct from 0 or -1 and let H be the divisor of G. Then λ appears in the spectrum of H.*

Proof It is known that the multiplicity of the eigenvalue 0 (resp. -1) in the spectrum of a co-graph G which does not contain isolated vertices (and therefore in the spectrum of a connected NSG) is equal to $\sum_{w \in V_0}(t_w - 1)$ (resp. $\sum_{w \in V_{-1}}(t_w - 1)$), where V_0 (resp. V_{-1}) is the set of interior vertices (in T_G) of $\oplus$-type (resp. $\otimes$-type) which have t_w direct successors as terminal vertices. In addition, 0 and -1 are non-main eigenvalues of G. Therefore, exactly n eigenvalues of $G = C(a_1, a_2, \ldots, a_n)$ are distinct from 0 or -1. The spectrum of H consists of exactly n eigenvalues and includes the main part of the spectrum of G, which completes the proof. □

Corollary 4.28 ([420]) *Let G be an arbitrary NSG and let H be its divisor. Then $\lambda_2(G) \leq 1$ if and only if $\lambda_2(H) \leq 1$.*

Proof The proof follows from the previous theorem. □

The remaining results are computational. By the use of Lemma 4.26 and Corollary 4.28, all possible candidates for NSGs are considered, along with the simultaneous computation of the second largest eigenvalues of their divisors. The maximal NSGs with $\lambda_2 < 1$ and $\lambda_2 = 1$ are given in Tables A.1 and A.2 of Section 4.3.

Remark 4.29 The star complement technique can produce many graphs with $\lambda_2 = 1$. By the rule, all these graphs have a small number of distinct eigenvalues and a large multiplicity of the second largest eigenvalue.

Some results on this subject were obtained by Stanić [419, 423, 424] and Stanić and Simić [425]. For example, it has been shown in [419, 423] that $K_{1,5}$, $K_{1,9}$, and $K_{1,10}$ are the only stars which can be star complements for $\lambda_2 = 1$ (for this result, see also [392]). In addition, the Clebsch graph is a unique maximal graph with star complement $K_{1,5}$. Next, there are exactly two complete graphs which can be star complements for the same eigenvalue: K_{10} and K_{11}. The former graph produces exactly two maximal graphs. The first of them is described in Example 2.9 and depicted in Fig. 2.1, while the second has 24 vertices and the spectrum $[11.28, 1^{14}, -1, -3^7, -3.28]$.

Cocktail party graphs and unicyclic graphs as star complements for $\lambda_2 = 1$ are considered in [424] and [425], respectively. □

4.2.4 Trees with $\lambda_2 \leq \sqrt{2}$

All trees with the property $\lambda_2 \leq \sqrt{2}$ were determined by Stanić [422]. The family of graphs illustrated in Fig. 4.8 played a crucial role in their determination. There, $Z_{k,l}$ (we assume that $k + l \geq 2$) is a graph with $3(k + l) + 1$ vertices

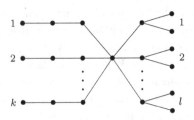

Figure 4.8 Trees $Z_{k,l}$.

obtained by joining a single vertex (this vertex will be referred to as the central vertex) with an end (resp. middle vertex) of each of k (resp. l) paths with 3 vertices. By removing the central vertex from $Z_{k,l}$ we get a disconnected graph, each component of which is P_3, and therefore its largest eigenvalue is equal to $\sqrt{2}$ and has multiplicity at least 2. By the interlacing theorem, we get $\lambda_2(Z_{k,l}) = \sqrt{2}$.

There follows a technical lemma.

Lemma 4.30 ([422]) *Let T be a tree whose diameter D is not 3 and whose second largest eigenvalue is at most $\sqrt{2}$. Then T is an induced subtree of $Z_{k,l}$.*

Considering the double star $DS(4,5)$, we get $\lambda_2(DS(4,5)) = \sqrt{2}$ and the second largest eigenvalue of any tree of diameter 3 which contains $DS(4,5)$ as an induced subtree is greater than $\sqrt{2}$. Therefore, using the previous lemma we get the following result.

Theorem 4.31 ([422]) *If T is a tree with the property $\lambda_2(T) \leq \sqrt{2}$, then T is an induced subtree of $Z_{k,l}$ or it is isomorphic to either $DS(4,4)$ or $DS(4,5)$.*

Proof T must be an induced subtree of $Z_{k,l}$ or $DS(4,5)$, but any proper induced subtree of $DS(4,5)$ distinct from $DS(4,4)$ is also a proper induced subtree of $Z_{k,l}$. $\square$

Remark 4.32 Let $Z^=$ consist of $DS(4,5)$ and all induced subtrees of $Z_{k,l}$ which contain two copies of P_3 as an induced subgraph. Let Z^- consist of $DS(4,4)$ and all induced subtrees of $Z_{k,l}$ which do not belong to $Z^=$. Finally, let Z^+ be the set of all remaining trees (i.e., those which do not belong to the union of the previous two sets). Then for any tree T we have:

(i) $\lambda_2(T) < \sqrt{2}$ if and only if $T \in Z^-$,
(ii) $\lambda_2(T) = \sqrt{2}$ if and only if $T \in Z^=$,
(iii) $\lambda_2(T) > \sqrt{2}$ if and only if $T \in Z^+$.

The previous classification, in fact, determines whether the second largest eigenvalue of a tree is less than, equal to or greater than $\sqrt{2}$ by considering its structure. To conclude where it belongs it is sufficient to compare it with $Z_{k,l}, DS(4,5), DS(4,4)$ or with some specific subtrees of $Z_{k,l}$. A similar result for 2 as the largest eigenvalue of a graph is given on page 63. □

4.2.5 Notes on reflexive cacti

Here we consider graphs with the property $\lambda_2 \le 2$. Following [329, 330], we observe that such graphs correspond to a set of vectors in the Lorentz space $\mathbb{R}^{p,1}$ having Gram matrix $2I - A$. They are Lorentzian counterparts of the spherical and Euclidean graphs which occur in the theory of reflection groups and they are often called *reflexive graphs*.

Most of the results on reflexive graphs are related to cacti or regular graphs. Here we give some short notes on reflexive cacti. Reflexive regular graphs are considered in the next subsection.

In all investigations on reflexive cacti the Smith graphs played a central role and some large classes of these graphs were constructed and described by means of Smith graphs. The following theorem explains their importance.

Theorem 4.33 (Radosavljević and Simić [377]) *Let G be a graph with a cut vertex u. Then*

(i) *if at least two components of $G - u$ are the supergraphs of Smith graphs, and if at least one of them is a proper supergraph, then $\lambda_2(G) > 2$;*

(ii) *if at least two components of $G - u$ are Smith graphs, while the rest are subgraphs of Smith graphs, then $\lambda_2(G) = 2$;*

(iii) *if at most one component of $G - u$ is a Smith graph, while the rest are proper subgraphs of Smith graphs, then $\lambda_2(G) < 2$.*

In relation to the previous theorem, if after removing a vertex u we get one proper supergraph, while the rest are proper subgraphs of Smith graphs, the theorem does not answer the question whether the graph is reflexive or not (we call such graphs undecidable) and such cases are interesting for investigation.

If all cycles of a cactus have a unique common vertex, we say that they form a *bundle*. It is clear that the number of cycles in a cactus with a bundle is not limited. On the contrary, it is proved by Radosavljević and Rašajski [376] that an undecidable reflexive cactus whose cycles do not form a bundle has at most five cycles. Such graphs with exactly five cycles belong to the four families of

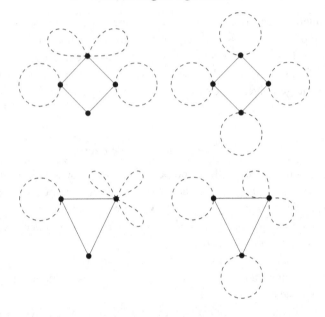

Figure 4.9 Maximal 5-cyclic reflexive cacti. Adapted from [376] with permission from Elsevier.

graphs illustrated in Fig. 4.9 (there, cycles depicted with closed dashed lines denote cycles of an arbitrary length).

Further investigation is focused on undecidable reflexive cacti with at most four cycles. Reflexive trees and unicyclic graphs are investigated in [308, 329] and [243, 244, 373]. Reflexive bicyclic graphs with an edge between the cycles are considered in [377]. For example, it is proved that an undecidable bicyclic graph with an edge between cycles is reflexive if and only if it is an induced subgraph of the 99 graphs or families of graphs given in the same reference. Various classes of multicyclic reflexive cacti whose cycles do not form a bundle are studied in [320, 374, 375, 376]. In particular, maximal reflexive cacti with four cycles are identified in [375, 376].

4.2.6 Regular graphs

In what follows, we reproduce some results of Stanić [421] and Koledin and Stanić [245, 246, 247] considering regular graphs with the property $\lambda_2 \leq 1$, regular triangle-free graphs with $\lambda_2 \leq \sqrt{2}$, and regular bipartite graphs with $\lambda_2 \leq 2$.

We start with the following simple but useful fact.

Theorem 4.34 ([247]) *Let G be a connected r-regular ($r \geq 3$) graph with diameter D satisfying $\lambda_2 < \sqrt{r}$. Then $D \leq 3$.*

Proof Considering two vertices u and v at a distance 4, we get the following inequalities: $\lambda_2(G) \geq \lambda_2(G[N[u] \cup N[v]]) \geq \min(\lambda_1(G[N[u]]), \lambda_1(G[N[v]])) \geq \lambda_1(K_{1,r}) = \sqrt{r}$. A contradiction. □

Regular graphs with $\lambda_2 \leq 1$

By Theorem 1.10, each connected regular graph G with the property $\lambda_2 \leq 1$ is a complement of a (not necessary connected) regular graph, each of whose components is one of the graphs listed in Theorem 3.15.

In what follows we give a closer characterization of regular graphs whose second largest eigenvalue does not exceed 1. In particular, we determine all connected regular graphs with small vertex degree. We first consider one special situation.

Theorem 4.35 ([421]) *Let G be a connected regular graph with n vertices and the properties $\lambda_2 \leq 1$ and $\lambda_n \geq -2$. Then one of the following holds:*

(i) G is a connected regular line graph satisfying $\lambda_2 \leq 1$ (these graphs are C_n ($3 \leq n \leq 6$), K_n ($n \geq 1$), the complement of the Petersen graph, and the graphs $\overline{C_6}, \overline{3K_2}, \mathrm{Line}(\overline{2K_3})$);
(ii) G is a cocktail party graph;
(iii) G is the Petersen graph.

Proof Since $\lambda_n(G) \geq -2$, we have that G is either a generalized line graph or an exceptional graph.

Recall that a connected regular generalized line graph is either a line graph or a cocktail party graph (if necessary, see [105, Proposition 1.1.9]). In addition, for each cocktail party graph we have $\lambda_2 \leq 1$. Further, by considering the graphs from Theorem 4.24, we identify all regular graphs with $\lambda_2 \leq 1$ which are generalized line graphs (it turns out that all of them are line graphs).

Finally, among the 187 regular exceptional graphs only the Petersen graph satisfies $\lambda_2 \leq 1$, and the proof follows. □

Recall that for any pair of complementary graphs G and $\overline{G}$, at least one is connected. The following corollary is an immediate consequence.

Corollary 4.36 ([421]) *Let G be a connected regular graph. Then both G and $\overline{G}$ have second largest eigenvalue not exceeding 1 if and only if G is one of the three types of graph listed in the previous theorem.*

We continue with regular graphs of fixed vertex degree. The following consideration is easily verified.

Each connected regular graph (with n vertices) of degree $n - 1$ ($n \geq 1$) or $n - 2$ ($n \geq 4$) has second largest eigenvalue not exceeding 1. A connected regular graph G (with $n \geq 3$ vertices) of degree 2 has second largest eigenvalue not exceeding 1 if and only if it is the cycle C_n ($3 \leq n \leq 6$). A connected regular graph (with $n \geq 5$ vertices) of degree $n - 3$ has second largest eigenvalue not exceeding 1 if and only if its complement $\overline{G}$ is either the cycle C_n ($n \geq 5$) or a disconnected graph, all of whose components are cycles.

In the following two theorems we determine all graphs in question with vertex degree 3 or 4.

Theorem 4.37 ([421]) *Let G be a cubic graph with the property $\lambda_2 \leq 1$. Then G is one of the following graphs: K_4, the Petersen graph, $\overline{C_6}$ or H_i ($1 \leq i \leq 3$) of Fig. 4.10.*

Proof The vertex degree of $\overline{G}$ is $r_{\overline{G}} = n - 4$. In addition, $\overline{G}$ belongs to either (i) or (iii) of Theorem 3.15.

By inspecting all cubic exceptional graphs, we find that exactly one of them has a complement whose second largest eigenvalue does not exceed 1. This is the graph H_1 of Fig. 4.10.

Assume now that $\overline{G}$ is a (possibly disconnected) line graph.

First let $\overline{G} \cong \text{Line}(H)$, where H is a regular graph with n edges. Since $r_{\overline{G}} = n - 4$, we get that each edge of H has degree $n - 4$, and therefore $r_H = \frac{n-4}{2} + 1$. Denote by N the number of vertices of H. We have $r_H N = 2n$, that is, we get $N = \frac{4(r_H + 1)}{r_H}$ (by setting $n = 2(r_H + 1)$), which provides $r_H \in \{1, 2, 4\}$.

For $r_H = 1$ we get $N = 8$, and therefore $H \cong 4K_2$. Further, $\overline{G} \cong \text{Line}(H) \cong 4K_1$ and $G \cong K_4$.

Similarly, for $r_H = 2$ we get $N = 6$, and therefore $H \cong C_6$ or $H \cong 2C_3$, which implies $\overline{G} \cong C_6$ or $\overline{G} \cong 2C_3$ and $G \cong \overline{C_6}$ or $G \cong H_2$.

Finally, for $r_H = 4$ we get $N = 5$, $H = K_5$, while $\overline{G}$ is the complement of the Petersen graph.

Now let $\overline{G} \cong \text{Line}(H)$, where H is a semiregular bipartite graph with n edges. Since each edge of H has degree $n - 4$, we have that H is (r, s)-semiregular bipartite with $r + s = n - 2$. With no loss of generality, we may assume that $r \geq s$. If s is equal to 1, we get that each component of H is a star, and then the only solution is $H \cong 2K_{1,3}$, which again provides $G \cong H_2$. Next, if $r \geq s \geq 2$, we easily get $r, s \leq \lfloor \frac{n}{2} \rfloor$. Thus, the only possibilities are: $r = \frac{n}{2}, s = \frac{n}{2} - 2$; $r = s = \frac{n}{2} - 1$ (where n is even); and $r = \frac{n-1}{2}, s = \frac{n-3}{2}$ (where n is odd). In the

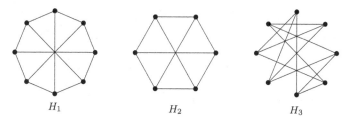

H_1 H_2 H_3

Figure 4.10 Graphs for Theorem 4.37.

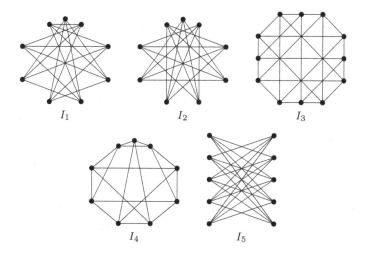

I_1 I_2 I_3

I_4 I_5

Figure 4.11 Graphs for Theorem 4.38. Adapted from [421] with permission from the publisher (Taylor & Francis Ltd, http://www.tandfonline.com).

second possibility, H reduces to an already considered regular graph. In the third possibility, we get that one of r and s is even, while n is odd, which is not possible, since both r and s must divide n.

Now, let $r = \frac{n}{2}, s = \frac{n}{2} - 2$ hold. Let further v_r (resp. v_s) denote the number of vertices of degree r (resp. s) in H. We have $n = rv_r = sv_s$, that is, $n = \frac{n}{2}v_r = (\frac{n}{2} - 2)v_s$. Hence, $v_r = 2$ holds and $v_s = \frac{2n}{n-4}$. The only solution is the graph H_3 which is obtained for $n = 8$. $\qquad\square$

The next result is proved in a similar way.

Theorem 4.38 ([421]) *Let G be a connected regular graph of degree $r_G = 4$ with the property $\lambda_2(G) \leq 1$. Then G is one of the following graphs: K_5, $\overline{C_7}$, $\overline{C_3 \cup C_4}$, $\overline{3K_2}$, Line($\overline{2K_3}$) or I_i ($1 \leq i \leq 5$) of Fig. 4.11.*

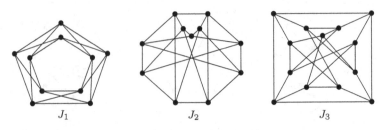

J_1 J_2 J_3

Figure 4.12 Graphs for Theorem 4.41.

Remark 4.39   Note that all r-regular graphs with the property $\lambda_2 \leq 1$, where $5 \leq r \leq 8$, are determined in [246].

Regular non-bipartite triangle-free graphs with $\lambda_2 \leq \sqrt{2}$

We present all the graphs described in the title above. The following lemma, whose proof can be found in [247], is needed.

Lemma 4.40   *The vertex degree of a regular non-bipartite triangle-free graph with the property $\lambda_2 \leq \sqrt{2}$ is at most 5.*

By the lemma, the degree r of regular non-bipartite triangle-free graphs with $\lambda_2 \leq \sqrt{2}$ satisfies $2 \leq r \leq 5$. Case $r = 2$ is simple, and the resulting graphs are C_5 and C_7. For the remaining cases we bound the number of vertices using the inequality (4.6). We also use the known fact that if G is a triangle-free graph with n vertices and minimal vertex degree δ satisfying $\delta > \frac{2}{5}n$, then G must be bipartite [140]. This gives the lower bound for the number of vertices in the non-bipartite case. Thus:

(1) if $r = 5$, then $14 \leq n \leq 22$;
(2) if $r = 4$, then $10 \leq n \leq 19$;
(3) if $r = 3$, then $8 \leq n \leq 18$.

The consideration is finished by use of a computer. All possible candidates are generated and those with $\lambda_2 > \sqrt{2}$ are eliminated. The result is summarized in the following theorem.

Theorem 4.41   ([247]) *There are exactly nine regular non-bipartite triangle-free graphs whose second largest eigenvalue does not exceed $\sqrt{2}$. These graphs are C_5, C_7, the Petersen graph, the Clebsch graph, H_1 of Fig. 4.10, I_4 of Fig. 4.11, and the graphs J_i ($1 \leq i \leq 3$) of Fig. 4.12.*

Regular bipartite reflexive graphs

Recall from Subsection 4.2.5 that a graph whose second largest eigenvalue does not exceed 2 is called reflexive. In what follows we present all regular bipartite reflexive (RBR) graphs.

Following [245], every regular bipartite graph (connected or not) of degree at most 2 is reflexive. These graphs are disjoint unions of complete graphs of order either 1 or 2, or disjoint unions of cycles of even orders. In addition, any disconnected RBR graph has degree at most 2 (since it is disconnected, its second largest eigenvalue is equal to its degree), so we can proceed to determine all connected RBR graphs. The set of all such graphs will be denoted by $\mathscr{R}$. Obviously, any graph in $\mathscr{R}$ must have an even order, and thus we may assume that any graph under consideration has $2n$ vertices. We prove the following result.

Theorem 4.42 ([245]) *Every connected regular bipartite graph with $2n$ vertices whose degree is at least $n-2$ is reflexive.*

Proof The degree r of a graph G described in the theorem is at most n, and then G is a bipartite complement of some regular bipartite graph of degree at most 2. Using the equality (4.7), we get $\lambda_2(G) = \lambda_2(\overline{G})$, and the result follows.
□

Next, if we denote

$$\mathscr{R}^* = \left\{ G \in \mathscr{R}, \, 3 \leq r_G \leq \frac{n}{2} \right\},$$

then any graph in $\mathscr{R} \backslash \mathscr{R}^*$ is either a bipartite complement of a graph from $\mathscr{R}^*$ or is necessarily reflexive. Therefore, it is sufficient to determine all graphs belonging to $\mathscr{R}^*$. We bound their vertex degree.

Theorem 4.43 ([245]) *The degree of any graph in $\mathscr{R}^*$ is at most 7.*

Proof Let G be a graph in $\mathscr{R}^*$ and assume that its vertex degree is greater than 7. Since the diameter of any connected r-regular triangle-free graph with the property $\lambda_2 \leq \sqrt{r}$ is at most 4, we have $D \leq 3$. Next, only connected regular bipartite graphs of diameter 2 are complete bipartite graphs, but they do not belong to $\mathscr{R}^*$, so it holds that $D = 3$. If G has $2n$ vertices, according to Theorem 4.8, $n \leq \frac{r^2 - \lambda_2^2}{r - \lambda_2^2}$. It is easily verified that for any fixed $r \geq 8$, the function $\frac{r^2 - \lambda_2^2}{r - \lambda_2^2}$ is positive and increases as λ_2 increases, and thus we get $2r \leq n \leq \frac{r^2 - 4}{r - 4}$. It follows that $r^2 - 8r + 4 \leq 0$ must hold, which is impossible since $r > 7$. A contradiction.
□

We now consider RBR graphs of degree 3.

Theorem 4.44 ([247]) *Every RBR graph of degree 3 has at most 30 vertices. If any such graph has exactly 30 vertices then its second largest eigenvalue must be equal to 2.*

Proof Consider an RBR graph of degree 3 with $2n \geq 30$ vertices, and let G be its bipartite complement. Then G is $(n-3)$-regular and, by Theorem 4.34, its diameter is 3. Let v be an arbitrary vertex of G, and consider the partition of the vertex set $V(G)$ with $A = \{v\}$, $B = \{u \in V(G), d(u,v) = 1\}$, $C = \{u \in V(G), d(u,v) = 2\}$, $D = \{u \in V(G), d(u,v) = 3\}$.

We have $|B| = n-3$, $|C| = n-1$, and $|D| = 3$. Also, each vertex in D has exactly $n-3$ neighbours in C, and simple counting shows that there are at least $n-7$ vertices in C adjacent to all three vertices in D. If C_1 is the set of such $n-7$ vertices of C, we denote $C_2 = C \backslash C_1$, so $|C_2| = 6$. Now the partition A, B, C_1, C_2, D induces the quotient matrix

$$\begin{pmatrix} 0 & n-3 & 0 & 0 & 0 \\ 1 & 0 & \frac{n^2-13n+42}{n-3} & \frac{6(n-5)}{n-3} & 0 \\ 0 & n-6 & 0 & 0 & 3 \\ 0 & n-5 & 0 & 0 & 2 \\ 0 & 0 & n-7 & 4 & 0 \end{pmatrix}.$$

The second largest eigenvalue of G is at least the second largest eigenvalue of the above matrix, that is, $\lambda_2(G) \geq \sqrt{\frac{5n-27}{n-3}}$. Thus, if $n > 15$ then $\lambda_2(G) > 2$, while if $n = 15$ then $\lambda_2(G) = 2$. $\qquad\square$

The determination of RBR graphs with vertex degree 3 is finished by use of a computer, and there are exactly 20 resulting graphs.

Note that the computer search can be reduced (which is not necessary) by further theoretical examinations like those in [245] (where it is proved that exactly one RBR graph of degree 3 has 30 vertices, and none of them has 28 vertices).

The remaining cases $r \in \{4,5,6,7\}$ are considered in a similar way [245]. The final result, which includes the above consideration, reads as follows.

    (i) All regular bipartite graphs satisfying either $r \leq 2$ or $r \geq n-2$ are reflexive. If an RBR graph is disconnected then its degree is at most 2.

    (ii) The set $\mathscr{R}^*$ consists of exactly 70 graphs. Precisely

       • 20 graphs of degree 3 (determined in [247]),

- 30 graphs of degree 4 (1 determined in [247], and 29 determined in [245]),
- 6 graphs of degree 5 (1 determined in [247], and 4 determined in [245]),
- 9 graphs of degree 6 (determined in [245]),
- 5 graphs of degree 7 (determined in [245]).

(iii) Inspecting the obtained graphs, we conclude that 53 bipartite complements of graphs in $\mathscr{R}^*$ do not belong to $\mathscr{R}^*$, which means that the set $\mathscr{R}$ consists of exactly 123 graphs. In other words, there is an infinite family of RBR graphs described in (i), and additional 123 individual graphs.

The graphs of $\mathscr{R}^*$ are given in Table A.3 of Section 4.3. Using this table, we single out all regular bipartite graphs with the property $\lambda_2 \leq \sqrt{2}$ and complete the search for regular triangle-free graphs with the same property started in Theorem 4.41.

Theorem 4.45 ([247]) *There are two infinite families of connected regular triangle-free graphs with the property $\lambda_2 \leq \sqrt{2}$ ($K_{r,r}$ ($r \geq 1$) and $\overline{rK_2}$ ($r \geq 3$)) and exactly 12 additional graphs: C_8, the Heawood graph, its bipartite complement, and all graphs from Theorem 4.41.*

We conclude with the following remark.

Remark 4.46 A graph is called *integral* if its spectrum consists entirely of integers. All connected cubic integral graphs are determined in [63] and separately in [396] (there are 13 such graphs), while all connected 4-regular integral graphs avoiding ± 3 in the spectrum are determined in [428] (there are 24 such graphs).

Since the Cartesian product of a regular non-bipartite integral graph and K_2 is a regular bipartite integral graph of the same vertex degree, the essential parts in [396, 428] were the determinations of regular bipartite integral graphs (of the corresponding vertex degree). Hence, exactly eight resulting connected cubic bipartite integral graphs and exactly 16 resulting connected 4-regular bipartite integral graphs avoiding ± 3 in the spectrum were determined.

We observe that all these regular bipartite integral graphs are reflexive. Since we have just determined all RBR graphs, they can be singled out from this set of graphs. $\square$

4.3 Appendix

This section contains the following tables of graphs.

Table A.1 – Graphs of $\mathcal{N}_1$.
Here $\mathcal{N}_1$ denotes the set of NSGs such that any NSG with $\lambda_2 < 1$ either belongs to $\mathcal{N}_1$ or is an induced subgraph of at least one NSG of $\mathcal{N}_1$. These graphs are considered in Subsection 4.2.3 (pages 121–123). We use the notation introduced on page 122. An asterisk stands for any positive integer, unless it denotes the last parameter which must be greater than 1.

Table A.2 – NSGs with $\lambda_2 = 1$.
These graphs are considered in the same subsection and we use the same notation.

Table A.3 – RBR graphs which belong to $\mathcal{R}^*$.
These graphs are considered in Subsection 4.2.6 (pages 131–133), while the set $\mathcal{R}^*$ is defined on page 131.

Each row of Table A.3 contains the order, the graph representation, and the non-negative part of its spectrum. We suppress the data of any graph whose bipartite complement also belongs to $\mathcal{R}^*$; the identifications of such graphs are given in the same row.

Following [245, 247], we use the LCF notation to represent all reflexive bipartite cubic graphs in this table, while for the other RBR graphs we use the hexadecimal notation. Here are the descriptions of these notations.

If G is a cubic Hamiltonian graph then its vertices can be arranged in a cycle C which accounts for two edges per arbitrary vertex v. The third edge vu can be described by the length l of the path vu in C with a plus sign if we turn clockwise or a minus sign if we turn counterclockwise along C. If the pattern of the LCF notation repeats, it is indicated by a superscript in the notation. If the second half of the numbers of the LCF notation is the reverse of the first half, but with all the signs changed, then it is replaced by a semicolon and a dash.

For the hexadecimal notation, we assume that the adjacency matrix of each graph obtained has the form (1.9) and then the binary integer obtained by concatenating the rows of the submatrix B is expressed as a hexadecimal integer (in some cases this integer is given in more than one line).

In Tables A.1 and A.2 the graphs are ordered lexicographically with respect to the number of vertices in the corresponding cells. In Table A.3 we use the lexicographical ordering with respect to the graph order and spectrum.

Table A.1 is given in a slightly different form in [420]. Table A.2 is taken from [321], while Table A.3 is taken from [245].

Table A.1 Graphs of $\mathcal{N}_1$. Adapted from [420] with permission from Elsevier.

G	a_1	a_2	a_3	a_4	a_5	a_6	a_7	a_8	a_9	a_{10}
1.	1	3	3							
2.	3	1	7							
3.	5	1	5							
4.	1	3	2	*						
5.	1	4	1	*						
6.	1	5	1	4						
7.	1	7	1	2						
8.	2	2	2	*						
9.	2	3	1	*						
10.	2	4	1	4						
11.	3	1	6	1						
12.	3	2	2	2						
13.	3	3	1	12						
14.	4	1	4	1						
15.	4	3	1	8						
16.	5	3	1	7						
17.	6	1	3	1						
18.	7	1	3	8						
19.	8	1	3	4						
20.	9	1	3	3						
21.	9	3	1	6						
22.	1	1	1	1	7					
23.	1	1	3	1	5					
24.	1	1	*	1	4					
25.	1	1	*	2	3					
26.	1	2	*	1	3					
27.	1	3	1	*	2					
28.	2	1	*	1	3					
29.	3	1	3	1	3					
30.	3	1	5	*	2					
31.	5	1	1	1	3					
32.	5	1	3	*	2					
33.	1	1	1	1	6	*				
34.	1	1	1	2	2	2				
35.	1	1	1	3	1	12				
36.	1	1	2	1	4	*				
37.	1	1	2	3	1	8				
38.	1	1	3	3	1	7				
39.	1	1	4	1	3	*				
40.	1	1	5	1	3	8				
41.	1	1	6	1	3	4				
42.	1	1	7	1	3	3				
43.	1	1	7	3	1	6				
44.	1	1	11	1	3	2				
45.	1	1	*	3	1	5				
46.	1	1	*	4	1	3				
47.	1	1	*	6	1	2				
48.	1	2	1	1	2	*				
49.	1	2	*	2	1	*				
50.	1	2	*	3	1	4				
51.	1	2	*	5	1	2				
52.	1	3	1	1	1	4				

(continued on next page)

Table A.1 (continued).

G	a_1	a_2	a_3	a_4	a_5	a_6	a_7	a_8	a_9	a_{10}
53.	1	3	1	3	1	2				
54.	2	1	*	1	2	*				
55.	2	1	*	2	1	*				
56.	2	1	*	3	1	4				
57.	2	1	*	5	1	2				
58.	2	2	1	1	1	4				
59.	2	2	1	3	1	2				
60.	3	1	1	1	2	2				
61.	3	1	1	2	1	10				
62.	3	1	2	2	1	8				
63.	3	1	3	2	1	6				
64.	3	1	3	3	1	3				
65.	3	1	3	5	1	2				
66.	3	1	4	1	1	*				
67.	3	1	4	2	1	4				
68.	3	1	4	4	1	2				
69.	3	1	5	3	1	2				
70.	3	2	1	1	1	2				
71.	4	1	1	2	1	6				
72.	4	1	2	1	1	*				
73.	4	1	3	1	1	4				
74.	4	1	3	3	1	2				
75.	5	1	1	2	1	5				
76.	5	1	1	3	1	3				
77.	5	1	1	5	1	2				
78.	5	1	2	1	1	8				
79.	5	1	2	2	1	3				
80.	5	1	2	4	1	2				
81.	6	1	1	1	1	*				
82.	6	1	2	1	1	4				
83.	7	1	1	1	1	36				
84.	7	1	2	1	1	3				
85.	7	1	2	3	1	2				
86.	8	1	1	1	1	20				
87.	9	1	1	1	1	15				
88.	9	1	1	2	1	4				
89.	9	1	2	2	1	2				
90.	10	1	1	1	1	12				
91.	11	1	1	1	1	11				
92.	11	1	2	1	1	2				
93.	12	1	1	1	1	10				
94.	13	1	1	1	1	9				
95.	16	1	1	1	1	8				
96.	21	1	1	1	1	7				
97.	37	1	1	1	1	6				
98.	1	1	1	1	3	1	3			
99.	1	1	1	1	5	*	2			
100.	1	1	3	1	1	1	3			
101.	1	1	3	1	3	*	2			
102.	1	1	*	1	2	*	2			
103.	1	1	*	2	1	*	2			
104.	1	2	*	1	1	*	2			

(continued on next page)

Table A.1 (continued).

G	a_1	a_2	a_3	a_4	a_5	a_6	a_7	a_8	a_9	a_{10}
105.	2	1	*	1	1	*	2			
106.	3	1	3	1	1	*	2			
107.	5	1	1	1	1	*	2			
108.	1	1	1	1	1	1	2	2		
109.	1	1	1	1	1	2	1	10		
110.	1	1	1	1	2	2	1	8		
111.	1	1	1	1	3	2	1	6		
112.	1	1	1	1	3	3	1	3		
113.	1	1	1	1	3	5	1	2		
114.	1	1	1	1	4	1	1	*		
115.	1	1	1	1	4	2	1	4		
116.	1	1	1	1	4	4	1	2		
117.	1	1	1	1	5	1	1	4		
118.	1	1	1	1	5	3	1	2		
119.	1	1	1	2	1	1	1	2		
120.	1	1	2	1	1	2	1	6		
121.	1	1	2	1	2	1	1	*		
122.	1	1	2	1	2	2	1	4		
123.	1	1	2	1	3	1	1	4		
124.	1	1	2	1	3	3	1	2		
125.	1	1	3	1	1	2	1	5		
126.	1	1	3	1	1	3	1	3		
127.	1	1	3	1	1	5	1	2		
128.	1	1	3	1	2	1	1	8		
129.	1	1	3	1	2	2	1	3		
130.	1	1	3	1	2	4	1	2		
131.	1	1	4	1	1	1	1	*		
132.	1	1	4	1	2	1	1	4		
133.	1	1	5	1	1	1	1	36		
134.	1	1	5	1	2	1	1	3		
135.	1	1	5	1	2	3	1	2		
136.	1	1	6	1	1	1	1	20		
137.	1	1	7	1	1	1	1	15		
138.	1	1	7	1	1	2	1	4		
139.	1	1	7	1	2	2	1	2		
140.	1	1	8	1	1	1	1	12		
141.	1	1	9	1	1	1	1	11		
142.	1	1	9	1	2	1	1	2		
143.	1	1	10	1	1	1	1	10		
144.	1	1	11	1	1	1	1	9		
145.	1	1	14	1	1	1	1	8		
146.	1	1	19	1	1	1	1	7		
147.	1	1	35	1	1	1	1	6		
148.	1	1	*	1	1	1	1	5		
149.	1	1	*	1	1	2	1	3		
150.	1	1	*	1	1	4	1	2		
151.	1	2	*	1	1	1	1	4		
152.	1	2	*	1	1	3	1	2		
153.	2	1	*	1	1	3	1	2		
154.	3	1	1	1	1	1	1	2		
155.	1	1	1	1	3	1	1	*	2	
156.	1	1	3	1	1	1	1	*	2	
157.	1	1	1	1	1	1	1	1	1	2

Table A.2 NSGs with $\lambda_2 = 1$ (cf. [321]).

G	a_1	a_2	a_3	a_4	a_5	a_6	a_7	a_8	a_9
1.	3	1	8						
2.	1	2	7						
3.	4	1	6						
4.	6	1	5						
5.	1	3	4						
6.	2	2	4						
7.	1	4	3						
8.	2	3	3						
9.	14	1	3	2					
10.	10	1	3	3					
11.	8	1	3	5					
12.	7	1	3	9					
13.	5	1	4	2					
14.	4	2	2	2					
15.	3	2	2	3					
16.	1	8	1	2					
17.	2	7	1	2					
18.	1	6	1	3					
19.	1	5	1	5					
20.	2	5	1	3					
21.	2	4	1	5					
22.	3	4	1	4					
23.	3	3	1	13					
24.	4	3	1	9					
25.	6	3	1	7					
26.	10	3	1	6					
27.	1	4	1	*	2				
28.	1	3	2	*	2				
29.	2	3	1	*	2				
30.	1	2	*	1	4				
31.	1	2	*	2	3				
32.	2	2	2	*	2				
33.	1	1	1	1	8				
34.	1	1	2	1	6				
35.	1	1	4	1	5				
36.	2	1	*	1	4				
37.	2	1	*	2	3				
38.	6	1	1	1	3				
39.	4	1	2	1	3				
40.	3	1	4	1	3				
41.	3	1	6	*	2				
42.	4	1	4	*	2				
43.	6	1	3	*	2				
44.	1	3	1	4	1	2			
45.	1	3	1	2	1	3			
46.	1	3	1	1	1	5			
47.	3	2	1	2	1	2			
48.	2	2	1	4	1	2			
49.	2	2	1	2	1	3			
50.	2	2	1	1	1	5			
51.	1	2	*	6	1	2			

(continued on next page)

Table A.2 (continued).

G	a_1	a_2	a_3	a_4	a_5	a_6	a_7	a_8	a_9
52.	1	2	*	4	1	3			
53.	1	2	*	3	1	5			
54.	2	1	*	6	1	2			
55.	6	1	1	5	1	2			
56.	4	1	2	5	1	2			
57.	3	1	4	5	1	2			
58.	1	1	1	4	1	4			
59.	2	1	*	4	1	3			
60.	3	1	5	4	1	2			
61.	4	1	3	4	1	2			
62.	6	1	2	4	1	2			
63.	1	1	1	3	1	13			
64.	1	1	2	3	1	9			
65.	1	1	4	3	1	7			
66.	1	1	8	3	1	6			
67.	2	1	*	3	1	5			
68.	6	1	1	3	1	3			
69.	4	1	2	3	1	3			
70.	3	1	4	3	1	3			
71.	8	1	2	3	1	2			
72.	1	1	2	2	2	2			
73.	1	1	1	2	2	3			
74.	3	1	1	2	1	11			
75.	3	1	2	2	1	9			
76.	3	1	3	2	1	7			
77.	3	1	4	2	1	5			
78.	3	1	5	2	1	3			
79.	4	1	1	2	1	7			
80.	4	1	2	2	1	5			
81.	4	1	3	2	1	3			
82.	6	1	1	2	1	5			
83.	6	1	3	2	1	3			
84.	10	1	1	2	1	4			
85.	10	1	2	2	1	2			
86.	1	1	12	1	3	2			
87.	1	1	8	1	3	3			
88.	1	1	6	1	3	5			
89.	1	1	5	1	3	9			
90.	3	1	2	1	2	2			
91.	12	1	2	1	1	2			
92.	8	1	2	1	1	3			
93.	3	1	5	1	1	5			
94.	4	1	3	1	1	5			
95.	6	1	2	1	1	5			
96.	38	1	1	1	1	6			
97.	22	1	1	1	1	7			
98.	14	1	1	1	1	9			
99.	5	1	2	1	1	9			
100.	10	1	1	1	1	13			
101.	7	1	1	1	1	37			
102.	8	1	1	1	1	21			
103.	2	1	*	2	1	*	2		

(continued on next page)

Table A.2 (continued).

G	a_1	a_2	a_3	a_4	a_5	a_6	a_7	a_8	a_9
104.	1	2	*	2	1	*	2		
105.	1	2	*	1	2	*	2		
106.	1	1	4	1	1	1	3		
107.	1	1	2	1	2	1	3		
108.	1	1	1	1	4	1	3		
109.	6	1	1	1	1	1	2		
110.	4	1	2	1	1	*	2		
111.	3	1	4	1	1	*	2		
112.	2	1	*	1	2	*	2		
113.	1	1	1	1	6	*	2		
114.	1	1	2	1	4	*	2		
115.	1	1	4	1	3	*	2		
116.	1	2	*	1	1	4	1	2	
117.	1	2	*	1	1	2	1	3	
118.	1	2	*	1	1	1	1	5	
119.	1	1	1	2	1	2	1	2	
120.	1	1	1	1	2	1	2	2	
121.	2	1	*	1	1	4	1	2	
122.	2	1	*	1	1	2	1	3	
123.	2	1	*	1	1	1	1	5	
124.	1	1	4	1	1	5	1	2	
125.	1	1	2	1	2	5	1	2	
126.	1	1	1	1	4	5	1	2	
127.	1	1	1	1	5	4	1	2	
128.	1	1	2	1	3	4	1	2	
129.	1	1	4	1	2	4	1	2	
130.	1	1	1	1	4	3	1	3	
131.	1	1	2	1	2	3	1	3	
132.	1	1	6	1	2	3	1	2	
133.	1	1	4	1	1	3	1	3	
134.	1	1	1	1	5	2	1	3	
135.	1	1	1	1	4	2	1	5	
136.	1	1	2	1	3	2	1	5	
137.	1	1	1	1	3	2	1	7	
138.	1	1	8	1	2	2	1	2	
139.	1	1	4	1	2	2	1	3	
140.	1	1	2	1	2	2	1	5	
141.	1	1	1	1	2	2	1	9	
142.	1	1	1	1	1	2	1	11	
143.	1	1	2	1	1	2	1	7	
144.	1	1	4	1	1	2	1	5	
145.	1	1	8	1	1	2	1	4	
146.	1	1	1	1	5	1	1	8	
147.	1	1	2	1	3	1	1	5	
148.	1	1	3	1	2	1	1	9	
149.	1	1	4	1	2	1	1	5	
150.	1	1	6	1	2	1	1	3	
151.	1	1	10	1	2	1	1	2	
152.	1	1	5	1	1	1	1	37	
153.	1	1	6	1	1	1	1	21	
154.	1	1	8	1	1	1	1	13	

(continued on next page)

Table A.2 (continued).

G	a_1	a_2	a_3	a_4	a_5	a_6	a_7	a_8	a_9
155.	1	1	12	1	1	1	1	9	
156.	1	1	20	1	1	1	1	7	
157.	1	1	36	1	1	1	1	6	
158.	1	1	4	1	1	1	1	1	2
159.	1	1	2	1	2	1	1	1	2
160.	1	1	1	1	4	1	1	1	2

Table A.3 RBR graphs belonging to $\mathscr{R}^*$. Reprinted from [245] with permission from Elsevier.

Gr.	$2n$	Notation	Spectrum (non-negative part)
G_1	12	$[5,-5]^6$	$3, 1.73^2, 1^3$
G_2	12	$[-5,-5,3,-5,-5,-3;-]$	$3, 2, 1.41^2, 1, 0^2$
G_3	12	$[-3,3]^6$	$3, 2^2, 1, 0^4$
G_4	14	$[5,-5]^7$	$3, 1.41^6$
G_5	14	$[7,7,-5,3,5,7,-3]^2$	$3, 1.93^2, 1.41^2, 0.52^2$
G_6	14	$[5,-3,5,7,5,-5,5;-]$	$3, 2, 1.41^4, 0^2$
G_7	16	$[5,-5]^8$	$3, 1.73^4, 1^3$
G_8	16	$[-5,7,-5,7,-5,7,3,7;-]$	$3, 2, 1.73^2, 1.41^2, 1, 0^2$
G_9	16	$[7,-5,3,-5;-]^2$	$3, 2^2, 1.73^2, 1, 0^4$
G_{10}	18	$[5,7,-7;-]^3$	$3, 1.73^6, 0^4$
G_{11}	18	$[5,-5]^9$	$3, 1.97^2, 1.73^2, 1.29^2, 0.68^2$
G_{12}	18	$[5,7,-5,7,9,-5,5,7,-7;-]$	$3, 2, 1.73^4, 1^2, 0^2$
G_{13}	20	$[-9,5,-7,7,-7,7,-5,7,-9,7;-]$	$3, 2^2, 1.88^2, 1.53^2, 1, 0.35^2$
G_{14}	20	$[5,9,-9,5,-9,-5,-9,5,-5,9;-]$	$3, 2^2, 1.88^2, 1.53^2, 1, 0.35^2$
G_{15}	20	$[-5,9,-9,5,-5,7,-9,7,-5,9;-]$	$3, 2^3, 1.73^2, 1^3, 0^2$
G_{16}	20	$[5,9,-5,9,-9,-5,5,9,5,9;-]$	$3, 2^3, 1.73^2, 1^3, 0^2$
G_{17}	20	$[5,-5,9,-9]^5$	$3, 2^4, 1^5$
G_{18}	20	$[5,-9,-7,7,-7,-5,5,-9,5,7;-]$	$3, 2^4, 1^5$
G_{19}	24	$[5,-9,7;-]^4$	$3, 2^6, 1^3, 0^4$
G_{20}	30	$[-13,-9,7;-]^5$	$3, 2^9, 0^{10}$
G_{21}	16	F03A2DA3CC954B56	$4, 2, 1.85^2, 1.41^2, 0.77^2$
G_{22}	16	F01E99536CC52BA6	$4, 2^2, 1.41^4, 0^2$
G_{23}	16	F08E552D66A3995A	$4, 2^2, 1.73^2, 1^2, 0^2$
G_{24}	16	F02D1E6A93C559A6	$4, 2^2, 1.73^2, 1^2, 0^2$
G_{25}	16	F00F9C35A95366CA	$4, 2^3, 1.41^2, 0^4$
G_{26}	16	F02B1E6CC59359A6	$4, 2^3, 1.41^2, 0^4$
G_{27}	16	F00FAA63C9559C36	$4, 2^4, 0^6$
G_{28}	16	F03C0F5AC3A56996	$4, 2^4, 0^6$
G_{29}	18	1E03A50D0EA9171663594	$4, 2^2, 1.88^2, 1.53^2, 0.35^2$
G_{30}	18	1E03A48D0F19270E99394	$4, 2^3, 1.73^2, 1^2, 0^2$
G_{31}	18	1E0662CA172517560A695	$4, 2^3, 1.73^2, 1^2, 0^2$
G_{32}	18	1E092E609C781B44D8A8B	$4, 2^4, 1^4$
G_{33}	18	1E019D15B0A51E42C9D92	$4, 2^4, 1^4$
G_{34}	18	1E02C8B93847A9495638A	$4, 2^4, 1^4$
G_{35}	20	F00B1239544AA9211E251E309	$4, 2^5, 0^4$
G_{36}	20	F00B1239524969411E2A1E309	$4, 2^5, 0^4$
G_{37}	20	F007111F14A0CD82B10D4CA2A	$4, 2^5, 0^4$
G_{38}	20	F030A170A925A19899855C063	$4, 2^5, 0^4$
G_{39}	24	F0024C43448A13829154180D2231868528E0	$4, 2^5, 1.41^6$
G_{40}	24	F0084903518393029860514C22A8864524E0	$4, 2^6, 1.41^4, 0^2$
G_{41}	24	F0004D43448A138291541A0C2231868528E0	$4, 2^6, 1.41^4, 0^2$
G_{42}	24	F008C423480B8311A82C150519246245820E	$4, 2^6, 1.41^4, 0^2$
G_{43}	24	F0014903518393029860584C22A8864524E0	$4, 2^7, 1.41^2, 0^4$
G_{44}	24	F000C523480B8311A8AC050519246245820E	$4, 2^7, 1.41^2, 0^4$
G_{45}	24	F0032229814544AC841892540A6831413868	$4, 2^8, 0^6$
G_{46}	24	F00322294149446C8811C2C10AA831413864	$4, 2^8, 0^6$

(continued on next page)

Table A.3 (continued)

Gr.	$2n$	Notation	Spectrum
G_{47}	26	1281CA03280CA03280CA0 3280CB03240CD03140C503	$4, 1.73^{12}$
G_{48}	28	F0009424C408A411C05884130819 88A0606408E10582482C1	$4, 2^7, 1.41^6$
G_{49}	30	1E000501844641A02610851262024 2900B060C4340421512048248131	$4, 2^{10}, 1^4$
G_{50}	32	F000005C818812094025108684112212 09038062489025041C4006A028284340	$4, 2^{12}, 0^6$
G_{51}	22	189789389B88B88B8CB84B84B84BC4B	$5, 1.73^{10}$
G_{52}	24	F808A94A62CA19554C61313A943A1CC70265	$5, 2^8, 1^3$
G_{53}	24	F80C52A499158AA1CC3230F148766453821E	$5, 2^8, 1^3$
G_{54}	24	F80C52A499158AA1CC1632B148766453821E	$5, 2^8, 1^3$
G_{55}	24	F80C0EA158C39381CC1274B22A955166425A	$5, 2^8, 1^3$
G_{56}	42	1F0000802C1400836090A10B020410A22158 01000662831022094040B80821284844110825 0490604C1021C0061061210A8091111011901	$5, 2^{20}$
$G_{57}(\overline{G_{58}})$	24	FC065615DA2738E6693B18F44BA8CBD2CD13	$6, 2^9, 0^4$
$G_{59}(\overline{G_{60}})$	24	FC0C5C88F2751E6969B13595E266A953A2DA	$6, 2^9, 0^4$
$G_{61}(\overline{G_{62}})$	24	FC0D1C15BA2D3876553B84E9D234B68CEA72	$6, 2^9, 0^4$
G_{63}	32	FC00303A0C6318D48A8A492C160D2B1 1A26446B0258691A18558C017534260C9	$6, 2^{15}$
G_{64}	32	FC00302B17A0921C18C55152A1C82474 2B118926450DC0B10C9A4A6862868643	$6, 2^{15}$
G_{65}	32	FC00300F1B48C22C907151A2256428B8 48C5451916940E23891684CAA3816252	$6, 2^{15}$
G_{66}	30	1FC02A4CD8744DE03658B270D69C8 4EC4E98E2AA1D87590D4CD51E8CA	$7, 2^{14}$
G_{67}	30	1FC02525AAB1263B32729532656C3 8B14CCCE91C65525C1FA383C348B	$7, 2^{14}$
G_{68}	30	1FC0154D1A719878C47562399A6C5 4B85CAA4987D40E57272692734E8	$7, 2^{14}$
G_{69}	30	1FC00AE932B8B0368C5CCD2B44D8 79314F29330E711D1D928D4B568D4	$7, 2^{14}$
G_{70}	30	1FC014EA292B878657A26534C594D 629938D39433CE0CA6A34A8BF09A	$7, 2^{14}$

Some of the graphs presented in Table A.3 appear in the literature. For example, G_1, G_4, G_7, G_{10}, G_{17}, G_{19}, and G_{20} are respectively known as the Franklin, Heawood, Möbius–Kantor, Pappus, Desargues, Nauru, and Tutte–Coxeter graphs.

Exercises

4.1 ([248]) Let G be a connected r-regular bipartite graph with n vertices whose second largest eigenvalue satisfies the property $\lambda_2^4 < r$. Prove that $2(r + \lambda_2) \leq n \leq 2(r + \lambda_2^2)$. If $n = 2(r + \lambda_2^2)$, then prove $r \leq \lambda_2^2(\lambda_2^2 - 1)^2$.

4.2 ([200]) Given an r-regular graph G with n vertices, if H denotes its induced subgraph with s vertices and the average vertex degree $\overline{d}_H$, then show that

$$\overline{d}_H - \frac{rs}{n} \leq \lambda_2(G)\frac{n-s}{n}. \tag{4.18}$$

Hint: Use Lemma 4.12.

4.3 Show that $\lambda_2(T) \leq \sqrt{\frac{n-3}{2}}$ does not hold for every tree T of even order n.

4.4 Prove Lemma 4.30.

4.5 ([246]) Let G be a connected r-regular ($r \geq 4$) graph on $n = 2r$ vertices with the property $\lambda_2 \leq 1$. Prove that

(i) if $r \in \{6,7\}$, then $G \cong \overline{2K_r}$,
(ii) if $r = 6$, then $G \cong \overline{2K_6}$ or $G \cong \overline{\text{Line}(K_{3,4})}$,
(iii) if $r = 7$, then $G \cong \overline{2K_7}$ or $G \cong \overline{\text{Line}(H)}$, where H is any of the two 4-regular graphs with 7 vertices.

4.6 Prove Theorem 4.38.

4.7 [246] Show that $\overline{\text{Line}(K_{2,r+1})}$ is the unique r-regular graph with n vertices and the property $\lambda_2 \leq 1$, whenever $n > 2r > 34$.

Hint: Show that $\overline{G}$ is connected and that cannot be either a cocktail party graph, a regular exceptional graph or a line graph of a regular graph. By Theorem 3.15, it remains to consider the line graphs of semiregular bipartite graphs as possible candidates for $\overline{G}$.

Notes

Alon and Milman [10] have proved several inequalities for λ_2 which are related to the girth, connectivity and diameter of a graph.

In relation to regular graphs, we note a result of Nilli [348]. If G is a connected r-regular graph which contains two edges whose distance apart is at least $2k + 2$, then

$$\lambda_2(G) \geq 2\sqrt{r-1}\left(1 - \frac{1}{k+1}\right) + \frac{1}{k+1}. \tag{4.19}$$

It is also worthwhile adding that the cubic graphs for which the second largest eigenvalue is maximal are identified in [53]. For each even n, there is a unique such graph described in the corresponding reference and also in [102, p. 66].

The second largest eigenvalue (in modulus) of regular graphs is investigated in [6, 82, 130]. An upper bound for the diameter of a regular graph in terms of the second largest eigenvalue in modulus was obtained by Chung [82]. The relevance of expanders to theoretical computer science is discussed in [6, 7]. In the latter paper they are put in connection with parallel sorting networks.

A sequence of upper bounds for the second largest eigenvalue of symmetric matrices with non-negative entries was obtained recently by Kolotilina [249]. Each of them can be considered as the upper bound for the second largest eigenvalue of a graph.

All double nested graphs with $\lambda_2 \leq \sqrt{2}$ are determined in [21].

The second largest eigenvalue plays an important role in the ordering of regular graphs (of the same vertex degree). For example, a classical result is the ordering of cubic graphs given in [62].

5

Other eigenvalues of the adjacency matrix

In Section 5.1 we give the upper and lower bounds that apply to any eigenvalue of the adjacency matrix. Subsequently, we consider two additional topics. In Section 5.2 we determine all graphs with negative third largest eigenvalue and give some inequalities for the remaining eigenvalues of these graphs. In Section 5.3 we determine all graphs G such that both G and its complementary graph $\overline{G}$ have second least eigenvalue at least -1.

5.1 Bounds for λ_i

We do not recommend the inequalities obtained in this section be applied to λ_1 or λ_n. One can find better results in Chapters 2 and 3.

We start with the following result.

Theorem 5.1 (Brigham and Dutton [56]) *For a graph G with n vertices let n^+ and n^- be the number of positive and negative eigenvalues, respectively. Then, for $1 \le i \le n$,*

$$-\sqrt{\frac{2mn^+}{(n-i+1)(n-i+n^++1)}} \le \lambda_i \le \sqrt{\frac{2mn^-}{i(i+n^-)}}. \quad (5.1)$$

Proof The right inequality is valid for any $\lambda_i \le 0$. From $\sum_{j=1}^{n} \lambda_j^2 = 2m$, we get

$$\lambda_i^2 = 2m - \sum_{\lambda_j>0, j \ne i} \lambda_j^2 - \sum_{\lambda_j<0} \lambda_j^2.$$

146

Since $\sum_{j=1}^{n} \lambda_j = 0$, the right-hand side is maximized for $\lambda_1 = \lambda_2 = \cdots = \lambda_i$, and $\lambda_j < 0$, $\lambda_j = -\frac{i\lambda_i}{n^-}$, so

$$\lambda_i^2 \leq 2m - (i-1)\lambda_i^2 - \frac{i\lambda_i^2}{n^-},$$

which gives the assertion.

The left inequality is considered in a similar way. □

Example 5.2 For any path P_n, we have $m = n - 1$ and $n^+ = n^- = \lfloor \frac{n}{2} \rfloor$. We compute the eigenvalues of P_8 and the sequences of lower and upper bounds of (5.1):

i	1	2	3	4	5	6	7	8
Lower bound	-0.775	-0.845	-0.930	-1.035	-1.168	-1.342	-1.581	-1.937
λ_i	1.879	1.532	1.000	0.347	-0.347	-1.000	-1.532	-1.879
Upper bound	3.873	2.536	1.937	1.581	1.342	1.168	1.035	0.930

□

From the last theorem we get that $\lambda_i > 0$ (resp. $\lambda_i < 0$) implies $n^- \leq n - i$ (resp. $n^+ \leq i - 1$). Thus we have the following corollary.

Corollary 5.3 ([56]) *For a graph G with n vertices and $1 \leq i \leq n$,*

$$-\sqrt{\frac{2m(i-1)}{n(n-i+1)}} \leq \lambda_i \leq \sqrt{\frac{2m(n-i)}{in}}. \tag{5.2}$$

If a connected graph G is not a tree then we have $n \leq m$, and since $\frac{n-i}{n}$ increases as n increases, using the upper bound of (5.1) we get

$$\lambda_i < \sqrt{2\frac{m-i}{i}}, \quad \text{for } i \leq \frac{n}{2} \text{ and } n \leq m. \tag{5.3}$$

The upper bound of (5.2) gives a better estimate than (5.3), but in the latter λ_i is controlled only by m.

Remark 5.4 In 1989, Powers [370] gave a similar upper bound for λ_i ($1 \leq i \leq \frac{n}{2}$) of connected graphs, just in terms of the graph order:

$$\lambda_i \leq \begin{cases} \dfrac{n-i}{i}, & \text{if } i \text{ divides } n, \\[2mm] \left\lfloor \dfrac{n}{i} \right\rfloor, & \text{otherwise.} \end{cases}$$

This inequality evidently holds for $i \leq 2$ (if necessary, for λ_2 see (4.1)). Recently, Nikiforov (private communication) obtained a sequence of graphs with the property $\lambda_i > \left\lfloor \frac{n}{i} \right\rfloor$ for all $i \geq 5$, and disproved the above result. At this moment, it is not known if the upper bound of Powers holds for $i \in \{3,4\}$. $\square$

Considering spectral moments, we get another pair of bounds for λ_i. Since for $\lambda_i \leq 0$

$$M_{2k} = \sum_{j=1}^{n} \lambda_j^{2k} \geq \lambda_1^{2k} + \sum_{j=i}^{n} \lambda_j^{2k} \geq (n-i+2)\lambda_i^{2k},$$

and similarly for $\lambda_i \geq 0$

$$M_{2k} = \sum_{j=1}^{n} \lambda_j^{2k} \geq \sum_{j=1}^{i} \lambda_j^{2k} \geq i\lambda_i^{2k},$$

we get the following result.

Theorem 5.5 ([56]) *For a graph G with spectral moments M_k ($k \geq 1$),*

$$-\sqrt[2k]{\frac{M_{2k}}{(n-i+2)}} \leq \lambda_i \leq \sqrt[2k]{\frac{M_{2k}}{i}}. \tag{5.4}$$

There are two results concerning trees. The first was obtained by Shao [401],

$$\lambda_i(T) \leq \sqrt{\left\lfloor \frac{n}{i} \right\rfloor - 1} \quad \left(1 \leq i \leq \frac{n}{2}\right). \tag{5.5}$$

The second result is due to Hong [217],

$$\lambda_i(T) \leq \sqrt{\left\lfloor \frac{n-2}{i} \right\rfloor} \quad \left(2 \leq i \leq \frac{n}{2}\right). \tag{5.6}$$

Observe that for $i \geq 3$, the former result gives at least an equal estimate.

5.2 Graphs with $\lambda_3 < 0$

Let $DNG(m_1, m_2, \ldots, m_h; n_1, n_2, \ldots, n_h)$ be a double nested graph as defined on page 7.

The hereditary property $\lambda_3 < 0$ yields the existence of minimal graphs that do not satisfy this condition. By direct computation we check that there are exactly three such graphs with at most five vertices: $3K_1, C_4$, and C_5. Bearing in mind that any DNG is a bipartite graph which does not contain any of the graphs P_4, C_4 or $2K_2$ as an induced subgraph, we easily verify the following fact. If a graph G does not contain as an induced subgraph any of the graphs $3K_1, C_4$ or C_5, then its complement is either $DNG(m_1, m_2, \ldots, m_h; n_1, n_2, \ldots, n_h)$ or $DNG(m_1, m_2, \ldots, m_h; n_1, n_2, \ldots, n_h) \cup pK_1$, where $h, p \geq 1$. Denote by DNG_h the set of DNGs with $2h$ cells and by DNG_h^+ the set of graphs obtained as a union of a DNG with $2h$ cells and a non-empty set of isolated vertices.

Thus, to determine all the graphs in the title, we need to check whether the complement of a graph belonging to $DNG_h \cup DNG_h^+$ has the property $\lambda_3 < 0$. This is done by means of a computer search and the result reads as follows.

Theorem 5.6 (Petrović [357]) *A graph G has the property* $\lambda_3 < 0$ *if and only if* $\overline{G}$ *is an induced subgraph of one of the graphs listed below or* $\overline{G}$ *is an induced subgraph of one of the 323 graphs with 12 vertices belonging respectively to* DNG_2 *(there are 10 such graphs),* DNG_2^+ *(25 graphs),* DNG_3 *(69 graphs),* DNG_3^+ *(74 graphs),* DNG_4 *(80 graphs),* DNG_4^+ *(40 graphs),* DNG_5 *(20 graphs),* DNG_5^+ *(4 graphs), and* DNG_6 *(1 graph).*

    (i) $DNG(1, m_2, 1, 1; 1, n_2, 1, 1) \cup pK_1$,

    (ii) $DNG(m_1, 1, 1, m_4; 1, 1, n_3, 1)$,

    (iii) $DNG(2, 2, 1; 1, n_2, 1) \cup pK_1$,

    (iv) $DNG(m_1, 1, 1; 1, 1, 2) \cup K_1$,

    (v) $DNG(m_1, 1, 2; 1, n_2, 1) \cup K_1$,

    (vi) $DNG(2, 1, 1; 2, 1, 1) \cup pK_1$,

    (vii) $DNG(2, 2, 1; 1, 1, n_3)$,

    (viii) $DNG(1, 2, 2; 1, n_2, 1)$,

    (ix) $DNG(2, 1, m_3; 2, 1, n_3)$,

    (x) $DNG(m_1, 1, m_3; 1, 2, 2)$,

    (xi) $DNG(m_1, 2; 1, 2) \cup 2K_1$,

    (xii) $DNG(3, 2; 2, 1) \cup pK_1$,

    (xiii) $DNG(m_1, 1; 2, 3) \cup K_1$,

    (xiv) $DNG(1, m_2; 2, 3) \cup K_1$,

    (xv) $DNG(3, 2; 2, n_2)$.

Consider now the behaviour of the remaining negative eigenvalues of these graphs.

Theorem 5.7 (Liu and Bo [276]) *For a graph G with at least five vertices and $\lambda_3(G) < 0$,*

$$
\begin{aligned}
-1 \leq \lambda_i(G) < 0, & \quad \textit{for } 3 \leq i \leq \tfrac{n+1}{2}, \\
\lambda_i(G) \leq -1, & \quad \textit{for } \tfrac{n+3}{2} \leq i \leq n.
\end{aligned}
\tag{5.7}
$$

In addition, $\lambda_n \geq -1 - \sqrt{\binom{n}{2} - m}$, with the last equality if G is a complete graph.

Proof Set $\lambda_i(G) = 1 + k$, for $i \geq 3$. Since $\lambda_3 < 0$, we have $k < 1$. Using the Courant–Weyl inequalities, we get $\lambda_{n-i+2}(\overline{G}) \leq -k \leq \lambda_{n-i+1}(\overline{G})$. By the previous theorem, $\overline{G}$ is bipartite and thus

$$
-k \leq \lambda_{n-i+1}(\overline{G}) = -\lambda_i(\overline{G})
\tag{5.8}
$$

and

$$
-k \leq \lambda_{n-i+2}(\overline{G}) = -\lambda_{i-1}(\overline{G}).
\tag{5.9}
$$

Now the inequalities (5.7) follow from (5.8) and (5.9), respectively.

Considering $\lambda_n(G)$, we have $\lambda_n(\overline{G}) \leq \sqrt{\binom{n}{2} - m}$ (otherwise, by Theorem 2.38, $\overline{G}$ would contain at least one triangle, which is not possible since it is bipartite). The application of the Courant–Weyl inequalities gives the result. $\square$

5.3 Graphs G with $\lambda_{n-1}(G)$ and $\lambda_{n-1}(\overline{G}) \geq -1$

Graphs of the form $G_1 \nabla (G_2 \cup G_3 \cup \cdots \cup G_k)$ play a crucial role in this section. We prove the following lemma.

Lemma 5.8 *The graph $\overline{K_r} \nabla (\overline{K_s} \cup \overline{K_p} \cup \overline{K_q})$ has exactly one eigenvalue less than -1 if and only if $rspq - (rs + sp + pq) + 1 \leq 0$.*

Proof Computing the characteristic polynomial, we get that the non-zero eigenvalues of this graph are determined by

$$
f(x) = x^4 - (rs + sp + pq)x^2 + rspq = 0.
$$

Since this equation has exactly one root less than -1 if and only if $f(-1) \leq 0$ we get the assertion.   □

Now we prove the following result.

Theorem 5.9   (Petrović [360]) *Both graphs G and $\overline{G}$ have exactly one eigenvalue less than -1 if and only if either of them belongs to one of the following families of graphs:*

(i) $K_r \nabla (\overline{K_s} \cup \overline{K_p} \cup K_q)$, *for* $r, s, p, q \geq 1$;

(ii) $\overline{K_r} \nabla (K_s \cup K_p \cup \overline{K_q})$, *for* $r = s = 1$, $p, q \geq 1$ *or* $r = p = 1$, $s, q \geq 1$ *or* $r = 1$, $s = p = 2$, $q \geq 1$ *or* $r = q = 2$, $s = 1$, $p \geq 1$ *or* $r = s = p = q = 2$;

(iii) $\overline{K_r} \nabla (K_s \cup \overline{K_p} \cup K_q)$, *for* $1 \leq r \leq 2$, $s, p, q \geq 1$ *or* $r = 3$, $s \geq 1$, $1 \leq p \leq 2$, $q \geq 1$ *or* $r \geq 4$, $s \geq 1$, $p = 1$, $q \geq 1$;

(iv) $\overline{K_r} \nabla (\overline{K_s} \cup \overline{K_p} \cup K_q)$, *for* $r = s = 2$, $p = 3$, $q \geq 1$ *or* $r = p = 2$, $s, q \geq 1$;

(v) $K_r \nabla (\overline{K_s} \cup \overline{K_p})$, *for* $r \geq 2$, $s, p \geq 1$;

(vi) $K_r \nabla (K_s \cup \overline{K_p})$, *for* $r = 2$, $s, p \geq 1$;

(vii) $K_r \nabla (\overline{K_s} \cup K_p)$, *for* $r \geq 2$, $s, p \geq 1$;

(viii) $K_r \nabla (K_s \cup K_p)$, *for* $r \geq 2$, $s, p \geq 1$.

Proof   The fact that any of the graphs described in (i)–(viii) has exactly one eigenvalue less than -1 is checked directly.

Assume that both G and $\overline{G}$ have n vertices and exactly one eigenvalue less than -1. Since at least one of them must be connected, without loss of generality, we may assume that G is connected. Now, G is not a complete multipartite graph (since then $\lambda_n(\overline{G}) = -1$), and thus it must contain at least one of the graphs P_4 or $K_1 \nabla (K_1 \cup K_2)$ as an induced subgraph.

Let, in the first case, G contain P_4 as an induced subgraph and let the vertices of P_4 be labelled $1, 2, 3, 4$ in natural order. Note that, unless $G \cong P_4$, each vertex of G is adjacent to at least one vertex of P_4 (otherwise, G contains $K_1 \cup P_4$ as an induced subgraph and thus $\lambda_{n-1}(\overline{G}) < -1$).

Let further $N_{i_1, i_2, \ldots, i_k}$ $(1 \leq i_1 < i_2 < \cdots < i_k \leq 4;\ 1 \leq k \leq 4)$ denote the set of vertices in $V(G) \backslash V(P_4)$ that are adjacent exactly to the vertices $i_1, i_2, \ldots, i_k$ of P_4. Considering these sets, and using the interlacing theorem, we get $N_1 = N_4 = N_{1,4} = N_{2,3} = N_{1,2,4} = N_{1,3,4} = N_{1,2,3,4} = \emptyset$.

The adjacency relations between the remaining sets are considered by direct checking. It turns out that any two of these sets must be either completely

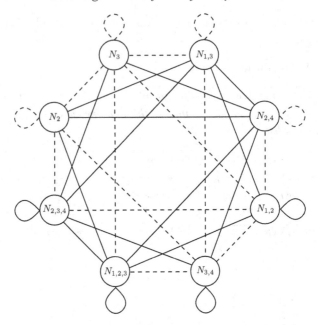

Figure 5.1 Adjacency relations.

adjacent, completely non-adjacent or non-coexistent (in the sense that the existence of one implies the non-existence of the other). They are presented by a graph-like diagram illustrated in Fig. 5.1. There, the sets are completely adjacent if they are joined by a full line, completely non-adjacent if they are joined by a dashed line and non-coexistent if there is no line between them.

Considering the adjacency relations of Fig. 5.1 and bearing in mind that both G and $\overline{G}$ must have exactly one eigenvalue less than -1, we get that G must belong to one of the following families

$$K_r \nabla (\overline{K_s} \cup \overline{K_p} \cup K_q), \quad \overline{K_r} \nabla (K_s \cup K_p \cup \overline{K_q}),$$
$$\overline{K_r} \nabla (K_s \cup \overline{K_p} \cup K_q), \quad \overline{K_r} \nabla (\overline{K_s} \cup K_p \cup \overline{K_q}),$$
$$\overline{K_r} \nabla (\overline{K_s} \cup \overline{K_p} \cup K_q), \quad \overline{K_r} \nabla (\overline{K_s} \cup \overline{K_p} \cup \overline{K_q}).$$

It remains to obtain the explicit values of parameters r, s, p, q for which G and $\overline{G}$ satisfy the condition of the theorem. This is done by using the previous lemma. It arises that if $\lambda_n(G) < -1 \leq \lambda_{n-1}(G)$ and $\lambda_n(\overline{G}) < -1 \leq \lambda_{n-1}(\overline{G})$ then G belongs to one of the families (i)–(iv).

The second case in which G contains $K_1 \nabla (K_1 \cup K_2)$ but does not contain P_4

as an induced subgraph is considered in the same way. As a result, G belongs to one of the families (v)–(viii), and the proof follows. □

Here is the main result of this section.

Theorem 5.10 ([360]) *The second smallest eigenvalue of both graphs G and $\overline{G}$ is at least -1 if and only if each of them belongs to one of the families (i)–(viii) listed in the previous theorem or one of the following families:*

> *(ix) $K_r \nabla \overline{K_s}$, for $r,s \geq 1$;*
>
> *(x) $\overline{K_r \nabla K_s}$, for $r,s \geq 1$.*

Proof From the previous theorem we have $\lambda_{n-1}(G), \lambda_{n-1}(\overline{G}) \geq -1$ whenever any of these graphs belongs to one of the families (i)–(viii). It remains to consider the possibility which we excluded in the previous proof: G is a complete multipartite graph. Since $\lambda_{n-1}(G) \geq -1$, we get $G \cong K_{1,1,\ldots,1,s}$ or $G \cong K_{r,s}$, which completes the proof. □

Exercises

5.1 Show that if G does not contain any of the graphs $3K_1, C_4$ or C_5 as an induced subgraph then its complement is a DNG.

5.2 Prove that if $\overline{G}$ is non-bipartite then $\lambda_3(G) \geq 0$.

Hint: This follows directly from Theorem 5.6. To prove it without using this theorem, show that G contains an induced subgraph with the property $\lambda_3 \geq 0$ whenever its complement is non-bipartite.

5.3 Prove that $\lambda_3(G) < -1$ if and only if $G \cong P_3$ and that no graphs with $-1 \leq \lambda_3 \leq \frac{1-\sqrt{5}}{2}$ exist.

Hint: Apply similar reasoning to the previous exercise.

5.4 Prove that for a complete multipartite graph $K_{n_1, n_2, \ldots, n_k}$ ($n_1, n_2, \ldots, n_k \geq 1$, $\sum_{i=1}^{k} n_i = n$),

$$\lambda_{n-1}(K_{n_1, n_2, \ldots, n_k}) \geq -1$$

holds if and only if $k \leq 2$ or at most one of n_i's ($1 \leq i \leq k$) is greater than 1.

Notes

All minimal graphs with exactly three non-negative eigenvalues (i.e., those with $\lambda_4 < 0$) are determined in [357]. There are 285 such graphs.

Graphs with exactly two positive eigenvalues (i.e., those with $\lambda_2 > 0 \geq \lambda_3$) are considered in [50, 356, 362]. For example, all bipartite and all line graphs with this property are identified.

6

Laplacian eigenvalues

The first five sections are devoted to the L-spectral radius μ_1 of a graph. In Section 6.1 we give general upper and lower bounds for μ_1. Section 6.2 deals with the particular classes of graph. In Section 6.3 we consider the graphs with small values of μ_1. The graphs with maximal value of μ_1 and the ordering of graphs with respect to μ_1 are respectively considered in Sections 6.4 and 6.5.

The next four sections are devoted to the second smallest L-eigenvalue. Similarly to above, in Section 6.6 we give general upper and lower bounds for this L-eigenvalue. Trees are considered in Section 6.7. Extremal graphs and the ordering of graphs are considered in Sections 6.8 and 6.9.

Some less investigated L-eigenvalues are considered in Section 6.10.

6.1 General inequalities for L-spectral radius

In 1967 Kel'mans (cf. [325]) gave the upper bound of (6.1). Twenty years later, Grone and Merris [175] gave the lower bound of (6.1) completing the possible range for the L-spectral radius of an arbitrary graph containing at least one edge,

$$\Delta + 1 \leq \mu_1 \leq n. \tag{6.1}$$

Both inequalities are based on the main properties of the L-spectrum. The left equality holds if and only if $\Delta = n - 1$, while the right equality holds if and only if the complement of a considered graph is disconnected.

In what follows we consider some upper bounds for μ_1 depending on vertex degrees or average 2-degrees and some lower bounds depending on the graph diameter or domination number.

155

6.1.1 Upper bounds

All the results considered in this subsection are proved either by considering the coordinates of the eigenvector corresponding to μ_1 or by using the inequalities obtained in Chapter 2 and the relation (1.10) that makes a connection between the L-spectral radius and the spectral radius of a graph.

We start with the following result.

Theorem 6.1   (Guo [177]) *For a graph G,*

$$\mu_1 \leq \max\left\{ \frac{d_i + \sqrt{d_i^2 + 8d_i m_i'}}{2} : 1 \leq i \leq n \right\}, \tag{6.2}$$

where $m_i' = \frac{\sum_{i \sim j}(d_i - |N(i) \cap N(j)|)}{d_i}$. Equality holds if and only if G is a regular bipartite graph.

Proof Let $\mathbf{x} = (x_1, x_2, \ldots, x_n)^T$ denote the corresponding eigenvector, and assume that one of its coordinates, say x_i, is positive and largest in modulus (i.e., $|x_j| \leq x_i$ for all $1 \leq j \leq n$).

Since $L\mathbf{x} = (D - A)\mathbf{x} = \mu_1\mathbf{x}$, following [177] we have

$$(\mu_1 - d_i)x_i = -\sum_{j \sim i} x_j \tag{6.3}$$

and $(D - A)^2\mathbf{x} = \mu_1^2\mathbf{x}$. Next, since $(D - A)^2\mathbf{x} = D^2\mathbf{x} - DA\mathbf{x} - AD\mathbf{x} + A^2\mathbf{x}$, we have

$$\mu_1 x_i = d_i^2 x_i - d_i \sum_{j \sim i} x_j - \sum_{j \sim i} d_j x_j + \sum_{j \sim i} \sum_{k \sim j} x_k. \tag{6.4}$$

Substituting (6.4) into (6.3), and after simple transformations, we get $\mu_1^2 - d_i\mu_1 - 2d_i m_i' \leq 0$; thus we have the inequality (6.2).

The second part of the theorem is proved directly.   □

An immediate consequence of the previous theorem is the following inequality:

$$\mu_1 \leq \max\left\{ \frac{d_i + \sqrt{d_i^2 + 8d_i m_i}}{2} : 1 \leq i \leq n \right\}. \tag{6.5}$$

Here is another bound.

Theorem 6.2 (Das [122]) *For a graph G,*

$$\mu_1 \leq \max\{d_i + d_j - |N(i) \cap N(j)| \ : \ i \sim j\}. \tag{6.6}$$

Proof Similarly to the previous theorem, assume that the coordinate x_i of the corresponding eigenvector $\mathbf{x}$ is the largest in modulus, say $x_i = 1$. Let $x_j = \min\{x_k : k \sim i\}$ and $c_{ij} = |N(i) \cap N(j)|$. Since $x_j \leq x_k$ for all k such that $k \sim i$, following [122], we get

$$\sum_{k \sim i, k \not\sim j} x_k \geq (d_i - c_{ij})x_j,$$

and since $x_k \leq 1$ for all k,

$$\sum_{k \not\sim i, k \sim j} x_k \geq (d_i - c_{ij}).$$

Since $L\mathbf{x} = \mu_1 \mathbf{x}$, we get

$$\mu_1 x_i = d_i x_i - \sum_{k \sim i} x_k, \tag{6.7}$$

or, since $x_i = 1$,

$$\mu_1 = d_i - \sum_{k \sim i, k \sim j} x_k - \sum_{k \sim i, k \not\sim j} x_k. \tag{6.8}$$

In a similar way, we get

$$\mu_1 x_j = d_j x_j - \sum_{k \sim j, k \sim i} x_k - \sum_{k \sim j, k \not\sim i} x_k. \tag{6.9}$$

Subtracting (6.9) from (6.8), we get

$$\mu_1(1 - x_j) \leq d_i - d_j x_j - (d_i - c_{ij})x_j + (d_j - c_{ij}) \leq (d_i + d_j - c_{ij})(1 - x_j).$$

Using (6.7), we eliminate the possibility $x_j = 1$, and then, by the last inequality, we have $\mu_1 \leq d_i + d_j - c_{ij}$ for $i \sim j$ and the proof follows. $\square$

Remark 6.3 The whole paper [480] is devoted to graphs attaining the upper bound (6.6). It is proved that equality is attained exactly for the graphs G^+

described as follows. If G is a semiregular bipartite graph, then G^+ is the supergraph of G with the following property. If $ij \in E(G^+)$ then either $ij \in E(G)$ or i and j belong to the same colour class and have equal number of neighbours in the other class. □

The upper bound (6.6) obviously improves the upper bound obtained by Rojo et al. [386]:

$$\mu_1 \leq \max\{d_i + d_j - |N(i) \cap N(j)| \ : \ 1 \leq i < j \leq n\}. \qquad (6.10)$$

Moreover, it is proved in [386] that the latter bound (and consequently, the bound (6.6)) does not exceed the order of any graph G, which makes them non-trivial for any graph. Next, the upper bound (6.6) also improves two classical upper bounds for μ_1:

$$\mu_1 \leq \max\{d_i + d_j \ : \ i \sim j\} \qquad (6.11)$$

and

$$\mu_1 \leq \max\{d_i + m_i \ : \ 1 \leq i \leq n\}. \qquad (6.12)$$

The inequality (6.11) was obtained by Anderson and Morley [16] and its improvement (6.12) by Merris [315]. In [124], Das showed that $\max\{d_i + m_i \ : \ 1 \leq i \leq n\} \leq \frac{2m}{n-1} + n - 2$. Using this result we get another consequence:

$$\mu_1 \leq \frac{2m}{n-1} + n - 2. \qquad (6.13)$$

This bound is exact for a star, its complement, and a complete graph. There is another upper bound obtained by the same author. The proof is very similar to that of Theorem 6.2.

Theorem 6.4   ([122]) *For a graph G,*

$$\mu_1 \leq \max\left\{ \sqrt{2(d_i^2 + d_i m_i')} \ : \ 1 \leq i \leq n \right\}, \qquad (6.14)$$

where m_i' is the same as in Theorem 6.1.

Although the upper bounds (6.2) and (6.14) have similar expressions, they are incomparable. On the contrary, (6.14) is better than the upper bound of Li and Pan [265]:

$$\mu_1 \le \max\{\sqrt{2d_i(d_i + m_i)} \ : \ 1 \le i \le n\}. \tag{6.15}$$

We list two similar upper bounds.

Theorem 6.5 (Zhu [527]) *For a graph G,*

$$\mu_1 \le \max\left\{ \frac{d_i(d_i + m_i) + d_j(d_j + m_j)}{d_i + d_j} - 2\frac{\sum_{k \sim i, k \sim j} d_k}{d_i + d_j} \ : \ i \sim j \right\} \tag{6.16}$$

and

$$\mu_1 \le \max\left\{ \frac{d_i m_i}{d_j} + \frac{d_j m_j}{d_i} \ : \ i \sim j \right\}. \tag{6.17}$$

The upper bound (6.16) is an improvement on the result of Li and Zhang:

$$\mu_1 \le \max\left\{ \frac{d_i(d_i + m_i) + d_j(d_j + m_j)}{d_i + d_j} \ : \ i \sim j \right\} \tag{6.18}$$

(so, with $-2\frac{\sum_{k \sim i, k \sim j} d_k}{d_i + d_j}$ omitted). Both results improve the mentioned upper bound (6.12).

The application of the inequality (1.10) together with an additional upper bound for λ_1 is demonstrated in the next theorem.

Theorem 6.6 (Shu et al. [402]) *For a connected graph G,*

$$\mu_1 \le \delta + \frac{1}{2} + \sqrt{\left(\delta - \frac{1}{2}\right)^2 + \sum_{i=1}^{n} d_i(d_i - \delta)}. \tag{6.19}$$

Equality holds if and only if G is a regular bipartite graph.

Proof Let the line graph Line(G) have n' vertices, m' edges, and minimal vertex degree δ'. Using the upper bound (2.35), we get

$$\lambda_1(\text{Line}(G)) \le \frac{\delta' - 1 + \sqrt{(\delta' + 1)^2 + 4(2m' - \delta'n')}}{2}.$$

We next compute

$$n' = m = \frac{1}{2}\sum_{i=1}^{n} d_i, \quad 2m' = \sum_{i=1}^{n} d_i(d_i - 1), \quad \text{and } \delta' \geq 2\delta - 2.$$

Since $\delta' - 1 + \sqrt{(\delta' + 1)^2 + 4(2m' - \delta'n')}$ decreases as δ' increases, we have

$$\lambda_1(\text{Line}(G)) \leq \delta - \frac{3}{2} + \sqrt{\left(\delta - \frac{1}{2}\right)^2 + \sum_{i=1}^{n} d_i(d_i - \delta)}.$$

Using (1.10) we get (6.19).

If the equality holds in (6.19) then it can easily be seen (using Theorem 2.19) that G must be regular bipartite. Next, if G is regular bipartite then $d_i = \delta = r$ and the right-hand side of (6.19) is equal to $2r$. Next, Line(G) is regular and $\delta' = 2r - 2$, which yields $\lambda_1(\text{Line}(G)) = 2r - 2$ and (by Theorem 1.10) $\mu_1 = 2r$. $\square$

Using similar reasoning, we prove the following result.

Theorem 6.7   (Zhang [512]) *For a connected graph G,*

$$\mu_1 \leq \max\{d_i + \sqrt{d_i m_i} : 1 \leq i \leq n\}, \tag{6.20}$$

with equality if and only if G is regular bipartite.

There are some consequences of the previous theorem. First, bearing in mind that $\sum_{j\sim i} d_j = d_i m_i$, it is easy to see that (6.20) is better than the upper bound obtained by Shi [400]:

$$\mu_1 \leq \sqrt{2}\max\left\{\sqrt{d_i^2 + \sum_{j\sim i} d_j} : 1 \leq i \leq n\right\}. \tag{6.21}$$

It is also better than the upper bound for connected graphs obtained by Zhang and Luo [516]:

$$\mu_1 \leq \Delta + \sqrt{2m + \Delta(\delta - 1) - \delta(n - 1)}. \tag{6.22}$$

Namely, for fixed i we have

$$d_i m_i = \sum_{j \sim i} d_j + \sum_{j \nsim i} d_j - \sum_{j \nsim i} d_j = 2m - \sum_{j \nsim i} d_j$$

$$= 2m - d_i - \sum_{j \nsim i, j \neq i} d_j \le 2m - d_i - (n - 1 - d_i)\delta$$

$$= 2m - (n - 1)\delta + d_i(\delta - 1) \le 2m - (n - 1)\delta + (\delta - 1)\Delta,$$

giving the assertion.

Finally, (6.20) also improves the mentioned upper bound (6.15).

Example 6.8   Let G be a (δ, Δ)-semiregular bipartite graph. We get the following computational results for some of the upper bounds mentioned:

μ_1	(6.2)	(6.14)	(6.19)	(6.20)
$\delta + \Delta$	$\frac{\Delta + \sqrt{\Delta^2 + 8\delta\Delta}}{2}$	$\sqrt{2(\Delta^2 + \delta\Delta)}$	$\delta + \frac{1}{2} + \sqrt{(\delta - 1)^2 + (\Delta - \delta)m}$	$\Delta + \sqrt{\delta\Delta}$

Note that the upper bounds (6.6) and (6.17) are attained for any semiregular bipartite graph.   □

6.1.2 Lower bounds

The lower bounds for μ_1 have been investigated less. We point out one such bound expressed in terms of order and size.

Theorem 6.9   (Zhang and Li [514]) *For a graph G,*

$$\mu_1 \ge \frac{1}{n - 1}\left(2m + \sqrt{\frac{2m(n(n - 1) - 2m)}{n(n - 2)}}\right), \tag{6.23}$$

with equality if and only if G is a complete graph.

Proof   We have

$$\left((n - 1)\mu_1 - \mathrm{tr}(L(G))\right)^2 \ge \sum_{i=1}^{n-1}(\mu_1 - \mu_i)^2$$

$$= \mathrm{tr}(L(G)^2) - 2\mu_1\mathrm{tr}(L(G)) + (n - 1)\mu_1^2.$$

Since $\mathrm{tr}(L(G)) = 2m$ and $\mathrm{tr}(L(G)^2) \ge 2m + \frac{4m^2}{n}$, we get the inequality.

Next, equality holds if and only if there are equalities in the above computation, that is, if and only if G is a complete graph.   □

We may consider induced paths, and consequently the diameter of a graph.

Theorem 6.10   (Lu et al. [301]) *Let P_{s+1} be an induced path of a graph G and let the vertices of P_{s+1} be labelled $1, 2, \ldots, s + 1$. Then*

$$\mu_1 \geq \frac{2s + \sum_{i=1}^{s+1} d_i}{s+1}. \tag{6.24}$$

Proof   The proof follows from the Rayleigh principle applied to the vector $\mathbf{y} = (y_1, y_2, \ldots, y_n)^T$ with coordinates

$$y_i = \begin{cases} 1, & \text{if } i \text{ is odd and } i \leq s+1, \\ -1, & \text{if } i \text{ is even and } i \leq s+1, \\ 0, & \text{otherwise.} \end{cases}$$

□

The lower bound of (6.1) can easily be derived from (6.24) whenever G contains an edge uv such that $d_u = d_v = \Delta$. Moreover, if the vertices of G have degrees $\delta = d_1 \leq d_2 \leq \cdots \leq d_n = \Delta$, by setting

$$e_r = \frac{1}{r} \sum_{i=1}^{r} d_i \ (1 \leq r \leq n) \tag{6.25}$$

we immediately get the following corollary.

Corollary 6.11   ([301]) *For a connected graph with diameter D,*

$$\mu_1 \geq \frac{(D+1)e_{D+1} + 2D}{D+1}. \tag{6.26}$$

Since $d_i \geq 2$ for $2 \leq i \leq D$ and $d_1, d_{D+1} \geq 1$, using the previous corollary we get another consequence.

Corollary 6.12   ([301]) *For a connected graph with diameter D,*

$$\mu_1 \geq \frac{4D}{D+1}. \tag{6.27}$$

Example 6.13 Here are the computational results for the previous lower bounds applied to double kite graphs:

Graph	(6.27)	(6.26)	(6.24)	μ_1
$DK(5,3)$	3.429	4.857	5.143	6.069
$DK(10,20)$	3.833	5.250	5.333	11.011

$\square$

We mention without proof another result concerning the L-spectral radius and the domination number of a graph.

Theorem 6.14 (Nikiforov [331]) *For a non-trivial graph with domination number $\varphi \geq 2$,*

$$\mu_1 \geq \left\lceil \frac{n}{\varphi} \right\rceil. \qquad (6.28)$$

Equality holds if and only if $G = G_1 \cup G_2$, where $|V(G_1)| = \left\lceil \frac{n}{\varphi} \right\rceil$, $\varphi(G_1) = 1$ and $\varphi(G_2) = \varphi - 1$, $\mu_1(G_2) \leq \left\lceil \frac{n}{\varphi} \right\rceil$.

6.2 Bounding L-spectral radius of particular types of graph

Here we consider triangle-free graphs, so-called triangulation graphs, bipartite graphs, and trees.

6.2.1 Triangle-free graphs

These graphs were considered by Zhang and Luo [517].

Theorem 6.15 ([517]) *For a triangle-free graph G,*

$$\mu_1 \geq \max \left\{ \frac{16m^2}{n^3}, \frac{2m}{n} + \frac{m^{\frac{3}{4}}}{2\sqrt{2n}} \right\}. \qquad (6.29)$$

Equality holds if G is the complete bipartite graph with equal number of vertices in each part.

Proof Following [517], we use a simple combinatorial result [141]. If G is a triangle-free graph, there exists a bipartite subgraph H such that

$$m(H) \geq \max\left\{\frac{4m(G)^2}{n(G)^2}, \frac{m(G)}{2} + \frac{\sum_{u \in V(G)} \sqrt{d_u}}{8\sqrt{2}}\right\}$$

$$\geq \max\left\{\frac{4m(G)^2}{n(G)^2}, \frac{m(G)}{2} + \frac{m(G)^{\frac{3}{4}}}{8\sqrt{2}}\right\}.$$

Now let H be a bipartite spanning subgraph of G with maximal number of edges. By Theorems 1.14 and 1.19, we have

$$\mu_1(G) \geq \kappa_1(G) \geq \frac{\mathbf{j}^T Q(H) \mathbf{j}}{\mathbf{j}\mathbf{j}^T} = \frac{4m(H)}{n(G)}$$

$$\geq \max\left\{\frac{16m(G)^2}{n(G)^3}, \frac{2m(G)}{n(G)} + \frac{m(G)^{\frac{3}{4}}}{2\sqrt{2}n(G)}\right\}.$$

If $G = K_{\frac{n}{2},\frac{n}{2}}$ (with n even), the inequality (6.29) gives $\mu_1 \geq n$, but by (6.1) we have $\mu_1 \leq n$, and thus equality holds. □

From this theorem it is easy to get a known result, in some literature named after Turán.

Corollary 6.16 *A connected graph with $4m > n^2$ contains at least one triangle.*

Proof If it does not contain any triangle, by the previous theorem $\frac{16m^2}{n^3} \leq \mu_1 \leq n$ and thus $4m \leq n^2$. □

We have a lower bound for μ_1 of triangle-free graphs expressed in terms of vertex degrees and average 2-degrees.

Theorem 6.17 ([517]) *For a triangle-free graph G,*

$$\mu_1 \geq \max\left\{\frac{d_i + m_i + \sqrt{(d_i - m_i)^2 + 4d_i}}{2} : 1 \leq i \leq n\right\}. \tag{6.30}$$

Proof According to [517], let L' be the principal submatrix of L corresponding to the closed neighbourhood of a vertex u, $N[u] = \{u, u_1, u_2, \ldots, u_k\}$ (where $d_u = k$). Obviously, $\mu_1(L) \geq \mu_1(L')$, and since G is triangle-free we compute the characteristic polynomial of L' as follows:

$$\det(xI - L') = \left(x - d_u - \sum_{i=1}^{k} \frac{1}{x - d_{u_i}}\right) \prod_{i=1}^{k} (x - d_{u_i}).$$

Since $\mu_1(L') > d_{u_i}$ for each $1 \le i \le k$, $\mu_1 = \mu_1(L)$ satisfies

$$\mu_1 - d_u \ge \sum_{i=1}^{k} \frac{1}{\mu_1 - d_{u_i}}.$$

By the Cauchy–Schwarz inequality, we have

$$\sum_{i=1}^{k} (\mu_1 - d_{u_i}) \cdot \sum_{i=1}^{k} \frac{1}{\mu_1 - d_{u_i}} \ge \left(\sum_{i=1}^{k} \frac{\sqrt{\mu_1 - d_{u_i}}}{\sqrt{\mu_1 - d_{u_i}}}\right)^2 = k^2.$$

Hence,

$$\mu_1 - d_u \ge \frac{k^2}{\sum_{i=1}^{k} (\mu_1 - d_{u_i})} = \frac{d_u}{\mu_1 - m_u},$$

and the result follows. $\qquad\square$

In addition, if G is regular then we have the following result.

Corollary 6.18 ([517]) *For an r-regular triangle-free graph,*

$$\mu_1 \ge \max\left\{\frac{4r^2}{n}, r + \sqrt{r}\right\}. \tag{6.31}$$

Proof Since $\frac{16m^2}{n^3} = \frac{4r^2}{n}$ and $\frac{d_u + m_u + \sqrt{(d_u - m_u)^2 + 4d_u}}{2} = r + \sqrt{r}$, the inequality follows from Theorems 6.15 and 6.17. $\qquad\square$

6.2.2 Triangulation graphs

A graph G is called a *triangulation* if every pair of adjacent vertices in G has at least one common neighbour. The upper bounds for triangulation graphs were considered by Guo et al. [185].

Theorem 6.19 ([185]) *For any triangulation G,*

$$\mu_1 \le \max\left\{\frac{2d_i - 1 + \sqrt{4d_i m_i - 4d_i + 1}}{2} : 1 \le i \le n\right\}. \tag{6.32}$$

Proof For a vertex i we have $(d_i - \mu_1)x_i = \sum_{j \sim i} x_j$, where x_i, x_j are the coordinates of the corresponding eigenvector. Following [185], we use the Lagrange identity to get

$$(d_i - \mu_1)^2 x_i^2 = d_i \sum_{j \sim i} x_j^2 - \sum_{j \sim i, k \sim i} (x_j - x_k)^2.$$

Summing the above equality over all vertices i and bearing in mind that

$$\mu_1 = \sum_{j \sim k} (x_j - x_k)^2 \leq \sum_{i=1}^{n} \left(\sum_{j \sim i, k \sim i} (x_j - x_k)^2 \right) \qquad (6.33)$$

(which holds since G is a triangulation), we get

$$\sum_{i=1}^{n} \left((d_i - \mu_1)^2 - d_i m_i + \mu_1 \right) x_i^2 \leq 0.$$

This yields $(d_i - \mu_1)^2 - d_i m_i + \mu_1 \leq 0$, for some vertex i, and the inequality (6.32) follows.   □

Furthermore we have the following more general result.

Theorem 6.20 ([185]) *If each edge of a graph G belongs to $t \geq 1$ triangles, then*

$$\mu_1 \leq \max \left\{ \frac{2d_i - t + \sqrt{4d_i m_i - 4t d_i + t^2}}{2} : 1 \leq i \leq n \right\}. \qquad (6.34)$$

Equality occurs if G is the complete graph K_{t+2}.

Proof Replacing (6.33) with

$$t\mu_1 = t \sum_{j \sim k} (x_j - x_k)^2 \leq t \sum_{i=1}^{n} \left(\sum_{j \sim i, k \sim i} (x_j - x_k)^2 \right),$$

we get the result. For K_{t+2} the equality is verified by direct computation.   □

Remark 6.21 In the previous theorem, the expression 'each edge belongs to $t \geq 1$ triangles' means precisely that there exists an integer $t \geq 1$ such that each edge belongs to t triangles, while some edges (but not all) may belong to more than t triangles.

Observe that the result of Theorem 6.20, and those of Theorems 6.15 and 6.17 cannot be applied to the same graph.

It has been shown in [185] that the upper bound (6.34) improves the upper bound obtained by Lu et al. [298] (concerning the same class):

$$\mu_1 \leq \frac{2\Delta - t + \sqrt{(2\Delta - t)^2 + 8m - 4\delta(n-1) - 4\delta^2 + 4(\delta - 1)\Delta}}{2}. \quad (6.35)$$

Moreover, the same bound gives a better estimate than the general upper bound of Zhang (6.20) and its consequences (6.21), (6.22) or (6.15) when they are applied to the graphs specified in Theorem 6.20.   □

6.2.3 Bipartite graphs and trees

Many upper and lower bounds for μ_1 for bipartite graphs can be obtained by using the connection between λ_1 and μ_1 given in (1.10) and an appropriately chosen bound for λ_1. For example, using the simple lower bound (2.4), we get

$$\mu_1 \geq 2 + \sqrt{\frac{1}{m}\sum_{i \sim j}(d_i + d_j - 2)^2} \quad (6.36)$$

for any bipartite graph.

We proceed with more specific bounds by giving a result concerning trees with given matching number. Let S_n^l ($2l \leq n+1$) denote the starlike tree obtained from the star $K_{1,n-1}$ by attaching a pendant edge at $l-1$ vertices of degree 1.

Theorem 6.22 (Guo [179]) *Let T be a tree with n vertices and matching number μ. Then*

$$\mu_1(T) \leq t, \quad (6.37)$$

where t is the largest root of the equation

$$x^3 - (n - \mu + 4)x^2 + (3n - 3\mu + 4)x - n = 0.$$

Equality holds if and only if $T = S_n^\mu$.

The proof is derived by considering the Laplacian characteristic polynomial of S_n^μ:

$$L_{S_n^\mu}(x) = x(x-1)^{n-2\mu}(x^2 - 3x + 1)^{\mu-2}\big(x^3 - (n - \mu - 4)x^2$$
$$+ (3n - 3\mu + 4)x - n\big).$$

There is a simple corollary.

Corollary 6.23 ([179]) *Let T be a tree with $2n$ vertices with perfect matching, then*

$$\mu_1 \leq \frac{n + 2 + \sqrt{n^2 + 4}}{2}. \tag{6.38}$$

Equality holds if and only if $T \cong S_{2n}^n$.

Proof The largest root of the above polynomial (with $2n$ instead of n) is $\frac{n+2+\sqrt{n^2+4}}{2}$. $\qquad\square$

Trees are considered in the context of extremal graphs (Section 6.4) and ordering of graphs (Section 6.5).

6.3 Graphs with small L-spectral radius

We characterize all graphs whose L-spectral radius does not exceed 4.5. As in Subsection 2.3.1, we start with the Hoffman program with respect to the Laplacian matrix. Recall from page 21 that this program concerns precisely the determination of graphs whose L-spectral radius does not exceed $2 + \varepsilon$, where ε is the real root of $x^3 - 4x - 4 = 0$ ($\varepsilon \approx 2.3830$).

The graphs whose L-spectral radius is at most 4 were determined by Omidi [353], while the remaining graphs were determined by Wang et al. [450]. The result reads as follows.

Theorem 6.24 (Laplacian Hoffman Program) *If G is connected and $\mu_1(G) \leq 2 + \varepsilon$ (where ε is the real root of $x^3 - 4x - 4 = 0$), then G is one of the following graphs:*

    (i) P_n,
    (ii) C_n,
    (iii) K_4,
    (iv) $K_{1,3}$,
    (v) $K_4 - e$,
    (vi) $K_{1,3} + e$,

(vii) T_n^2 $(n \geq 5)$,

(viii) $K(3,l)$ $(l \geq 2)$,

(ix) the graph obtained by attaching a pendant vertex at any two vertices of K_3,

(x) the graph obtained by attaching a pendant vertex at each of the vertices of K_3,

(xi) T_n^i $(3 \leq i \leq n-3)$,

(xii) the graph obtained by attaching a pendant vertex at a vertex of an odd cycle of length at least five,

(xiii) the graph obtained by inserting an edge between two pendant vertices at distance 2 in $T_n^{2,n-3}$ $(n \geq 8)$,

(xiv) $DK(3,l)$ $(l \geq 2)$,

(xv) $S_{i,j,k}$ $(i \geq j+k+1)$.

Prior to the previous theorem, for graphs (i)–(vi), $\mu_1 \in [0,4]$; for graphs (vii) and (viii), $\mu_1 \in (4, 2+\sqrt{5})$; and for graphs (ix)–(xv), $\mu_1 \in (2+\sqrt{5}, 2+\varepsilon)$. There is no graph whose L-spectral radius is equal to $2+\sqrt{5}$ or $2+\varepsilon$.

Graphs with the property $\mu_1 \leq 4.5$ are considered in the following theorem. The theorem itself is similar to Theorem 2.59.

Theorem 6.25 ([450]) *Any connected graph whose L-spectral radius does not exceed* 4.5 *has a spanning tree which is an open quipu. Consequently, any such graph is obtained by adding a set (possibly empty) of edges to an open quipu.*

6.4 Graphs with maximal L-spectral radius

We consider graphs of fixed order whose L-spectral radius attains its absolute maximum given in (6.1), that is, graphs with the property $\mu_1 = n$. In the sequel, we consider graphs of fixed diameter or some other structural invariant with maximal L-spectral radius and conclude the section with the particular types of tree having the same spectral property.

6.4.1 Graphs with $\mu_1 = n$

Here we consider the determination of graphs whose L-spectral radius attains its maximal value. According to Liu et al. [277], we first consider the following theorem. By $\text{mult}(\mu_i)$ we denote the multiplicity of the eigenvalue μ_i.

Theorem 6.26 ([277]) *Given a graph G, for $1 \le k \le n-1$, mult(n) = k if and only if G has a spanning subgraph which is complete $(k+1)$-partite.*

Proof Assume that mult$(n) = k$. By Theorem 1.12, $\overline{G}$ has $k+1$ components, say $\overline{G}_1, \overline{G}_2, \ldots, \overline{G}_{k+1}$. But then G contains a subgraph H, where $V(H) = V(\overline{G}_1) \cup V(\overline{G}_2) \cup \cdots \cup V(\overline{G}_{k+1})$ and $E(H) = \{uv : u \in V(\overline{G}_i), v \in V(\overline{G}_j), i \ne j\}$. Clearly, H is a spanning subgraph that is complete $(k+1)$-partite.

Conversely, if G has a complete $(k+1)$-partite graph as a spanning subgraph then its complement contains $k+1$ components and then, by Theorem 1.12, mult$(n) = k$. $\square$

The following corollary is an immediate consequence.

Corollary 6.27 ([277]) *For a graph G, the following statements are equivalent.*

> *(i) $\mu_1 = n$.*
> *(ii) G has a complete bipartite spanning subgraph.*
> *(iii) G is a join of two graphs.*

Applying the previous statements, we get some graphs attaining $\mu_1 = n$.

Corollary 6.28 ([277]) *Given a graph G, then*

> *(i) if G is bipartite we have mult(n) ≤ 1 with equality if and only if G is complete bipartite,*
> *(ii) if G is a tree we have $\mu_1 = n$ if and only if G is a star,*
> *(iii) if G is a cycle we have $\mu_1 = n$ if and only if $G \cong C_3$ or $G \cong C_4$,*
> *(iv) if G is a unicyclic graph we have $\mu_1 = n$ if and only if $G \cong C_4$ or $G \cong K_{1,n-1} + e$,*
> *(v) if G is a bicyclic graph we have $\mu_1 = n$ if and only if $G \cong K_{2,3}$ or G is obtained by inserting two edges into $K_{1,n-1}$.*

Consider now regular graphs. By Corollary 6.27, G is an r-regular graph with the property $\mu_1 = n$ if and only if $G \cong G_1 \triangledown G_2$, where G_i is an r_i-regular graph of order n_i $(1 \le i \le 2)$ and $n_1 + r_2 = n_2 + r_1 = r$. Thus, we have $r \ge \frac{n}{2}$. Is it true that there exists an r-regular graph with $\mu_1 = n$ whenever $\frac{n}{2} \le r \le n-1$? The answer is given in the following theorem.

Theorem 6.29 ([277]) *There is a connected r-regular graph G with $\mu_1 = n$ if and only if nr is even and $\frac{n}{2} \le r \le n-1$.*

Proof If $\mu_1 = n$, by Corollary 6.27 the maximal vertex degree in G satisfies $\Delta \ge \frac{n}{2}$, and so the necessity follows.

Assume that nr is even and $\frac{n}{2} \le r \le n-1$. Following [277], we distinguish two cases depending on the parity of r.

Case 1: r is even. Since $0 \le 2r-n \le r-1$, there must be a $(2r-n)$-regular graph, say G_1, of order r. Let G_2 be the graph with $n-r$ isolated vertices. Then $G = G_1 \triangledown G_2$ is r-regular and the result follows from Corollary 6.27.

Case 2: r is odd. Since nr is even, n must be even. If $r = \frac{n}{2}$, $K_{r,r}$ is the desired graph. So, we may assume that $r \ge \frac{n}{2}+1$. Then there is a $(2r-n-1)$-regular graph, say again G_1, of order $r-1$. For G_2 we may take $\frac{1}{2}(n-r+1)K_2$, and again $G = G_1 \triangledown G_2$ is the desired graph. □

The graphs with a given number of cut edges are considered in the next theorem.

Theorem 6.30 (Li et al. [261]) *The L-spectral radius of a connected graph G with exactly k $(1 \le k \le n-1)$ cut edges is equal to n if and only if G has k pendant edges and its maximal vertex degree is $n-1$.*

Proof By Theorem 1.12, $\overline{G}$ is disconnected. Let $\widetilde{E} = \{e_1, e_2, \ldots, e_k\}$ be the set of cut edges of G. We first note that each cut edge must be a pendant edge (otherwise $\overline{G}$ is connected). It is also easy to verify that all k cut edges are attached at a common vertex, say u, with $d_u = n-1$ (otherwise, by inequality (6.10), $\mu_1 < n$), and the proof follows. □

6.4.2 Various graphs

The graphs considered here are precisely those of fixed order and diameter, bipartite graphs, and (in particular) trees.

Order, diameter, and maximal L-spectral radius

There is an interesting property of clique chain graphs, defined on page 5.

Theorem 6.31 (Wang et al. [446]) *The maximum of any L-eigenvalue μ_i $(1 \le i \le n)$ within the set of all graphs with n vertices and diameter D is attained for some graph $G^D(1, n_2, n_3, \ldots, n_D, 1)$.*

Proof The proof follows directly from Theorem 1.14. □

The last theorem gives the form of a graph of fixed order and diameter with maximal L-eigenvalue μ_1. In particular, for μ_1 we have the following results. We first consider bipartite graphs, for which we need co-clique chain graphs instead.

Theorem 6.32 (Zhai et al. [497]) *The graph* $G_*^3\left(1,\left\lceil\frac{n-2}{2}\right\rceil,\left\lfloor\frac{n-2}{2}\right\rfloor,\ 1\right)$ *is the unique graph with maximal L-spectral radius within the set of bipartite graphs with* n *vertices and diameter* 3. *Moreover,* $\max\left\{n-\frac{9}{2n},n-1\right\}<\mu_1(G_*^3)<n-\left\lceil\frac{n}{2}\right\rceil^{-1}$.

For $D\geq 4$, $G_*^D\left(\left\lceil\frac{D}{2}\right\rceil\cdot 1,n-D,\left\lfloor\frac{D}{2}\right\rfloor\cdot 1\right)$ *(where* $\left\lceil\frac{D}{2}\right\rceil\cdot 1$ *denotes that there are* $\left\lceil\frac{D}{2}\right\rceil$ *consecutive* V_i's *with one vertex) is the unique graph with maximal L-spectral radius within the set of bipartite graphs with* n *vertices and diameter* D. *Moreover,* $n-D+3-\frac{1}{n-D}<\mu_1(G_*^D)<n-D+3$.

We conclude with the following result.

Theorem 6.33 ([497]) *A graph* G *has maximal L-spectral radius within the set of graphs with* n *vertices and diameter* 3 *if and only if* $G_*^D\left(1,\left\lceil\frac{n-2}{2}\right\rceil,\left\lfloor\frac{n-2}{2}\right\rfloor,1\right)$ *is a spanning subgraph of* G *and* G *is a spanning subgraph of* $G^D\left(1,\left\lceil\frac{n-2}{2}\right\rceil,\left\lfloor\frac{n-2}{2}\right\rfloor,1\right)$.

A graph G *has maximal L-spectral radius within the set of graphs with* n *vertices and diameter* $D\geq 4$ *if and only if* $G_*^D\left(\left\lceil\frac{D}{2}\right\rceil\cdot 1,n-D,\left\lfloor\frac{D}{2}\right\rfloor\cdot 1\right)$ *is a spanning subgraph of* G *and* G *is a spanning subgraph of* $G^D\left(\left\lceil\frac{D}{2}\right\rceil\cdot 1,\ n-D,\left\lfloor\frac{D}{2}\right\rfloor\cdot 1\right)$.

Bipartite graphs and trees

Li et al. [261] identified $K_{k,n-k}$ as the unique graph with maximal L-spectral radius within the set of bipartite graphs with vertex (and also, edge) connectivity at most k $(k\geq 1)$. Zhang and Zhang [504] proved that the unique graph with the same spectral property within the set of bipartite graphs with k $(1\leq k\leq n-4)$ cut edges is obtained by identifying the centre of the star $K_{1,k}$ with one vertex of degree $n-k-2$ in $K_{2,n-k-2}$. Some additional results concerning bipartite graphs are pointed out in the Notes (to this chapter).

We continue with trees. Gutman [196] proved that the star has the maximal L-spectral radius within the set of trees with n vertices (see also Corollary 6.28), while Petrović and Gutman [366] proved that the path has the minimal L-spectral radius within the same set. We present the following result.

Theorem 6.34 (Hong and Zhang [221]) *Let* T *be a tree with* n *vertices and* k *pendant edges. Then* $\mu_1(T)\leq\mu_1(\widehat{T})$, *where* $\widehat{T}$ *is a starlike tree obtained by attaching the paths* $P^1,P^2,\ldots,P^k$ *of almost equal lengths at all pendant vertices of the star* $K_{1,k}$.
Equality holds if and only if $T\cong\widehat{T}$.

Proof Let t denote the number of vertices in T whose degrees are at least 3. Following [221], we distinguish three cases.

Case 1: $t = 0$. In this case T is a path and has the desired form.

Case 2: $t = 1$. The line graph $\mathrm{Line}(T)$ is obtained by attaching the paths $P^1, P^2, \ldots, P^{n-k}$ at the vertices of the complete graph K_{n-k}. Since, by Theorem 1.21, $\mu_1(T) = 2 + \lambda_1(\mathrm{Line}(T))$, consecutive application of Lemma 1.29 gives the result.

Case 3: $t > 1$. Let $\mathbf{x} = (x_1, x_2, \ldots, x_n)^T$ be the Perron eigenvector of the signless Laplacian matrix of T. Assume that u and v are two vertices of T whose degrees are at least 3 and $x_u \geq x_v$. There is a unique path P between u and v and exactly one vertex adjacent to v, say w, lies on P. Assume that $\{v_1, v_2, \ldots, v_{d_v-2}\} \subset N(v)\backslash w$. Following the proof of Lemma 1.29, it can be seen that deleting the edge vv_i and adding the edge uv_i for $1 \leq i \leq d_v - 2$ is followed by a strict increasing of the L-spectral radius (since T is bipartite, the L-spectrum and Q-spectrum coincide). In this way, we get a tree with also k pendant vertices in which the number of vertices of degree at least 3 decreases to $t - 1$. Repeating this procedure until this number is 1, and referring to Case 2, we get the proof. □

We next present the unique tree with maximal L-spectral radius within the set of trees with n vertices and given degree sequence. This problem was considered by Zhang [511] and the resulting tree is considered as a rooted tree for which the depth-first search (described on page 72) gives the ordering of its vertices such that the degree d_i corresponds to the ith vertex in the ordering. An example is given in Fig. 6.1.

Zhang also gave two consequences of his result:

- The unique tree with n vertices, independence number α, and maximal L-spectral radius is the tree described above with degree sequence

$$\pi = (\alpha, \underbrace{2, 2, \ldots, 2}_{n-\alpha-1}, \underbrace{1, 1, \ldots, 1}_{\alpha})$$

(see also [509]).

- The unique tree with n vertices, matching number μ, and maximal L-spectral radius is the tree described above with degree sequence

$$\pi = (n - \mu, \underbrace{2, 2, \ldots, 2}_{\mu-1}, \underbrace{1, 1, \ldots, 1}_{n-\mu})$$

(see also [179]).

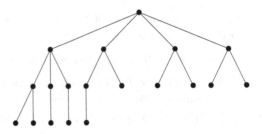

Figure 6.1 The tree of given degree sequence $\pi = (4,4,3,3,3,3,2,2,2,1,1,1,1,$
$1,1,1,1,1,1)$ with maximal L-spectral radius.

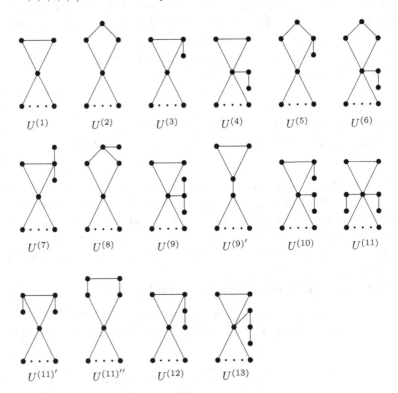

Figure 6.2 The first 16 unicyclic graphs ordered by their L-spectral radii.

6.5 Ordering graphs by L-spectral radius

We give partial orderings of trees, unicyclic graphs, bicyclic graphs, and cacti.

As mentioned in the previous section, if n is fixed, the star $K_{1,n-1}$ is the tree with maximal L-spectral radius. Considering the ordering of trees by their L-

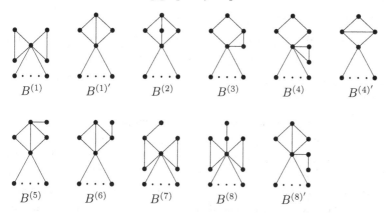

$B^{(1)}$ $\qquad$ $B^{(1)'}$ $\qquad$ $B^{(2)}$ $\qquad$ $B^{(3)}$ $\qquad$ $B^{(4)}$ $\qquad$ $B^{(4)'}$

$B^{(5)}$ $\qquad$ $B^{(6)}$ $\qquad$ $B^{(7)}$ $\qquad$ $B^{(8)}$ $\qquad$ $B^{(8)'}$

Figure 6.3 The first 11 bicyclic graphs ordered by their L-spectral radii. Adapted from [268] with permission from Elsevier.

spectral radii, Zhang and Li [514], and Guo [179] gave the second to the fourth tree ordered decreasingly, while Yu et al. [484] extended this result by giving the fifth to the eighth tree in the same ordering. The result reads as follows. For $n \geq 15$,[1] the first eight trees ordered by their L-spectral radii are:

(i) $T^{(1)}$ – the star $K_{1,n-1}$;

(ii) $T^{(2)}$ – the comet $C(n-3,3)$;

(iii) $T^{(3)}$ – obtained by attaching two pendant vertices at the same pendant vertex of the star $K_{1,n-3}$;

(iv) $T^{(4)}$ – obtained by attaching a pendant vertex at each of two different pendant vertices of the star $K_{1,n-3}$;

(v) $T^{(5)}$ – the comet $C(n-4,4)$;

(vi) $T^{(6)}$ – the double star $DS(4,n-4)$;

(vii) $T^{(7)}$ – obtained by attaching a pendant vertex at a pendant vertex of the star $K_{1,n-4}$ and by attaching two pendant vertices at another pendant vertex of the same star;

(viii) $T^{(8)}$ – obtained by attaching a pendant vertex at three different pendant vertices of the star $K_{1,n-4}$.

Let $T(i)$ denote the tree obtained from the path P_{D+1} whose vertices are labelled $v_1, v_2, \ldots, v_{D+1}$ by attaching $n-D-1$ pendant vertices at the vertex v_i $(1 \leq i \leq D+1)$. Concerning the ordering of trees of fixed diameter D, Guo [180] determined the first $\lfloor \frac{D}{2} \rfloor + 1$ trees ordered decreasingly. The result reads

---

[1] We bound the order to avoid overlapping between some trees, although most of them hold their positions for smaller values of n.

as follows. For $3 \leq D \leq n - 3$, the first $\lfloor \frac{D}{2} \rfloor + 1$ trees of diameter D ordered decreasingly by their L-spectral radii are: $T^{D(1)}, T^{D(2)}, \ldots, T^{D(\lfloor \frac{D}{2} \rfloor + 1)}$, where, for $1 \leq i \leq \lfloor \frac{D}{2} \rfloor$, $T^{D(i)} = T(\lfloor \frac{D}{2} \rfloor + 2 - i)$, while $T^{D(\lfloor \frac{D}{2} \rfloor + 1)}$ is obtained from $T(\lfloor \frac{D}{2} \rfloor + 1)$ by deleting a pendant vertex attached at $v_{\lfloor \frac{D}{2} \rfloor + 1}$ and attaching a pendant vertex at $v_{\lfloor \frac{D}{2} \rfloor + 2}$.

Guo [190] observed that the maximal L-spectral radius within the set of unicyclic graphs of order n is attained uniquely for the pineapple $P(3, n - 3)$. Further ordering of unicyclic graphs of fixed order is considered in several steps by Guo [190], Liu et al. [292], and Liu and Liu [291]. In this way the first 16 unicyclic graphs ordered by their L-spectral radii (for n sufficiently large) are obtained. These graphs are illustrated in Fig. 6.2. Graphs with equal L-spectral radii are enumerated by the same number.

Concerning the similar ordering of bicyclic graphs, He et al. [207] determined the first six such graphs, while Li et al. [268] determined the next five graphs, completing the list of all bicyclic graphs (11 in total) whose L-spectral radius lies in $[n - 1, n]$ (again, for n sufficiently large). This investigation was continued by Jin and Zuo, [236] who determined all 42 bicyclic graphs with $\mu_1 \in [n - 2, n - 1)$ along with their ordering. In Fig. 6.3 we give the first 11 bicyclic graphs, while the remaining ones can be found in the corresponding reference.

Guo and Wang [192] identified the cacti with n vertices and k cycles (with $k \geq 2$ and $n > 2k + 8$) satisfying the property $\mu_1 > n - 2$. There are exactly 20 such cacti, which are illustrated in Fig. 6.4.

Remark 6.35 Note that the values of the L-spectral radii of various graphs whose ordering is considered in the above are computed in the corresponding references. Note also that some bicyclic graphs from Fig. 6.3 can be identified in Fig. 6.4 (for $k = 2$). In addition, it appears that the L-spectral radii of some unicyclic graphs from Fig. 6.2 coincide with those of cacti from Fig. 6.4. For example, $\mu_1(U^{(i)}) = \mu_1(G^{(i)})$ ($1 \leq i \leq 4$). For more details see [192].

6.6 General inequalities for algebraic connectivity

Among all eigenvalues of different matrices associated with graphs, one of the most popular is the second smallest L-eigenvalue, called by Fiedler [160], the *algebraic connectivity*. Throughout this section and later, we denote the algebraic connectivity by a $(= a(G))$ (rather than μ_{n-1}), and call the corresponding eigenvector the *Fiedler eigenvector*.

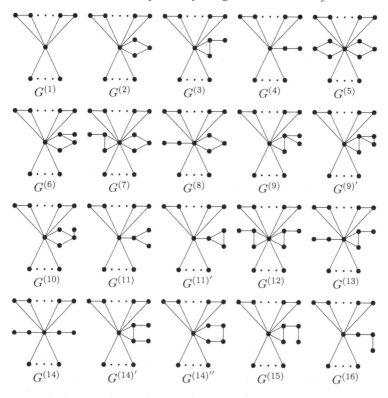

Figure 6.4 The first 20 k-cyclic cacti with $k \geq 2$ and $n > 2k + 8$ ordered by their L-spectral radii (cf. [192]).

By Theorem 1.22, the algebraic connectivity of a non-trivial complete graph is equal to its order. In the mentioned paper [160], the author established some classical results on this eigenvalue. First of all, he gave the possible range for the algebraic connectivity of any non-complete graph by

$$0 \leq a \leq n - \alpha, \tag{6.39}$$

where α denotes the independence number. By Theorem 1.13, the left inequality holds if and only if the graph is disconnected. From the right inequality we immediately obtain $a \leq n - 2$ for any non-complete graph. Another range is given in terms of the minimal vertex degree δ by

$$2\delta - n + 2 \leq a \leq \frac{n}{n-1}\delta. \tag{6.40}$$

If c_v and c_e respectively denote the vertex and the edge connectivity of a non-complete graph, then we have

$$a \leq c_v \leq c_e \leq \delta. \qquad (6.41)$$

The last inequalities explain the name of this eigenvalue: a larger a causes larger vertex and edge connectivity, and a smaller vertex connectivity causes smaller a. Apart from the original reference [160], the proof can be found in many books (e.g., [58, 98, 102]) or survey papers (e.g., [3, 315, 325]). Moreover, one can find a lower bound involving the edge connectivity:

$$a \geq \begin{cases} 2 \left(\cos \frac{\pi}{n} - \cos \frac{2\pi}{n} \right) c_e - 2 \cos \frac{\pi}{n} \left(1 - \cos \frac{\pi}{n} \right) \Delta, \\ 2 c_e \left(1 - \cos \frac{\pi}{n} \right). \end{cases} \qquad (6.42)$$

By Theorem 1.11, $\mu_1(\overline{G}) = n - a(G)$ and, by Theorem 1.22, for regular graphs, $\lambda_2(G) = r - a(G)$. These equalities enable us to observe that any result concerning μ_1 can be applied to the algebraic connectivity of the complementary graph. Also, any result concerning λ_2 of a regular graph can be transferred here directly. For example, a connection between a and the connectivity of a regular graph can be derived from Theorem 4.1.

The following two lemmas are used in the sequel. Both are proved using the Rayleigh principle, so we omit the proof of the first one.

Lemma 6.36 (Fiedler [161]) *For a graph G,*

$$a = 2n \min \frac{\sum_{i \sim j} (y_i - y_j)^2}{\sum_{i=1}^{n} \left(\sum_{j=1}^{n} (y_i - y_j)^2 \right)}, \qquad (6.43)$$

where the minimum is taken over all non-constant vectors $\mathbf{y} = (y_1, y_2, \ldots, y_n)^T$.

Lemma 6.37 (Fiedler [160]) *Let G_1 and G_2 be two edge-disjoint spanning subgraphs of a graph G such that $m(G_1) + m(G_2) = m(G)$. Then $a(G_1) + a(G_2) \leq a(G)$.*

Proof Using the Rayleigh principle, we get

$$a(G) = \min_{\mathbf{y} \in \mathbb{R}^n, \|\mathbf{y}\|=1, \mathbf{y} \perp \mathbf{j}} \mathbf{y}^T L(G) \mathbf{y} = \min_{\mathbf{y} \in \mathbb{R}^n, \|\mathbf{y}\|=1, \mathbf{y} \perp \mathbf{j}} \left(\mathbf{y}^T L(G_1) \mathbf{y} + \mathbf{y}^T L(G_2) \mathbf{y} \right)$$

$$\geq \min_{\mathbf{y_1} \in \mathbb{R}^n, \|\mathbf{y_1}\|=1, \mathbf{y_1} \perp \mathbf{j}} \mathbf{y_1}^T L(G_1) \mathbf{y_1} + \min_{\mathbf{y_2} \in \mathbb{R}^n, \|\mathbf{y_2}\|=1, \mathbf{y_2} \perp \mathbf{j}} \mathbf{y_2}^T L(G_2) \mathbf{y_2}$$

$$= a(G_1) + a(G_2).$$

$\square$

We proceed with the upper and lower bounds for the algebraic connectivity, continue with the bounds for some graph invariants expressed in terms of the algebraic connectivity, and finally consider the isoperimetric problem and expander graphs.

6.6.1 Upper and lower bounds

We give four upper and three lower bounds for the algebraic connectivity.

Upper bounds

We start with an upper bound expressed just in terms of the number of edges in a graph.

Theorem 6.38 (Belhaiza et al. [34]) *For a connected non-complete graph G,*

$$a \leq \lfloor \sqrt{1 + 2m} - 1 \rfloor. \tag{6.44}$$

Proof Let G have at least $\binom{n}{2} - \lfloor \frac{n}{2} \rfloor$ edges. In this case $2m + 1 \geq (n-1)^2$, and thus we have $n - 2 \leq \lfloor \sqrt{1 + 2m} - 1 \rfloor$. Since G is non-complete, the inequality (6.39) gives $a(G) \leq n - 2$, and we get the assertion.

Denote by H a graph with n vertices and $m(H)$ edges, such that $\binom{n-1}{2} \leq m(H) \leq \binom{n}{2} - \lfloor \frac{n}{2} \rfloor - 1$. In this case $2m(H) + 1 \geq (n-2)^2$, and so $n - 3 \leq \lfloor \sqrt{1 + 2m(H)} - 1 \rfloor$.

Observe that there is at least one vertex of degree at most $n - 3$ in H (since it has at most $\binom{n}{2} - \lfloor \frac{n}{2} \rfloor - 1$ edges). By the inequality (6.41), we have $a(H) \leq n - 3$. Since any graph G with at most $\binom{n}{2} - \lfloor \frac{n}{2} \rfloor - 1$ edges is a spanning subgraph of some graph H specified above, the assertion follows from Lemma 6.37. $\square$

We now give three upper bounds for the algebraic connectivity that involve the domination number φ of a graph.

Theorem 6.39 (Nikiforov [331]) *For a non-trivial graph G with domination number* φ,

$$a \leq \begin{cases} n, & \text{if } \varphi = 1, \\ n - \varphi, & \text{if } \varphi \geq 2. \end{cases} \tag{6.45}$$

If $\varphi = 1$, *equality holds if and only if* $G \cong K_n$. *If* $\varphi = 2$, *equality holds if and only if G is the complement of* $\frac{n}{2}K_2$. *If* $\varphi > 2$, *(6.45) is always a strict inequality.*

Proof The inequality $a \leq n$ holds for any graph, and it is attained only for the complete graph K_n (see the beginning of this section). For $\varphi \geq 2$, the inequality follows from a simple combinatorial result $\delta \leq n - \varphi$ and the inequality (6.41).

Let us determine when the equality holds for $\varphi \geq 2$. If $G \cong \overline{\frac{n}{2}K_2}$ then $a = n - \mu_1(\overline{G}) = n - 2$, so (6.45) is an equality.

Assume now that the equality holds. Then we have $a = \delta = n - \varphi$, and thus $\mu_1(\overline{G}) = \varphi = \Delta(\overline{G}) + 1$, that is, the equality in (6.1) is attained, which yields that $\overline{G}$ has a component G_1 with φ vertices and maximal vertex degree $\varphi - 1$. Since G_1 is a component in $\overline{G}$, each vertex of G_1 (considered in G) is adjacent to all vertices of $G - V(G_1)$, and therefore $\varphi = 2$, which gives $\Delta(\overline{G}) = 1$, and the proof follows.   □

Lu et al. [299] gave the second inequality:

$$a \leq n - \varphi + \frac{n - \varphi^2}{n - \varphi}, \tag{6.46}$$

which holds for $n \geq 2$. Its proof and a considerably long proof of the third inequality obtained by using a computer search can be found in the corresponding references.

Theorem 6.40 (Aouchiche et al. [15]) *Let G be a connected graph with odd order* $n = 2k + 1 \geq 9$, *algebraic connectivity a, minimal vertex degree* δ, *and domination number* $\varphi \geq 3$. *If* $\delta \in \{1, 3, 5\}$ *or* δ *is even and G is not isomorphic to any of the three regular graphs depicted in Fig. 6.5, then*

$$a \leq 2(k - \varphi) + \frac{k + 3 - \sqrt{(k+1)^2 + 4}}{2}. \tag{6.47}$$

It can be verified that the bound (6.47) is attained for any of the following graphs obtained from the complete graph K_{k+1} with the vertex set $\{v_1, v_2, \ldots, v_{k+1}\}$:

Figure 6.5 Three forbidden graphs from Theorem 6.40.

- a graph H_1 obtained by adding the vertices $v'_1, v'_2, \ldots, v'_k$ and the edges $v_i v'_i$, for $1 \leq i \leq k$;
- a graph H_2 obtained from H_1 by adding the edge $v_{k+1} v'_k$;
- a graph H_3 obtained from H_2 by deleting the edge $v_k v_{k+1}$.

Remark 6.41 An analogous bound for n even is given in Exercise 6.9. Comparing the inequalities (6.45)–(6.47), we can observe that (6.45) is finer than (6.46) whenever $n > \varphi^2$. As a matter of fact, (6.47) with notable restrictions on n, δ, and φ gives a better estimate than either of (6.45) and (6.46), whenever it can be applied. □

Example 6.42 We compute the upper bounds (6.44)–(6.46) for the complements of unions of equal paths:

Graph	a	(6.44)	(6.45)	(6.46)
$\overline{3P_3}$	6	6	7	7.714
$\overline{5P_3}$	12	12	13	13.846
$\overline{3P_5}$	11.382	12	13	13.846
$\overline{3P_5}$	21.382	22	23	23.913

□

Lower bounds

To present a lower bound for the algebraic connectivity we need some definitions taken from [372]. The *connection-graph-stability score* for an edge e of a graph G is defined as the sum of lengths of all paths that contain the edge e, that is

$$C_e = \frac{1}{2} \sum_{i=1}^{n} \sum_{j=1}^{n} \mathbf{I}_{ij}(e) |P_{ij}|,$$

where P_{ij} is a path connecting the vertices i and j, $|P_{ij}|$ its length, and

$$
\mathbf{I_{ij}}(e) = \begin{cases} 1, & \text{if } e \in P_{ij}, \\ 0, & \text{otherwise.} \end{cases}
$$

The *maximum connection-graph-stability score* assigned to the edges of G is denoted by $C_{\max}$.

Let $\mathscr{P}_{ij} = \left\{ P_{ij}^{(1)}, P_{ij}^{(2)}, \ldots, P_{ij}^{(n_{ij})} \right\}$ denote a non-empty set of paths connecting the vertices i and j in a graph G, and let $\mathbf{y_{ij}} = \left(y_{ij}^{(1)}, y_{ij}^{(2)}, \ldots, y_{ij}^{(n_{ij})} \right)$ be a unit non-negative vector. The set of all vectors $\mathbf{y_{ij}}$ is denoted by Y and called the *path weighting strategy*. The *extended connection-graph-stability score* for an edge e is defined as

$$
C_e(Y) = \frac{1}{2} \sum_{i=1}^{n} \sum_{j=1}^{n} \sum_{q=1}^{n_{ij}} \mathbf{I_{ij}^{(q)}}(e) |P_{ij}^{(q)}| y_{ij}^{(q)},
$$

where

$$
\mathbf{I_{ij}^{(q)}}(e) = \begin{cases} 1, & \text{if } e \in P_{ij}^{(q)}, \\ 0, & \text{otherwise.} \end{cases}
$$

The *maximum extended connection-graph-stability score* assigned to the edges of G is denoted by $C_{\max}(Y)$.

We prove the following theorem.

Theorem 6.43   (Rad et al. [372]) *For a connected graph,*

$$
a \geq \frac{n}{C_{\max}(Y)}. \tag{6.48}
$$

Proof   Let $P_{ij}^{(q)} = i w_1 w_2 \cdots w_k j$ be the qth path that connects i to j, and let $\mathbf{y} \in \mathbb{R}^n$ denote a unit vector orthogonal to $\mathbf{j}$. The difference between the coordinates y_i and y_j can be expressed as

$$
y_i - y_j = (y_i - y_{w_1}) + (y_{w_1} - y_{w_2}) + \cdots + (y_{w_k} - y_j) = \sum_{e \in P_{ij}^{(q)}} r_e,
$$

where r_e stands for the difference between the coordinates of $\mathbf{y}$ that correspond to the vertices incident with e. Using the Cauchy–Schwarz inequality, we get

$$(y_i - y_j)^2 = \left(\sum_{e \in P_{ij}^{(q)}} r_e \right)^2 \leq \left| P_{ij}^{(q)} \right| \sum_{e \in P_{ij}^{(q)}} r_e^2 = \left| P_{ij}^{(q)} \right| \sum_{e \in E(G)} \mathbf{I}_{ij}^{(q)}(e) r_e^2.$$

Next, $(y_i - y_j)$ can be expressed as a weighted average of its alternative expansions, as follows:

$$(y_i - y_j)^2 = \sum_{q=1}^{n_{ij}} y_{ij}^{(q)} (y_i - y_j)^2 \leq \sum_{q=1}^{n_{ij}} y_{ij}^{(q)} \left| P_{ij}^{(q)} \right| \sum_{e \in E(G)} \mathbf{I}_{ij}^{(q)}(e) r_e^2.$$

Hence,

$$\sum_{i=1}^{n} \sum_{j=1}^{n} (y_i - y_j)^2 \leq \left(\sum_{i=1}^{n} \sum_{j=1}^{n} \sum_{q=1}^{n_{ij}} \left| P_{ij}^{(q)} \right| y_{ij}^{(q)} \right) \sum_{e \in E(G)} \mathbf{I}_{ij}^{(q)}(e) r_e^2$$

$$= \sum_{e \in E(G)} 2C_e(Y) r_e^2 \leq 2C_{\max}(Y) \sum_{e \in E(G)} r_e^2.$$

Using (6.43), we get

$$a = 2n \min \frac{\sum_{i \sim j} (y_i - y_j)^2}{\sum_{i=1}^{n} \left(\sum_{j=1}^{n} (y_i - y_j)^2 \right)} \geq 2n \min \frac{\sum_{i \sim j} (y_i - y_j)^2}{2C_{\max}(Y) \sum_{i \sim j} (y_i - y_j)^2},$$

and the proof follows. $\qquad \square$

Following [372], we confirm that the inequality (6.48) is better than the inequality of Mohar [323]:

$$a \geq \frac{4}{nD}, \tag{6.49}$$

where D is the graph diameter. Namely, it is sufficient to prove

$$\frac{n}{C_{\max}(Y)} \geq \frac{4}{nD},$$

and, since $C_{\max} \geq C_{\max}(Y)$, it is sufficient to prove

$$\frac{n}{C_{\max}} \geq \frac{4}{nD}.$$

Consider the set of all shortest paths passing through an edge $e = u_1 u_2$, and let $V_1 = \{v \in V(G)$: there is a shortest path from u_2 to v containing $u_1\}$ and $V_2 = \{v \in V(G)$: there is a shortest path from u_1 to v containing $u_2\}$. Note that $u_i \in V_i$ $(1 \le i \le 2)$, and also $V_1 \cap V_2 = \emptyset$.

If $n_i = |V_i|$ $(1 \le i \le 2)$, then there are at most $n_1 n_2$ shortest paths between V_1 and V_2 passing through e, hence $C_e \le n_1 n_2 D \le \left(\frac{n}{2}\right)^2 D$. Since the last inequality holds for any edge e, it also holds for the edge that has the maximum connection-graph-stability score. Consequently, we have

$$C_{\max} \le \left(\frac{n}{2}\right)^2 D,$$

and thus

$$\frac{n}{C_{\max}} \ge \frac{n}{\left(\frac{n}{2}\right)^2 D} = \frac{4}{nD}.$$

It can be proved in a similar way (which we leave to the reader) that the inequality (6.48) improves the inequality of Lu et al. [301]:

$$a \ge \frac{2n}{2 + (n(n-1) - 2m)D}. \tag{6.50}$$

6.6.2 Bounding graph invariants by algebraic connectivity

An upper bound for the diameter was obtained by Mohar [323], who determined the limit points for the number of vertices at a fixed distance from a given vertex:

$$D \le 2 \left\lceil \frac{\Delta + a}{4a} \ln(n-1) \right\rceil. \tag{6.51}$$

As was shown in the same paper, this inequality improves the similar result obtained by Alon and Milman [10]:

$$D \le 2 \left\lceil \sqrt{\frac{2\Delta}{a}} \log_2 n \right\rceil. \tag{6.52}$$

The *mean distance* $\overline{\rho}(G)$ of a graph G is equal to the average of all distances between distinct vertices of G, that is,

$$\overline{\rho}(G) = \frac{2}{n(n-1)} \sum_{i \neq j} d(i,j).$$

An upper bound for $\overline{\rho}(G)$ was obtained by the same author [323]:

$$\overline{\rho}(G) \leq \frac{n}{n-1} \left(\left\lceil \frac{\Delta+a}{4a} \ln(n-1) \right\rceil + \frac{1}{2} \right). \tag{6.53}$$

6.6.3 Isoperimetric problem and graph expansion

According to [102, p. 205], the classical *isoperimetric problem* is to determine the maximum area with given perimeter or the maximum volume with given surface area. In graph theory, an analogue is to determine the maximum number of vertices in a set S with edge boundary ∂S of prescribed cardinality. (The edge boundary is defined in Subsection 4.1.1 to be the number of edges with exactly one end in S.) Recall that the expansion parameter is also defined in the same subsection as

$$h(G) = \min_{1 \leq |S| \leq \frac{n}{2}} \frac{|\partial S|}{|S|},$$

where $S \subset V(G)$. As we will see, this quotient is closely related to all L-eigenvalues (and especially to the algebraic connectivity). Note also that, when considered in relation to the L-eigenvalues, the expansion parameter is often called the *isoperimetric number* (sometimes the *conductance*) and denoted by $i(G)$. To be consistent we use the terminology and notation introduced earlier in Subsection 4.1.1. Notice that, in the same section, only regular graphs were considered in the context of graph expansion. Here we take into account all connected graphs (since $h(G) = 0$ if and only if G is disconnected, disconnected graphs are not of interest).

A small value of $h(G)$ causes a comparatively large set of vertices to be separated by a comparatively small number of edges, and so the expansion parameter can be seen as a measure of graph connectivity. If $\mathbf{y}$ denotes the characteristic vector of S then $\mathbf{y}^T L(G)\mathbf{y} = \sum_{i \sim j}(y_i - y_j)^2$, and thus

$$\frac{\mathbf{y}^T L(G)\mathbf{y}}{\mathbf{y}^T \mathbf{y}} = \frac{|\partial S|}{|S|},$$

which gives a connection between the expansion parameter and the L-eigen-values. Moreover, in view of the last equality we have

$$h(G) = \min\left\{\frac{\mathbf{y}^T L(G)\mathbf{y}}{\mathbf{y}^T\mathbf{y}} \; : \; \mathbf{y} \in \{0,1\}^n, \; 1 \leq \mathbf{y}^T\mathbf{j} \leq \frac{n}{2}\right\}.$$

Thus, $a(G)$ and $h(G)$ are obtained by optimizing the same function on different subsets of $\mathbb{R}^n$.

We present a lower and an upper bound for $h(G)$ that involve the algebraic connectivity. Both were obtained by Mohar [324]. The proof of the second bound can be found in [102] as well.

Theorem 6.44   *For any non-trivial graph,*

$$h(G) \geq \frac{a}{2}. \tag{6.54}$$

Proof   Let $\mathbf{y}$ be the characteristic vector of S. Since $|\partial S| = \sum_{i \sim j}(y_i - y_j)^2$ and $2|S||\partial S| = \sum_{i=1}^n \sum_{j=1}^n (y_i - y_j)^2$, using (6.43) we get

$$a\frac{|\overline{S}|}{n} \leq \frac{|\partial S|}{|S|}.$$

Since $h(G) \geq \frac{|\partial S|}{|S|}$ by definition, the proof follows from $\lceil\frac{n}{2}\rceil \leq |\overline{S}|$. $\square$

Theorem 6.45   *If $G \notin \{K_1, K_2, K_3\}$, then*

$$h(G) \leq \sqrt{a(2\Delta - a)}. \tag{6.55}$$

In Subection 4.1.1 we defined the vertex expansion parameter to be equal to $\min_{1 \leq |S| \leq \frac{n}{2}} \frac{|N(S)|}{|S|}$, where $N(S)$ stands for the set of vertices that have at least one neighbour in S. By its definition, $N(S)$ may include some vertices belonging to S. Here we use the *vertex boundary* of S, ∂S, defined to be the set of vertices outside S which are adjacent to some vertex inside S. In this way we get an alternative definition of the vertex expansion parameter:

$$\check{h}(G) = \min_{1 \leq |S| \leq \frac{n}{2}} \frac{|\partial S|}{|S|}.$$

According to [102, p. 208], the differences between the various measures of expansion used are largely superficial in that they all conform to the general principle that expansion in graphs of bounded vertex degree is controlled by the algebraic connectivity. In the case of the edge expansion parameter $h(G)$, this is confirmed by the previous two theorems. In the following two theorems we establish an analogous property for $\dot{h}(G)$.

Theorem 6.46 (Alon [6]) *Let G be a non-trivial graph. If $0 \leq \varepsilon \leq a$, then*

$$\dot{h}(G) \geq \frac{2\varepsilon}{\Delta + 2\varepsilon}. \tag{6.56}$$

Proof The inequality holds for any disconnected graph, so we may take $\varepsilon > 0$. Let $S \subset V(G)$ and $1 \leq |S| \leq \frac{n}{2}$. If $|S| + |\dot{\partial}S| = n$ (where n is the number of vertices in G), then

$$\frac{|\dot{\partial}S|}{|S|} \geq 1 > \frac{2\varepsilon}{\Delta} + 2\varepsilon.$$

Otherwise, we apply the inequality (6.78) of Exercise 6.14 with $V(G) \backslash (S \cup \dot{\partial}S)$ in the role of T. Since $d(S, T) = 2$ and $a \geq \varepsilon$, we have

$$\frac{n - |S| - |\dot{\partial}S|}{n} \leq \frac{1 - \frac{|S|}{n}}{1 + \frac{4\varepsilon|S|}{n\Delta}},$$

or

$$\frac{|\dot{\partial}S|}{n} \geq \left(1 - \frac{|\dot{\partial}S|}{n}\right)\left(1 - \frac{1}{1 + \frac{4\varepsilon|S|}{n\Delta}}\right).$$

Since $|S| \leq \frac{n}{2}$, we conclude that

$$\frac{|\dot{\partial}S|}{|S|} \geq \frac{\frac{2\varepsilon}{\Delta}}{1 + \frac{4\varepsilon|S|}{n\Delta}} \geq \frac{2\varepsilon}{\Delta + 2\varepsilon},$$

and the result follows. $\qquad\square$

Theorem 6.47 (Alon [6], cf. [102, Theorem 7.6.2]) *If G is a non-trivial graph with $0 < c \leq \dot{h}(G)$, then*

$$a \geq \frac{c^2}{2(c^2 + 2)}. \tag{6.57}$$

6.7 Notes on algebraic connectivity of trees

We first give some inequalities for trees and then consider trees with large values of the algebraic connectivity.

Inequalities for trees

Among all trees of fixed order, the maximal and the minimal algebraic connectivity are attained for the star $K_{1,n-1}$ and the path P_n, respectively. This fact appears in many references and, for example, it can be verified by Lemma 1.29. In addition, for $n \geq 1$ we have $a(P_n) = 2\left(1 - \cos\frac{\pi}{n}\right)$, while for $n \geq 3$, we have $a(K_{1,n-1}) = 1$. Grone et al. [176] proved that $a(T) \leq 0.49$ holds for any tree T with at least six vertices and distinct from the star. In other words, there is a considerably large gap between the algebraic connectivity of a star and that of any other tree with $n \geq 6$ vertices. Moreover, the same authors obtained a lower bound for a tree as a function of its diameter:

$$a(T) \leq \left(1 - \cos\left(\frac{\pi}{D+1}\right)\right). \tag{6.58}$$

We may consider the algebraic connectivity of rooted Bethe trees $T_{\Delta,k}$, defined on page 54.

Theorem 6.48 (Rojo and Medina [382]) *For a rooted Bethe tree $T_{\Delta,k}$, let*
$f(\Delta,k) = \frac{(\Delta-2)^2}{\Delta-2\sqrt{\Delta-1}\cos\frac{\pi}{2k-1}}$ *then*

$$\frac{(\Delta-2)^2}{\left((\Delta-1)^k - (2k-1)(\Delta-2)\left(1 - \frac{1}{(\Delta-1)^{k-\frac{1}{2}}+1}\right) - \sqrt{(\Delta-1)}\right) + f(\Delta,k)}$$

$$\leq a \leq \frac{(\Delta-2)^2}{(\Delta-1)^k - 2(k-1)\left((\Delta-1) + \frac{(\Delta-2)}{(\Delta-1)^{k-1}-1}\right) + 2k-1}. \tag{6.59}$$

The algebraic connectivity of generalized Bethe trees was considered in [380]. Binary trees are special case of Bethe trees (obtained for $\Delta = 3$). The analogous inequalities for these trees were given by Molitierno et al. [327]:

$$\frac{1}{(2^k - 2k + 2) - \frac{2k-\sqrt{2}(2k-1-2^{k-1})}{2^k-1-\sqrt{2(2^{k-1}-1)}} + \frac{1}{3-2\sqrt{2}\cos\frac{\pi}{2k-1}}}$$

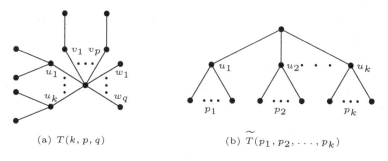

Figure 6.6 Trees $T(k,p,q)$ and $\widetilde{T}(p_1,p_2,\ldots,p_k)$.

$$\leq a \leq \frac{1}{(2^k - 2k + 3) - \frac{2(k-1)}{2^{k-1}-1}}. \tag{6.60}$$

We comment that for large k the difference between the denominators in the lower bound and the upper bound of (6.60) is close to $1 + \sqrt{2}$.

Trees with large algebraic connectivity

We determine all trees with suitably large order whose algebraic connectivity is at least $\frac{5-\sqrt{21}}{2}$. Let $T(k,p,q)$ denote a tree with n vertices obtained by taking k paths P_3, p paths P_2, and $q+1$ single vertices and by inserting q edges joining a fixed single vertex, say v, to the remaining single vertices, p edges joining v to an end of each path P_2, and k edges joining v to the middle vertex of each path P_3 (see Fig. 6.6(a)). We have $3k + 2p + q + 1 = n$.

Theorem 6.49 (Yuan et al. [493]) *If T is a tree with at least* 15 *vertices and $a(T) \geq 2 - \sqrt{3}$ then $T \in \bigcup_{i=1}^{6} \mathscr{T}_i$, where*

- $\mathscr{T}_1 = \{K_{1,n-1}\}$,
- $\mathscr{T}_2 = \{DS(2,n-2)\}$,
- $\mathscr{T}_3 = \{T(0,p,q) : p \geq 2\}$,
- $\mathscr{T}_4 = \{DS(3,n-3)\}$,
- $\mathscr{T}_5 = \{T(1,p,q) : p \geq 1\}$,
- $\mathscr{T}_6 = \{T(k,p,q) : k \geq 2\}$.

Proof If the diameter D of T is at least 5, then T contains at least one of the path P_7 or the caterpillars T_7^2 and T_7^3 as an induced subgraph. Since the algebraic connectivity of any of these graphs is less than $2 - \sqrt{3}$, by Theorem 1.14, $a(T) < 2 - \sqrt{3}$.

For $D = 4$, using similar reasoning we get $T \in \mathscr{T}_3 \cup \mathscr{T}_5 \cup \mathscr{T}_6$.

If $D = 3$ or $D = 2$, T must be a double star or a star, and the proof follows by direct computation. □

An extension of the previous result is given in the following theorem.

Theorem 6.50 (Wang and Tan [454]) *If T is a tree with at least 45 vertices and $\frac{5-\sqrt{21}}{2} \le a(T) < 2 - \sqrt{3}$ then $T \in \bigcup_{i=7}^{10} \mathscr{T}_i$, where*

- $\mathscr{T}_7 = \{DS(4, n-4)\}$,
- $\mathscr{T}_8 = \{\widetilde{T}(3, 1, p_3, p_4, \ldots, p_k)\}$,
- $\mathscr{T}_9 = \{\widetilde{T}(3, 2, p_3, p_4, \ldots, p_k)\}$,
- $\mathscr{T}_{10} = \{\widetilde{T}(3, 3, p_3, p_4, \ldots, p_k)\}$,

where $T(p_1, p_2, \ldots, p_k)$ is illustrated in Fig. 6.6(b).

The proof is similar. Furthermore, the graphs in $\bigcup_{i=1}^{6} \mathscr{T}_i$ can be ordered in the following way: if $T^{(i)} \in \mathscr{T}_i$ $(1 \le i \le 6)$, then $a(T^{(i)}) > a(T^{(j)})$ whenever $i < j$.

6.8 Graphs with extremal algebraic connectivity

We consider graphs of fixed order and diameter that attain the maximal value of algebraic connectivity. In the sequel, we give a number of particular graphs that attain the minimal value of the same invariant.

Order, diameter, and maximal algebraic connectivity

We consider the clique chain graphs denoted by $G^D(n_1, n_2, \ldots, n_{D+1})$. By Theorem 6.31, within the set of graphs of fixed order and diameter, an arbitrary L-eigenvalue attains its maximal value for some of these graphs. In particular, following Wang et al. [446], we reinterpret the following discussion for algebraic connectivity.

The Laplacian characteristic polynomial C_{G^D} of $G^D(n_1, n_2, \ldots, n_{D+1})$ is computed in [319]:

$$C_{G^D}(x) = p_D(x) \prod_{i=1}^{D+1} (\bar{d}_i + 1 - x)^{n_i - 1}, \tag{6.61}$$

where $\bar{d}_i$ denotes the degree of a vertex in the clique i. The polynomial $p_D(x) = \prod_{i=1}^{D+1} \theta_i$ is of degree $D + 1$ in x and the function $\theta_i = \theta_i(D, x)$ obeys the recursion

$$\theta_i = (\overline{d}_i + 1 - x) - \left(\frac{n_{i-1}}{\theta_{i-1}} + 1\right) n_i,$$

with initial condition $\theta_0 = 1$ and with the convention that $n_0 = n_{D+2} = 0$.

From (6.61) we see that the Laplacian C_{G^D} has roots at $n_{i-1} + n_i + n_{i+1} = \overline{d}_i + 1$ with multiplicity $n_i - 1$, for $1 \le i \le D + 1$ (with $n_0 = n_{D+2} = 0$). The remaining roots are the solutions of the polynomial p_D. Since the explicit L-eigenvalues $\mu_i = \overline{d}_i + 1$ are larger than zero, we must have $p_D(0) = 0$ and thus the polynomial of interest is

$$p_D(x) = \prod_{i=1}^{D+1} \theta_i(D,x) = \sum_{j=0}^{D+1} c_j(D)x^j = \prod_{j=1}^{D+1} (z_j - x).$$

Since the algebraic connectivity a of any non-complete graph satisfies $a \le \delta$, we have $a = z_D$ (i.e., the algebraic connectivity is always equal to the second smallest eigenvalue of the polynomial p_D).

In what follows we consider graphs of fixed order and diameter that maximize the algebraic connectivity when $D \le 4$.

For $D = 1$, K_n is the unique solution. For $D = 2$, from Theorem 6.31 we have $n_1 = n_3 = 1$, and thus the unique maximizer is $G^2(1, n - 2, 1)$ – that is, the graph obtained by deleting an edge from K_n. In addition, $a(G^2(1, n - 2, 1)) = n - 2$.

For $D = 3$, from (6.61) we get that the algebraic connectivity of the graph $G^3(1, n_2, n_3, 1)$ is the smallest root of

$$p_3(x) = x^3 - 2(n-1)x^2 + ((n-2)(n+1) + (n_2 - 1)(n - n_2 - 3))x \\ - nn_2(n - n_2 - 2).$$

Maximizing the smallest root of the above polynomial we get the unique solution $G^3(1, \lfloor \frac{n-2}{2} \rfloor, n - 2 - \lfloor \frac{n-2}{2} \rfloor, 1)$, with $\lfloor \frac{n-2}{2} \rfloor - 1 \le a(G^3(1, \lfloor \frac{n-2}{2} \rfloor, n - 2 - \lfloor \frac{n-2}{2} \rfloor, 1)) \le \lfloor \frac{n-2}{2} \rfloor$.

For $D = 4$, in a similar way (the details can be seen in [446]) we get the maximizing graph $G^4(1, n_2, n - n_2 - n_4, n_4, 1)$ with $(n_2, n_4) = (\lfloor k \rfloor, \lfloor k \rfloor)$, $(n_2, n_4) = (\lceil k \rceil, \lceil k \rceil)$ or $(n_2, n_4) = (\lfloor k \rfloor, \lceil k \rceil)$, where $k = \frac{n}{3} - \frac{5}{9} - \frac{\sqrt{2(3n-4)}}{18}$. Furthermore, the algebraic connectivity of this graph does not exceed $\frac{n}{3} - \frac{2}{9} - \frac{2\sqrt{2(3n-4)}}{9}$.

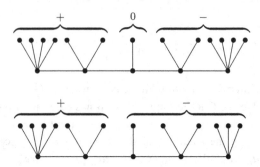

Figure 6.7 Trees with given degree sequence with minimal algebraic connectivity. The sign patterns of the Fiedler eigenvector are associated with the vertices (cf. [44]).

Graphs with minimal algebraic connectivity

Concerning the graphs with small algebraic connectivity, Godsil and Royle [170] assumed that they tend to be elongated graphs of large diameter with cut edges.

The problem of determining a graph of fixed order and diameter with minimal algebraic connectivity was resolved completely within the set of trees by Fallat and Kirkland [148]. The tree of fixed order and diameter that attains the minimal value of the algebraic connectivity is obtained by attaching an almost equal number of pendant edges at different ends of a path. Moreover, the same tree attains the minimal algebraic connectivity uniquely within the set of connected graphs of fixed order and number of pendant vertices k ($k \geq 2$) [255].

As we have seen, the graph of fixed order and diameter with maximal algebraic connectivity is a clique chain graph. In this context it is worth mentioning that Bıyıkoğlu and Leydold [45] proved that a connected graph of fixed order and size with minimal algebraic connectivity must consist of a chain of cliques. Moreover, it is conjectured in [34] that any connected graph of fixed order and size with minimal algebraic connectivity is a so-called path-complete graph. A *path-complete graph* consists of a clique, a path, and one or several edges joining an end of the path to one or several vertices of the clique. The conjecture is confirmed for graphs with at most 11 vertices.

Concerning trees of fixed degree sequence with minimal algebraic connectivity, Bıyıkoğlu and Leydold [44] proved that such a tree must be a caterpillar whose non-pendant vertices are ordered non-decreasingly in both subgraphs induced by the sign pattern of a Fiedler eigenvector. Two examples with de-

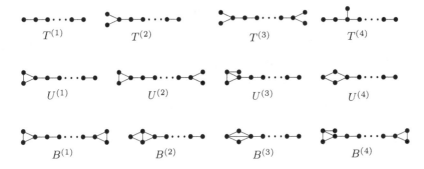

Figure 6.8 Trees, unicyclic graphs, and bicyclic graphs with smallest algebraic connectivity.

gree sequences $(5,5,4,4,3,\underbrace{1,1,\ldots,1}_{13})$ and $(5,4,4,4,3,\underbrace{1,1,\ldots,1}_{12})$ are depicted in Fig. 6.7.

6.9 Ordering graphs by algebraic connectivity

We consider the set of trees (with $n \geq 9$), unicyclic and bicyclic graphs (both with $n \geq 13$), and give the four graphs with the smallest algebraic connectivity ordered increasingly in each of these sets. The graphs are given in Fig. 6.8. For more details, see [258, 259, 398] (extended orderings, containing eight trees and seven unicyclic graphs, can be found in the first reference). Combining these results, we obtain the six connected graphs on at least nine vertices with the smallest algebraic connectivity. These graphs are $T^{(1)}, T^{(2)}, U^{(1)}, T^{(3)}$, $U^{(2)}$, and $B^{(1)}$. It holds that $a(T^{(1)}) < a(T^{(2)}) = a(U^{(1)}) < a(T^{(3)}) = a(U^{(2)}) = a(B^{(1)})$.

6.10 Other L-eigenvalues

We consider a sequence of bounds that concern any L-eigenvalue of a graph. In the sequel, we give more specific bounds for trees. At the end we consider the graphs with small L-eigenvalues μ_2 and μ_3, and some extremal graphs for μ_2.

6.10.1 Bounds for μ_i

We start with the following result.

Theorem 6.51 (Zhang and Li [513]) *For a connected graph G, denote by $t(G)$ the number of spanning trees of G and set*

$$\xi(G) = \min\{m((n-4)m + 2(n-1)), 2m((n(n-1) - 2m)\}.$$

Then, for $1 \le i \le n$,

$$\frac{1}{n-i}\left((n-1)(2^{n-i}nt(G))^{\frac{1}{n-1}} - 2m\right) \le \mu_i \le \frac{1}{n-1}\left(2m + \sqrt{\frac{n-i-1}{i}\xi(G)}\right).$$

$$(6.62)$$

The right equality holds if and only if G is a star or a complete graph.

Proof We prove only the right inequality, the proof of the left one is similar. We have

$$\mathrm{tr}(L(G)^2) = \sum_{k=1}^{i}\mu_k^2 + \sum_{k=i+1}^{n-1}\mu_k^2 \ge \frac{(\sum_{k=1}^{i}\mu_k)^2}{i} + \frac{(\sum_{k=i+1}^{n-1}\mu_k)^2}{n-i-1}$$

$$\ge \frac{(\sum_{k=1}^{i}\mu_k)^2}{i} + \frac{(2m - \sum_{k=1}^{i}\mu_k)^2}{n-i-1},$$

which gives

$$\mu_i \le \frac{\sum_{k=1}^{i}\mu_k}{i} \le \frac{1}{n-1}\left(2m + \sqrt{\frac{n-i-1}{i}((n-1)\mathrm{tr}(L(G)^2) - 4m^2)}\right),$$

and the inequality follows. The equality is easily verified. $\qquad\square$

We continue with the lower and upper bound given in the next two theorems. Both are expressed in terms of vertex degrees. After their presentation, we give some comments.

Theorem 6.52 (Brouwer and Haemers [57]) *Let $d_1 \ge d_2 \ge \cdots \ge d_n$ be the sequence of vertex degrees of a graph G which is not isomorphic to $K_i \cup (n - i)K_1$, then*

$$\mu_i \ge d_i - i + 2 \quad (1 \le i \le n-1). \tag{6.63}$$

Proof We use the following claim. Let S be the set of vertices in G such that each vertex in S has at least k neighbours outside S. If $m = |S| > 0$, then $\mu_m \geq k$. If S contains a vertex adjacent to all other vertices of S, then $\mu_m \geq k+1$.

The first part of the claim is proved by the L-interlacing theorem, and the second follows by Theorem 1.25.

Following [57], let the vertices of G be labelled $u_1, u_2, \ldots, u_n$ such that the degree of u_i is equal to d_i ($1 \leq i \leq n$). Set $S = \{u_1, u_2, \ldots, u_m\}$ and $k = d_m - m + 1$. We have to show $\mu_m \geq k+1$. If each vertex in S has at least $k+1$ neighbours outside S, then the proof follows by the above claim. If not then a vertex with only k neighbours outside S is adjacent to all vertices in S, and the proof follows unless $k = 0$.

For $k = 0$, let $T \subset S$ such that each vertex of T has at most $m-2$ neighbours in S. The case $T = \emptyset$ is easily resolved. Construct a new graph, say H. For all vertices of T with fewer than $m-2$ neighbours in S, delete the edges joining it with vertices outside S until each row in the Laplacian matrix of the graph induced by S gets row sum equal to 1. Delete any possible isolated vertices as well. It remains to show that $\mu_m(H) \geq 1$. Considering the partition of the vertex set $V(H)$ into the $m+1$ parts $\{v\}$ for $v \in S$ and $V \backslash S$, and the corresponding quotient matrix, we easily prove that the second smallest L-eigenvalue of H is at least 1, and the proof follows. $\qquad\square$

Now for the second result.

Theorem 6.53 *Let $d_1 \geq d_2 \geq \cdots \geq d_n$ be the sequence of vertex degrees of a graph G whose complement is not isomorphic to $K_i \cup (n-i)K_1$, then*

$$\mu_i \leq d_{i+1} + n - i - 1 \quad (1 \leq i \leq n-1). \tag{6.64}$$

Proof The proof follows directly from the previous theorem by considering the complementary graph $\overline{G}$ and taking into account Theorem 1.11. $\qquad\square$

Remark 6.54 The previous theorems give the magnitude for any single L-eigenvalue (recall that we have $\mu_n = 0$):

$$d_i - i + 2 \leq \mu_i \leq d_{i+1} + n - i - 1 \quad (1 \leq i \leq n-1).$$

Note also that the lower bound (6.63) for $i = 1$ coincides with the lower bound of (6.1). The lower bounds for $i = 2$ and $i = 3$ were also obtained separately by Li and Pan [264] and Guo [182]. $\qquad\square$

Example 6.55   There exists exactly one non-trivial connected graph G of order n with the property that for every pair of distinct vertices u and v, $d_u \neq d_v$, with exactly one exception [27]. In fact, the degree sequence of G is $n - 1, n-2, \ldots, \lfloor \frac{n}{2} \rfloor + 1, \lfloor \frac{n}{2} \rfloor, \lfloor \frac{n}{2} \rfloor, \lfloor \frac{n}{2} \rfloor - 1, \ldots, 2, 1$. For $n = 8$, we compute the L-eigenvalues of G together with the sequences of lower bounds (6.63) and upper bounds (6.64):

i	1	2	3	4	5	6	7	8
Lower bounds	8	6	4	2	1	-1	-3	
μ_i	8	7	6	5	3	2	1	0
Upper bounds	12	10	8	7	5	3	1	

Observe that the lower bounds less than 0 and the upper bounds greater than 8 are trivial.   □

We point out an extension of Theorem 6.52 for the second largest L-eigenvalue of a connected graph with at least three vertices. Due to Das [127],

$$\mu_2(G) \geq \begin{cases} \frac{d_2 + 2 + \sqrt{(d_2-2)^2 + 4c_{u_1 u_2}}}{2}, & \text{if } u_1 \sim u_2, \\ \frac{d_2 + 1 + \sqrt{(d_2+1)^2 - 4c_{u_1 u_2}}}{2}, & \text{otherwise,} \end{cases} \tag{6.65}$$

where $c_{u_1 u_2} = |N(u_1) \cap N(u_2)|$, while, as above, u_1 denotes a vertex with the largest degree d_1 and u_2 denotes a vertex with the second largest degree d_2.

Note that μ_2 and μ_{n-2} can be considered as more important than other L-eigenvalues between them since, for example, they may coincide respectively with μ_1 and a (for some graphs). In that case μ_1 and μ_2, as well as a and μ_{n-2}, give the same information about graph structure. For example, if $\overline{G}$ has at least three components then, by Theorem 1.11, $\mu_1(G) = \mu_2(G)$. Moreover, the same result gives another conclusion: $\mu_2(G) = n - \mu_{n-2}(\overline{G})$. In other words, any result for μ_2 implies the analogous result for μ_{n-2} of the complementary graph.

We mention two results concerning the L-eigenvalues of trees. The first, obtained by Guo [182], gives a simple upper bound. It can be proved in a way similar to the proof of Theorem 6.52. It reads

$$\mu_i(T) \leq \left\lceil \frac{n}{i} \right\rceil \quad (1 \leq i \leq n). \tag{6.66}$$

Equality holds if and only if $i < n$, i divides n, and T is spanned by i vertex-disjoint copies of $K_{1,\frac{n}{k}-1}$.

The second result is a curious property of $\mu_i(T)$ when T has a perfect matching. Namely, from Li et al. [263],

$$\mu_i(T) \begin{cases} = 2, & \text{if } n = 2i, \\ \leq \dfrac{\lceil\frac{n}{2i}\rceil + 2 + \sqrt{\lceil\frac{n}{2i}\rceil^2 + 4}}{2}, & \text{otherwise.} \end{cases} \tag{6.67}$$

Another upper bound for μ_2 of trees with a perfect matching is given in Exercise 6.17.

6.10.2 Graphs with small μ_2 or μ_3

In [198] the authors observed that the alkane energy levels are related to the L-eigenvalues of the corresponding molecular graphs. They also emphasized the relevance of graphs with a large number of small L-eigenvalues.

In what follows we identify all connected graphs whose second largest L-eigenvalue does not exceed a fixed constant ζ (≈ 3.2470). In particular, we present all connected graphs with $\mu_2 \leq 3$. In the sequel, we consider graphs with small values of μ_3.

The concept of determining graphs with bounded L-eigenvalues is very similar to that exploited in Section 4.2 or Chapter 5. There, we derived some structural properties of graphs we were looking for and then used these properties to identify the desired graphs by means of eigenvalue interlacing and forbidden subgraphs.

Let ζ denote the second largest L-eigenvalue of the path P_7, that is, the largest root of

$$x^3 - 5x^2 + 6x - 1 = 0.$$

We have $\zeta \approx 3.2470$. In order to determine connected graphs with $\mu_2 \leq \zeta$, we use the following two facts.

- By Theorem 6.52, the second largest degree of any such graph does not exceed three.
- Since $\mu_2(P_8) > \zeta$, the diameter of any such graph does not exceed six.

Now, it is a matter of routine to determine all the desired graphs. This was

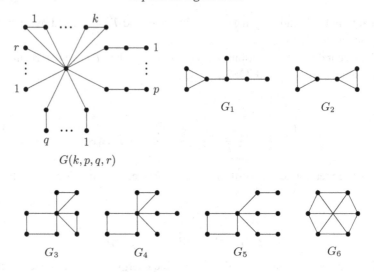

Figure 6.9 Graphs whose subgraphs have the property $\mu_2 \leq \zeta$. Adapted from [468] with permission from Elsevier.

done explicitly by Wu et al. [468] and the result reads as follows. For a connected graph G, $\mu_2(G) \leq \zeta$ if and only if G is a subgraph of one of the graphs $G(k,p,q,r)$ $(k,p,q,r \geq 0)$ or G_i $(1 \leq i \leq 6)$ illustrated in Fig. 6.9.

Connected graphs with the property $\mu_2 \leq 3$ can easily be singled out from those with $\mu_2 \leq \zeta$. These are precisely the subgraphs of $G(k,0,q,r)$ $(k,q,r \geq 0)$, G_2, G_6 or the additional two graphs, one of which is obtained by attaching two pendant edges at a vertex of C_4 and the other by attaching the path P_2 at the same vertex.

The investigation of graphs with bounded μ_i $(i \geq 3)$ is similar. Zhang and Luo [515] determined all connected non-bipartite graphs with $\mu_3 < 3$: there are 12 individual graphs and 6 infinite families of graphs whose subgraphs satisfy the given property. Connected bipartite graphs with the same property are considered by the first author in [505], and similarly there are 9 individual graphs and 10 infinite families of graphs whose bipartite subgraphs satisfy $\mu_3 < 3$.

Exercises

6.1 Prove the upper bound of Li and Zhang [266],

$$\mu_1 \le 2 + \sqrt{(c_1 - 2)(c_2 - 2)}, \qquad (6.68)$$

where $c_1 = \max\{d_i + d_j \ : \ ij \in E(G)\}$ and $c_2 = \max\{d_i + d_j \ : \ ij \in E(G) - k'k''\}$, with $k'k'' \in E(G)$ and $d_{k'} + d_{k''} = c_1$. Show that this bound improves the upper bound (6.11).

6.2 Prove Theorem 6.4.

Hint: Just follow the proof of Theorem 6.2.

6.3 Prove Theorem 6.5.

6.4 Prove the following two upper bounds of Zhang [512]:

$$\mu_1 \le \max\{\sqrt{d_i(d_i + m_i) + d_j(d_j + m_j)} \ : \ i \sim j\} \qquad (6.69)$$

and

$$\mu_1 \le \max\{2 + \sqrt{d_i(d_i + m_i - 4) + d_j(d_j + m_j - 4) + 4} \ : \ i \sim j\}. \qquad (6.70)$$

Hint: Consider the spectral radius of the line graph Line(G) and use the inequality (1.10).

6.5 Show that the upper bound for connected graphs obtained by Li and Pan [265],

$$\mu_1 \le \sqrt{2\Delta^2 + 4m + 2\Delta(\delta - 1) - 2\delta(n - 1)}, \qquad (6.71)$$

is a consequence of (6.22) and thus a consequence of (6.20).

6.6 Prove the following upper bound obtained by Lu et al. [300] and independently by Godsil and Newman [169]:

$$\mu_1 \le \frac{n\delta}{n - \alpha}. \qquad (6.72)$$

6.7 Show that the following inequality holds for any connected graph with diameter D:

$$\mu_1 \geq \frac{(D+1)\delta + 2D}{D+1}. \tag{6.73}$$

Hint: Use Corollary 6.11 and $e_r \leq e_{r+1}$, where e_r is defined by (6.25).

6.8 A graph is called maximal planar if it is planar and adding any edge makes it non-planar. If G is a maximal planar graph with at least four vertices, then show that

$$\mu_1 \leq \max\{d_i - 1 + \sqrt{d_i m_i - 2d_i + 1} \; : \; 1 \leq i \leq n\}. \tag{6.74}$$

Hint: Since G is maximal planar, each edge belongs to at least two triangles. The application of Theorem 6.20 will give the result.

6.9 ([15]) Prove the following upper bound:

$$a \leq 2(k - \varphi) + \frac{k + 2 - \sqrt{k^2 + 4}}{2} \tag{6.75}$$

for a connected graph with even order $n = 2k \geq 6$, algebraic connectivity a, and domination number $\varphi \geq 3$.

6.10 ([517]) Use the inequality (6.29) to derive the upper bound for the algebraic connectivity of an r-regular graph whose complement is triangle-free,

$$a \leq \min \left\{ \frac{(3n - 2r - 2)(2r + 2 - n)}{n}, r + 1 - \sqrt{n - r - 1} \right\}. \tag{6.76}$$

6.11 Prove that the inequality of Rad et al. (6.48) improves the inequality of Lu et al. (6.50).

6.12 Show that the bounds (6.49) and (6.50) are incomparable.

6.13 Prove the following property. Let $\mathbf{x} = (x_1, x_2, \ldots, x_n)$ denote the Fiedler eigenvector of a graph G and let G' be a graph obtained by deleting the edge uv and inserting the edge uw (with the assumption that u and w are not adjacent in G). If either $x_u \le x_w \le x_v$ or $x_u \ge x_w \ge x_v$, then $a(G') \le a(G)$. If at least one of these inequalities is strict, then $a(G') < a(G)$.

Hint: Follow the proof of Lemma 1.28.

6.14 Prove the following result of Alon and Milman [10]. Let G be a connected graph with n vertices and let S, T be disjoint non-empty subsets of $V(G)$ at distance $d(S,T) = k > 1$. If $|S| = c_1 n$ and $|T| = c_2 n$, then

$$k^2 \le \frac{\Delta}{a}\left(\frac{1}{c_1} + \frac{1}{c_2}\right)(1 - c_1 - c_2) \tag{6.77}$$

and

$$c_2 \le \frac{1 - c_1}{1 + \frac{a}{\Delta}c_1 k^2}. \tag{6.78}$$

6.15 ([183]) Prove the following result. Let $e = uv$ be a cut edge of a graph G and let G' be a graph obtained from G by deleting e and identifying the vertices u and v (this operation is called the *collapsion* of the edge e or the *contraction* of the vertices u and v). Then

$$a(G') \ge a(G) \tag{6.79}$$

and the inequality is strict if the coordinate of the Fiedler eigenvector corresponding to the new vertex is different from zero.

6.16 ([255]) Prove that a tree with maximal algebraic connectivity within the set of trees of fixed order n and number of pendant vertices k ($2 \le k \le n - 1$) is obtained by attaching r paths each of length q and $k - r$ paths each of length $q - 1$ at a single vertex, where $q = \lfloor \frac{n-1}{k} \rfloor$ and $kq + r = n - 1$.

6.17 ([181]) Prove that if T is a tree with a perfect matching then

$$\mu_2(T) \le \begin{cases} \frac{k+2+\sqrt{k^2+4}}{2}, & \text{if } n = 4k, \\ l, & \text{if } n = 4k+2, \end{cases} \tag{6.80}$$

where l is the largest root of the equation

$$x^3 - (k+5)x^2 + (3k+7)x - 2k - 1 = 0.$$

Notes

Some unmentioned results concerning inequalities for L-eigenvalues can be found in the survey papers [3, 315, 325]. A recent result is a decreasing sequence of upper bounds for μ_1, determined by Rojo and Rojo [387].

Additional results concerning the graphs with extremal L-spectral radius can be found in [186, 503].

Trees with given maximal degree that attain maximal L-spectral radius are determined in [491], while semiregular trees with minimal L-spectral radius are determined in [46]. Other extremal trees and their ordering are considered in [189, 483, 490].

More results on extremal unicyclic or bicyclic graphs can be found in [159, 498].

Isoperimetric inequalities and those for the algebraic connectivity of directed graphs are given in [10]. The algebraic connectivity of weighted graphs is considered in [26]. The interaction between the algebraic connectivity of a graph and its vertex-deleted subgraph, precisely the quantities $a(G) - a(G - u)$ and $\frac{a(G-u)}{a(G)}$, is considered by Kirkland in [238]. Some lower and upper bounds for both quantities are determined. The limit points for $a(G)$ are considered by the same author in [240]. It is proved that each non-negative real number is a limit point for this eigenvalue.

Let $W(k, n_1, n_2)$ denote the graph obtained by inserting an edge between the centre of the star K_{1,n_1-1} and one end of the path P_k and another edge between the centre of the star K_{1,n_2-1} and the other end of the same path. This graph has $n = k + n_1 + n_2$ vertices, and by the result of Fallat and Kirkland (page 192), if n and k are fixed then it attains the minimal algebraic connectivity for $|n_1 - n_2| \leq 1$. The same authors gave the ordering of these graphs with respect to the same eigenvalue. Namely, $a(W(k, n_1, n_2))$ is strictly decreasing for $1 \leq n_1 \leq \frac{n-k}{2}$. The same ordering is considered in [1].

The relevance of the algebraic connectivity to the combinatorial optimization, maximum cut problem, travelling salesman problem or theory of elasticity can be found in [3, 96, 162, 241, 304, 405].

More orderings of trees (in particular caterpillars) are considered in [1, 383, 453, 510]. The recent survey paper [2] concerns the ordering of trees and graphs with few cycles by their algebraic connectivity.

More results on L-eigenvalues distinct from μ_1 and a can be found in either of the survey papers [315, 325]. The corresponding eigenvectors are considered in [313].

The asymptotic behaviour of L-eigenvalues is considered in [237]. For example, it is proved that the ith smallest L-eigenvalue of a graph with bounded degree is $O\left(\frac{i}{n}\right)$ when $n \to \infty$.

Trees with maximal L-eigenvalues are considered in [381]. It is proved that $\mu_i \leq \frac{1}{2}\left(k+1+\sqrt{k^2+2k-3}\right)$, for $i \geq 2$ and any tree with $ik+1$ vertices.

The third smallest L-eigenvalue of graphs is considered in [354].

Graphs with small second largest L-eigenvalue are considered in [367]. All graphs with $\mu_4 < 2$ are identified in [507].

7

Signless Laplacian eigenvalues

The first five sections are devoted to the Q-spectral radius κ_1 of a graph. In Section 7.1 we give general upper and lower bounds for κ_1. In Section 7.2 we consider the bounds for κ_1 of nested graphs. In Section 6.3 we consider graphs with small values of κ_1. Graphs with maximal value of κ_1 and the ordering of graphs with respect to κ_1 are considered in Sections 7.4 and 7.5, respectively.

The least Q-eigenvalue is considered in Section 7.6. Less investigated Q-eigenvalues are considered in Section 7.7.

7.1 General inequalities for Q-spectral radius

Elementary calculus on the signless Laplacian matrix gives the following inequalities for the Q-spectral radius of a connected graph:

$$2\left(1+\cos\frac{\pi}{n}\right) \leq \kappa_1 \leq 2(n-1), \tag{7.1}$$

where the lower bound is attained for P_n and the upper for K_n. In addition, the largest eigenvalues of the matrices A, L, and Q satisfy $\mu_1 \leq \kappa_1$ (by Theorem 1.21) and $2\lambda_1 \leq \kappa_1$ (cf. [104]), with equality in the first place if and only if the graph is bipartite. These inequalities imply that any lower bound for μ_1 is also a lower bound for κ_1, and that doubling any lower bound for λ_1 also yields a valid lower bound for κ_1. Since, in the case of bipartite graphs, $\mu_1 = \kappa_1$, all results regarding these graphs remain valid for κ_1.

The additional two connections between the signless Laplacian characteristic polynomial Q_G of a graph and the characteristic polynomial of its line graph $P_{L(G)}$ or the characteristic polynomial of its subdivision $P_{S(G)}$ (respectively given in (1.7) and (1.8)) play an important role in research concerning Q-spectra, and especially their largest eigenvalue.

7.1.1 Transferring upper bounds for μ_1

Since $\mu_1 \leq \kappa_1$ a natural question that arises is which upper bound for μ_1 remains valid for κ_1 and which does not? In what follows, we consider this question and summarize the results in Table 7.1. We first prove a theorem.

Theorem 7.1 (Maden et al. [306]) *Given a graph G whose vertices are labelled* $1, 2, \ldots, n$, *with corresponding vertex degrees* $d_1, d_2, \ldots, d_n$, *let* $b_i \in \mathbb{R}^+$ $(1 \leq i \leq n)$, $b_i' = \frac{1}{b_i} \sum_{j \sim i} b_j$, $c_i = b_i(d_i + b_i')$, $c_i' = \frac{\sum_{j \sim i} c_j}{c_i}$, *and* $k_i = d_i + c_i'$. *Then*

(i)

$$\kappa_1 \leq \max_{1 \leq i \leq n} \left\{ \sqrt{d_i k_i + \frac{\sum_{j \sim i} \left(d_j c_j + \sum_{k \sim j} c_k \right)}{c_i}} \right\}. \quad (7.2)$$

If Q^2 *is irreducible then equality holds if and only if* $k_1 = k_2 = \cdots = k_n$, *while if* Q^2 *is reducible then equality holds if and only if there exists a permutation matrix P such that*

$$PQP^T = \begin{pmatrix} O_r & Q_1 \\ Q_2 & O_{n-r} \end{pmatrix}, \quad (7.3)$$

with $k_{\sigma(1)} = k_{\sigma(2)} = \cdots = k_{\sigma(r)}$ *and* $k_{\sigma(r+1)} = k_{\sigma(r+2)} = \cdots = k_{\sigma(n)}$, *where* σ *is a permutation on the set* $\{1, 2, \ldots, n\}$ *which corresponds to the permutation matrix P.*

(ii)

$$\kappa_1 \leq \max_{i \sim j} \left\{ \frac{d_i + d_j + \sqrt{(d_i - d_j)^2 + 4b_i'b_j'}}{2} \right\}, \quad (7.4)$$

with equality if and only if G is either regular or semiregular bipartite.

(iii)

$$\kappa_1 \leq \max_{1 \leq i \leq n} \{k_i\}, \quad (7.5)$$

with equality if and only if $k_1 = k_2 = \cdots = k_n$.

(iv)

$$\kappa_1 \leq \max_{1 \leq i \leq n} \left\{ \sqrt{\frac{c_i(d_i + c_i')}{b_i}} \right\}. \tag{7.6}$$

If Q^2 is irreducible then equality holds if and only if $d_1 + b_1' = d_2 + b_2' = \cdots = d_n + b_n'$, while if Q^2 is reducible then equality holds if and only if there exists a permutation matrix P such that (7.3) holds along with the same chains of equalities with $d_{\sigma(i)} + b_{\sigma(i)}'$ instead of $k_{\sigma(i)}$ $(1 \leq i \leq n)$.

Proof Let $B = \operatorname{diag}(b_1, b_2, \ldots, b_n)$ be an $n \times n$ diagonal matrix. Parts (i), (ii), and (iv) follow from Theorem 1.2 applied to the matrix $B^{-1}QB$.

Consider (ii) and let $\mathbf{x} = (x_1, x_2, \ldots, x_n)^T$ be the Perron eigenvector of $B^{-1}QB$. We may assume that $x_i = 1$ and the other coordinates are at most 1. Also, let $x_j = \max_{k \sim i} \{x_k\}$. Following [306], from

$$(B^{-1}QB)\mathbf{x} = \kappa_1 \mathbf{x},$$

we have

$$\kappa_1 = d_i + \frac{1}{b_i} \sum_{k \sim i} b_k x_k \leq d_i + \frac{1}{b_i} \sum_{k \sim i} b_k x_j$$

and

$$\kappa_1 x_j = d_j x_j + \frac{1}{b_j} \sum_{k \sim i} b_k x_k \leq d_j x_j + \frac{1}{b_j} \sum_{k \sim i} b_k.$$

Thus, we have

$$\kappa_1^2 - (d_i + d_j)\kappa + d_i d_j - \left(\frac{1}{b_i} \sum_{k \sim i} b_k \right) \left(\frac{1}{b_j} \sum_{k \sim i} b_k \right) \leq 0,$$

which leads to (7.4). Moreover, it can be checked that equality holds if and only if G is regular or semiregular bipartite. $\square$

Setting $b_i = 1$ for all i in (7.4), we get that the inequality (6.12) holds for κ_1. Since it is an improvement of (6.11) and (6.13), these inequalities remain valid for κ_1 as well. Setting $b_i = 1$ in (7.6), we get that (6.15) also holds for κ_1.

The following theorem of the same authors is proved in a similar way.

Table 7.1 *Upper bounds for* μ_1 *that hold and those that do not hold for* κ_1

Do hold		
(6.5)	$\kappa_1 \leq$	$\max\left\{ \frac{d_i + \sqrt{d_i^2 + 8 d_i m_i}}{2} : 1 \leq i \leq n \right\}$
(6.11)	$\kappa_1 \leq$	$\max\{ d_i + d_j : i \sim j \}$
(6.12)	$\kappa_1 \leq$	$\max\{ d_i + m_i : 1 \leq i \leq n \}$
(6.13)	$\kappa_1 \leq$	$\frac{2m}{n-1} + n - 2$
(6.15)	$\kappa_1 \leq$	$\max\{ \sqrt{2 d_i (d_i + m_i)} : 1 \leq i \leq n \}$
(6.18)	$\kappa_1 \leq$	$\max\left\{ \frac{d_i(d_i + m_i) + d_j(d_j + m_j)}{d_i + d_j} : i \sim j \right\}$
(6.69)	$\kappa_1 \leq$	$\max\{ \sqrt{d_i(d_i + m_i) + d_j(d_j + m_j)} : i \sim j \}$
(6.70)	$\kappa_1 \leq$	$\max\{ 2 + \sqrt{d_i(d_i + m_i - 4) + d_j(d_j + m_j - 4) + 4} : i \sim j \}$

Do not hold		
(6.2)	$\mu_1 \leq$	$\max\left\{ \frac{d_i + \sqrt{d_i^2 + 8 d_i m_i'}}{2} : 1 \leq i \leq n \right\}$
(6.6)	$\mu_1 \leq$	$\max\{ d_i + d_j - |N(i) \cap N(j)| : i \sim j \}$
(6.10)	$\mu_1 \leq$	$\max\{ d_i + d_j - |N(i) \cap N(j)| : 1 \leq i < j \leq n \}$
(6.14)	$\mu_1 \leq$	$\max\left\{ \sqrt{2(d_i^2 + d_i m_i')} : 1 \leq i \leq n \right\}$

Theorem 7.2 ([306]) *Given a graph G whose vertices are labelled $1, 2, \ldots, n$, with corresponding vertex degrees $d_1, d_2, \ldots, d_n$, let $b_i \in \mathbb{R}^+$ ($1 \leq i \leq n$), $b_i' = \frac{1}{b_i} \sum_{j \sim i} b_j$, $c_i = b_i(d_i + b_i')$, and $c_i' = \frac{\sum_{j \sim i} c_j}{c_i}$. Then*

$$\kappa_1 \leq \max_{1 \leq i \leq n} \{ d_i + b_i' \} \tag{7.7}$$

and

$$\kappa_1 \leq \max_{1 \leq i \leq n} \left\{ \frac{d_i + \sqrt{d_i^2 + \frac{4 c_i c_i'}{b_i}}}{2} \right\}. \tag{7.8}$$

Both equalities hold if and only if $d_1 + b_1' = d_2 + b_2' = \cdots = d_n + b_n'$.

Setting $b_i = d_i$ in (7.7), we again get (6.12). Moreover, with $b_i = 1$ in (7.8), we get (6.5). It can also be proved that the inequality of Li and Zhang (6.18) remains valid for κ_1. As a matter of fact, the authors proved this inequality by proving its validity for κ_1 and using $\mu_1 \leq \kappa_1$. Finally, both inequalities (6.69) and (6.70) mentioned in Exercise 6.4 also hold for κ_1.

In this way we get a list of upper bounds for μ_1 that can be used for κ_1 (see Table 7.1). The next question is: Which bounds for μ_1 do not hold for κ_1? A partial answer can be obtained by taking a simple graph and computing κ_1 and the corresponding bounds. For example, by taking the pineapple $P(3,1)$, we get $\kappa_1(P(3,1)) \approx 4.5616$, while the upper bounds of (6.2), (6.10), and (6.14) are 4.3723, 4.4721, and 4, respectively. So, none of these bounds hold for κ_1. Since (6.10) is improved by (6.6), it follows that (6.6) does not hold for κ_1 either.

Remark 7.3 It is interesting to note that two upper bounds that contain the value denoted by $m_i' = \frac{\sum_{i \sim j}(d_j - |N(i) \cap N(j)|)}{d_i}$ (i.e., those of (6.2) and (6.14)) are not valid for κ_1, but their consequences with m_i instead of m_i', (6.5) and (6.15), are.

Other bounds

The next result is the exact analogue to Theorem 2.19, and they have exactly the same extremal graphs.

Theorem 7.4 (Nikiforov [338]) *For a graph G,*

$$\kappa_1 \leq \min\left\{2\Delta, \frac{\Delta+2\delta-1+\sqrt{(\Delta+2\delta-1)^2+16m-8(n-1+\Delta)\delta}}{2}\right\}.$$
$$(7.9)$$

Equality holds if and only if G is regular or G has a component of order $\Delta+1$ in which every vertex is of degree δ or Δ, and all other components are δ-regular.

As a direct consequence of the above result, we get

$$\kappa_1 \leq \min\left\{2\Delta, \frac{\Delta-1+\sqrt{(\Delta-1)^2+16m}}{2}\right\}, \qquad (7.10)$$

with equality if and only if G is Δ-regular with possible isolated vertices [338]. Two more consequences are the upper bound obtained in [283]

$$\kappa_1 \leq \frac{\Delta+\delta-1+\sqrt{(\Delta+\delta-1)^2+8(2m-\delta(n-1))}}{2} \qquad (7.11)$$

and the upper bound obtained in [80]

$$\kappa_1 \leq \frac{\delta - 1 + \sqrt{(\delta - 1)^2 + 8(2m + \Delta^2 - \delta(n - 1))}}{2}. \tag{7.12}$$

Using similar arguments, we get the following lower bound (cf. [80]):

$$\kappa_1 \geq \frac{\Delta + \delta - 1 + \sqrt{(\Delta + \delta - 1)^2 + 8(2m + \Delta^2 - \Delta(n - 1))}}{2}. \tag{7.13}$$

For G connected, equality holds in (7.11), (7.12), and (7.13) if and only if G is regular.

In Subsection 2.1.2 we presented a sequence of upper and lower bounds for the spectral radius expressed in terms of the maximal vertex degree and diameter of a graph. The analogous result for κ_1 was obtained by Ning et al. [349], and the proof is similar to that of Theorem 2.12. Both inequalities hold for connected non-regular graphs:

$$\frac{4m}{n} + \frac{(\Delta - \delta)^2}{2n\Delta} < \kappa_1 < 2\Delta - \frac{1}{n\left(D - \frac{1}{4}\right)}. \tag{7.14}$$

Remark 7.5 Neither of the inequalities in (7.14) holds for regular graphs. For example, if G is complete or regular complete bipartite then the left-hand side is equal to $\kappa_1 \ (= 2\Delta)$, while the right-hand side is strictly less than κ_1.

We point out three results that bound κ_1 in terms of the clique number and consequently the chromatic number.

Theorem 7.6 (He et al. [204]) *For a connected graph with clique number $\omega \geq 2$,*

$$\kappa_1 \leq \frac{(3\omega - 4)k + 3l - 2 + \sqrt{k^2\omega^2 + ((2l + 4)\omega - 8l)k + (l - 2)^2}}{2}, \tag{7.15}$$

where $n = k\omega + l$ and $0 \leq l < \omega$. Moreover, equality holds if and only if G is a complete bipartite graph for $\omega = 2$ or a complete ω-bipartite graph with almost equal parts for $\omega \geq 3$.

The last result is slightly more precise than the following one [4]:

$$\kappa_1 \leq \frac{2(\omega - 1)}{\omega} n. \tag{7.16}$$

Using (2.11) and $\kappa_1 \leq \lambda_1 + \Delta$, where the last inequality follows directly from $Q = A + D$ and $d_{ii} \leq \Delta$ ($1 \leq i \leq n$), we get

$$\kappa_1 \leq \Delta + \sqrt{2m\left(1 - \frac{1}{\omega}\right)}. \tag{7.17}$$

Since, for the chromatic number, $\chi \geq \omega$, the above results remain valid if we replace ω with χ.

Inequalities involving the independence number and domination number are given in Exercise 7.3.

7.2 Bounds for Q-spectral radius of connected nested graphs

This subsection can be viewed as a counterpart to Subsection 2.2.4 since the bounds obtained for the spectral radius are reproduced for the Q-spectral radius. It turns out that the change in the graph matrix (i.e., the transition from A to Q) produces bounds with slightly modified expressions. Thus, we just present the final results, while proofs are left to the reader (they can also be found in the corresponding references [18, 19]).

Recall that $\overline{d}$ denotes the average vertex degree of a graph, while $\overline{d}_i$ stands for the average vertex degree in cell i of an NSG G. We use the notation introduced on page 6 and that of Subsection 2.2.4 to present one lower and two upper bounds for both NSGs and DNGs.

The lower bounds (7.18) and (7.21) are analogous to the lower bound obtained in Theorem 2.47, and therefore proved using the Rayleigh principle. Proofs of the remaining bounds are similar to that of Theorem 2.49.

Nested split graphs

We have the following bounds [19]:

$$\kappa_1 \geq \frac{1}{2}\left(\frac{\sum_{i=1}^{h} n_i\overline{d}_i}{N_h} + N_h - 1 + t + \sqrt{\left(\frac{\sum_{i=1}^{h} n_i\overline{d}_i}{N_h} + N_h - 1 + t\right)^2 + 4\widehat{e}_h^*}\right),$$
$$\tag{7.18}$$

where

$$t = \frac{\sum_{i=1}^{h} m_i N_i^3}{\sum_{i=1}^{h} m_i N_i^2} \text{ and } \widehat{e}_h^* = \sum_{i=1}^{h} \frac{N_i}{N_h} \widehat{e}_i;$$

$$\kappa_1 \le \frac{1}{2}\left(2N_h + n - 2 + \sqrt{(n-2)^2 + 4e_h'}\right), \tag{7.19}$$

where

$$e_h' = e_h - n_1 \left(\frac{\sum_{i=1}^{h} m_i \bar{e}_i + (M_h + \bar{d} - 1)\bar{e}_h}{(n - 1 + \bar{d} - n_1)(\bar{d} + 1)}\right);$$

$$\kappa_1 \le \frac{1}{2}\left(2N_h + n' - 2 + \sqrt{(n'-2)^2 + 4e_h''}\right), \tag{7.20}$$

where

$$n' = n - \frac{n_1 \bar{e}_h}{(n - 1 + \bar{d} - n_1)(\bar{d} + 1)} \text{ and } e_h'' = e_h - \frac{n_1 \sum_{i=1}^{h} m_i \bar{e}_i}{(n - 1 + \bar{d} - n_1)(\bar{d} + 1)}.$$

Example 7.7   For an NSG with parameters $(40, 20, 10, 10; 20, 10, 50, 40)$, we have

(7.18)	κ_1	(7.20)	(7.19)
277.454	284.920	329.782	330.250

For an NSG with parameters $(400, 200, 100, 100; 200, 100, 500, 400)$, we have

(7.18)	κ_1	(7.20)	(7.19)
2791.51	2866.13	3314.63	3319.32

□

Double nested graphs

Here are the bounds for DNGs [18]:

$$\kappa_1 \geq \frac{1}{2}\left(t + \frac{m}{N_h} + \sqrt{\left(t - \frac{m}{N_h}\right)^2 + 4\widehat{e}_h^*}\right), \tag{7.21}$$

where

$$t = \frac{\sum_{i=1}^{h} m_i N_{h+1-i}^3}{\sum_{i=1}^{h} m_i N_{h+1-i}^2} \text{ and } \widehat{e}_h^* = \sum_{i=1}^{h} m_i \frac{N_{h+1-i}^2}{N_h};$$

$$\kappa_1 \leq \frac{1}{2}\left(n + \sqrt{n^2 - 4(M_h N_h - m')}\right), \tag{7.22}$$

where

$$m' = m - \frac{n(n - N_h)}{(n - n_1)^2(n - m_1)} \sum_{i=1}^{h} m_i \overline{f}_{h-i+1}$$

for

$$\overline{f}_{h+1-i} = \sum_{i=1}^{h+1-i} n_i(M_h - M_{h+1-i});$$

$$\kappa_1 \leq \frac{1}{2}\left(n + \sqrt{n^2 - 4(M_h N_h - m'')}\right), \tag{7.23}$$

where

$$m'' = m - \frac{\kappa'(\kappa' - N_h)}{(\kappa' - n_1)^2(\kappa' - m_1)} \sum_{i=1}^{h} m_i f_{h-i+1}$$

for

$$\kappa' = \frac{1}{2}\left(n + \sqrt{n^2 - 4(M_h N_h - m')}\right).$$

Example 7.8 For a DNG with parameters $(2,2,5,3;2,3,1,1)$, we have

(7.21)	κ_1	(7.23)	(7.22)
13.6785	15.6451	17.0210	17.0550

For a DNG with parameters $(2000, 2, 5, 3; 2000, 3, 1, 1)$, we have

(7.21)	κ_1	(7.23)	(7.22)
4014.97	4014.97	4014.98	4014.98

$\square$

7.3 Graphs with small Q-spectral radius

We characterize all graphs whose L-spectral radius does not exceed 4.5, and simultaneously realize the Hoffman program for the signless Laplacian matrix. We start with the following result.

Theorem 7.9 *For any graph G:*

(i) $\kappa_1(G) = 0$ *if and only if G is totally disconnected;*

(ii) $0 < \kappa_1(G) < 4$ *if and only if all components of G are paths;*

(iii) *if G is connected and* $\kappa_1(G) = 4$, *then G is the cycle* C_n *or the star* $K_{1,3}$.

Proof (i) $\kappa_1(G) = 0$ if and only if Q_G is a zero-matrix (i.e., G is totally disconnected).

For (ii) and (iii), we use the connection between Q_G and $P_{S(G)}$ given in (1.8) and the corresponding results for λ_1. Hence, $\kappa_1(G) < 4$ for a connected graph G if its subdivision is a proper subgraph of at least one Smith graph, which gives (ii), and $\kappa_1(G) = 4$ if $S(G)$ is a Smith graph, which gives (iii). $\square$

Recall that all graphs whose spectral radius does not exceed $\frac{3\sqrt{2}}{2}$ are characterized in Theorem 2.59. To decide which graphs satisfy $\kappa_1 \leq 4.5$, it is sufficient to consider their subdivisions and use this result. In this way we get the following theorem.

Theorem 7.10 *Any connected graph whose Q-spectral radius satisfies* $4 < \kappa_1 \leq 4.5$ *is either an open quipu or a closed quipu.*

To realize the Hoffman program – that is, to determine all graphs with the property $\kappa_1 \le 2 + \varepsilon$ where ε is the real root of $x^3 - 4x - 4 = 0$ ($\varepsilon \approx 2.3830$) – we need to single out those graphs from the set of graphs described in the previous theorem. Inspecting all possibilities, Belardo et al. [33] obtained the following result.

Theorem 7.11 (Signless Laplacian Hoffman Program) *If G is connected and $\kappa_1(G) \le 2 + \varepsilon$ (where ε is the real root of $x^3 - 4x - 4 = 0$), then G is one of the connected graphs described in Theorem 7.9 or*

> *(i) T_n^2 ($n \ge 5$),*
> *(ii) T_n^i ($3 \le i \le n-3$, $n \ge 6$),*
> *(iii) $T_n^{i,j}$ ($2 \le i < j \le n-3$, $n-4 \le 2(j-i)$).*

7.4 Graphs with maximal Q-spectral radius

We consider graphs of fixed order and size whose Q-spectral radius attains its maximal value. In the sequel, we give a review of other results in this topic.

7.4.1 Order, size, and maximal Q-spectral radius

The following theorem can be proved in the same way as the corresponding result concerning λ_1 (Theorem 2.60).

Theorem 7.12 *Let G be a connected graph of fixed order and size with maximal Q-spectral radius. Then G does not contain as an induced subgraph any of the graphs P_4, C_4, and $2K_2$.*

Thus, both λ_1 and κ_1 of a connected graph of fixed order and size attain their maximal values for NSGs. The question arises whether these extremal NSGs are the same in both cases. For a small number of vertices this is true, as existing graph data shows. However, among the graphs with 5 vertices and 7 edges there are two graphs with maximal κ_1 (see Fig. 7.1), while only the first of them yields maximal λ_1. In fact, for any $n \ge 5$ and $m = n + 2$ there are two graphs with maximal κ_1 [439].

In the following theorem we consider trees, unicyclic graphs, and bicyclic graphs.

Theorem 7.13 *Let G be a graph with maximal Q-spectral radius within the set of connected graphs with n vertices and m edges.*

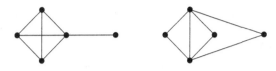

Figure 7.1 Two graphs with $n = 5$ and $m = 7$ with maximal κ_1 (≈ 6.3723).

(i) If $m = n - 1$, then G is the star $K_{1,n-1}$.
(ii) If $m = n$, then G is the graph obtained by adding an edge to the star $K_{1,n-1}$.
(iii) If $m = n + 1$, then G is the graph obtained by adding two adjacent edges to the star $K_{1,n-1}$.

Proof By Theorem 7.12, the resulting graph must be an NSG. The only tree which is an NSG is the star, and the only unicyclic graph which is an NSG is the star together with an additional edge, and there are just a few possibilities for bicyclic graphs. By inspecting them, we get the assertion. □

We proceed with disconnected graphs. Recall also from Subsection 2.2.4 that, if we do not restrict ourselves to connected graphs, then a graph of fixed order and size with maximal spectral radius is determined for any n and m. It turns out that there is no analogous result in the case of the Q-spectral radius, but there are some particular solutions. To present them we need the following discussion.

Reinterpreting Chang and Tam [77], we define the relations $\geq^G$ and $\sim^G$ on the set of vertices $V(G)$ by

$$u \geq^G v \text{ if and only if } N(u)\backslash\{v\} \supseteq N(u)\backslash\{v\}$$

and

$$u \sim^G v \text{ if and only if } N(u)\backslash\{v\} = N(u)\backslash\{v\}.$$

The relation $\geq^G$ is a pre-order – that is, it is reflexive and transitive – while $\sim^G$ is an equivalence relation.

If $(a_1, a_2, \ldots, a_r)$ and $(b_1, b_2, \ldots, b_s)$ are two finite sequences of real numbers then we say that the first sequence majorizes the second if $\sum_{i=1}^{r} a_i = \sum_{i=1}^{s} b_i$ and $\sum_{i=1}^{k} \bar{a}_i \geq \sum_{i=1}^{k} \bar{b}_i$ for $1 \leq k \leq \min\{r, s\}$, where $\bar{a}_1, \bar{a}_2, \ldots, \bar{a}_r$ is a rearrangement of $a_1, a_2, \ldots, a_r$ in non-increasing order. Following Merris [316], we call a graph *degree maximal* if it is connected and its degree sequence is not majorized by that of any other connected graph. It can be checked that a connected graph is degree maximal if and only if it is an NSG, and a disconnected NSG must be the disjoint union of a degree maximal graph and a set of isolated vertices. Among many equivalent conditions for a graph G to be an NSG, one is that the pre-order of $V(G)$ is total.

If G is an NSG then the total pre-order $\geq^G$ on $V(G)$ induces in a natural way a total partial order on the quotient set $V(G)/\sim^G$. Let $V_1, V_2, \ldots, V_r$ denote the neighbourhood equivalence classes of G arranged in strict ascending order with respect to the total partial order. It can be verified that, in this case, for any i, j $(1 \leq i, j \leq r)$, there are edges between V_i and V_j if and only if $i + j \geq r + 1$. So, the structure of an NSG G is completely determined once the cardinalities of the neighbourhood equivalence classes are specified. Denote by $F(n_1, n_2, \ldots, n_r)$ an NSG with neighbourhood equivalence classes $V_1, V_2, \ldots, V_r$ arranged in strict ascending order with respect to the total partial order such that $|V_i| = n_i$, for $1 \leq i \leq r$. Here, $n_1, n_2, \ldots, n_r$ can be any positive integers except that we require $n_{\lceil \frac{r}{2} \rceil} \geq 2$. The following result summarizes the investigation on graphs with maximal κ_1 within the set of graphs of fixed order n and size m, where $n \leq m - k$, $k \in \{0, 1, 2, 3\}$.

Theorem 7.14 ([77]) *Let G be a connected graph with maximal Q-spectral radius within the set of graphs of fixed order n and size m.*

- *If $n \leq m$ and $m \geq 4$, then $G \cong F(m - 3, 2, 1)$.*
- *Let $n \leq m - 1$. If $m = 5$ then $G \cong F(2, 2)$, if $m = 6$ then $G \cong K_4$, while if $m \geq 7$ then $G \cong F(m - 5, 2, 1, 1)$.*
- *Let $n \leq m - 2$. If $m = 6$ then $G \cong K_4$, if $m = 7$ then $G \in \{F(1, 3, 1), F(3, 2)\}$, while if $m \geq 8$ then $G \in \{F(m - 6, 3, 1), F(m - 7, 3, 1, 1)\}$.*
- *Let $n \leq m - 3$. If $m = 8$ then $G \cong F(1, 2, 2)$, if $m = 9$ then $G \cong F(4, 2)$, if $m = 10$ then $G \in \{F(1, 4, 1, 1), K_5\}$, while if $m \geq 11$ then $G \cong F(m - 9, 4, 1, 1)$.*

7.4.2 Other results

We give a brief review of other relevant results concerning graphs with maximal Q-spectral radius. There is a large similarity between these results and the corresponding results for λ_1 or μ_1. For example, according to [477, 526], the connected graph that maximizes κ_1 within the set of graphs with given number of cut vertices or given number of cut edges or given vertex (or edge) connectivity coincides with the graph that maximizes λ_1 within the same set. These graphs are considered in Subsection 2.3.5.

Li and Zhang [269] considered cacti with maximal κ_1. The result obtained is again similar to that for λ_1, which is mentioned in the same subsection. Other results concerning extremal values for κ_1 are obtained for graphs with given diameter [452], chromatic number [489], independence number [267], matching number [486] or degree sequence [227, 506].

7.5 Ordering graphs by Q-spectral radius

The ordering of graphs with respect to their Q-spectral radii for various classes of graph can be transferred from previous sections (e.g., trees with maximal μ_1 and κ_1 are ordered in the same way) or obtained in a similar way. We note that the first four smallest values for κ_1 among all connected graphs with given clique number are given in [502]. It is also noteworthy to add that Wei and Liu [459] determined all unicyclic graphs with $\kappa_1 \geq n-2$. For $n \geq 11$, this set of graphs coincides with the set of the first 16 unicyclic graphs ordered by their Laplacian spectral radii (see Fig. 6.2), but the orderings differ. The exact descending order, with respect to κ_1, of graphs from Fig. 6.2 follows (all inequalities between the corresponding Q-spectral radii are strict): $U^{(1)}$, $U^{(3)}$, $U^{(4)}$, $U^{(2)}$, $U^{(7)}$, $U^{(11)'}$, $U^{(10)}$, $U^{(9)}$, $U^{(11)}$, $U^{(12)}$, $U^{(13)}$, $U^{(5)}$, $U^{(6)}$, $U^{(9)'}$, $U^{(8)}$, $U^{(11)''}$.

7.6 Least Q-eigenvalue

The fact that $\kappa_n(G) = 0$ holds if and only if G has a bipartite component (Theorem 1.18) was first proved by Desai and Rao [133] in 1994. Using the result of Yan [475], we complete the range for κ_n by

$$0 \leq \kappa_n \leq n-2, \tag{7.24}$$

where the right equality holds if and only if G is a complete graph.

Another result of [133] deserves attention. Namely, if S and T are disjoint subsets of the vertex set V of a connected graph G, then

$$(|S|+|T|)\kappa_n \leq 4(m(S)+m(T)) + m(S \cup T, V \backslash (S \cup T)).$$

As a consequence, we have (see also [273])

$$\kappa_n \leq \frac{4(m(S)+m(V \backslash S))}{n}. \tag{7.25}$$

Moreover, let $m_{\min}(S)$ be the minimal number of edges whose removal from $G[S]$ results in a bipartite graph. Let also $\partial(S) = m(S, V \backslash S)$, and then $|\partial(S)| + m_{\min}(S)$ is the minimal number of edges whose removal disconnects $G[S]$ from $G[V \backslash S]$ and results in a bipartite subgraph induced by S. Let ψ be the minimum over all non-empty proper subsets S of $V(G)$ of

$$\frac{|\partial(S)| + m_{\min}(S)}{|S|}.$$

Then, for G connected,

$$\frac{\psi^2}{4\Delta} \leq \kappa_n \leq 4\psi, \tag{7.26}$$

where the left inequality is proved in [133], while the right follows from (7.25). The parameter ψ is used as a measure of the non-bipartiteness of a graph. Clearly, $\psi = 0$ if and only if $\kappa_n(G) = 0$ (i.e., if and only if G is bipartite).

Recall that the algebraic connectivity can be used as a measure of the connectedness of a graph G. A similar analogy between these eigenvalues follows from (6.41) and (7.50) (see Exercise 7.8).

Since, for G bipartite, $\kappa_n = 0$, a natural question that arises is what is the lower bound for κ_n in the case of non-bipartite graphs. The answer is provided by Cardoso et al. [71], who proved that the minimal value of κ_n of a connected non-bipartite graph with prescribed number of vertices is uniquely attained in the kite $K(3, n-3)$. Moreover, by direct computation we get that the least Q-eigenvalue of any such graph lies in $\left[\frac{1}{12n^2}, \frac{\pi^2}{n^2}\right)$.

7.6.1 Upper and lower bounds

We give a sequence of upper bounds for κ_n in terms of order, size, vertex degree, chromatic number or domination number and two lower bounds in terms of order and size.

Upper bounds

These bounds were considered by de Lima et al. [274].

Theorem 7.15 ([274]) *For a graph G,*

$$\kappa_n \leq \min\left\{\frac{d_i + d_j}{2} - \frac{\sqrt{(d_i - d_j)^2 + 4}}{2} : i \sim j\right\}. \tag{7.27}$$

Proof To apply the Rayleigh principle, assume that $d_j \geq d_i$, set $b = (d_i - d_j)^2 + 4$, and define the coordinates of a vector $\mathbf{y} = (y_1, y_2, \ldots, y_n)^T$ to be zero with the exceptions $y_i = \sqrt{\frac{b + \sqrt{b^2 - 4b}}{2b}}$ and $y_j = -\sqrt{\frac{b - \sqrt{b^2 - 4b}}{2b}}$. Clearly, $\mathbf{y}$ is a unit vector and then

$$\kappa_n \leq \mathbf{y}^T Q \mathbf{y} = \sum_{k \sim l} (y_k + y_l)^2 \leq \frac{d_i + d_j}{2} - \frac{\sqrt{(d_i - d_j)^2 + 4}}{2},$$

completing the proof. □

As a consequence, we have

$$\kappa_n \leq \frac{\delta + n - 1}{2} - \frac{\sqrt{(n - 1 - \delta)^2 + 4}}{2} < \delta. \tag{7.28}$$

Another result is provided by the same authors.

Theorem 7.16   ([274]) *For a graph G,*

$$\kappa_n \leq \frac{2m}{n} - 1. \tag{7.29}$$

Observe that the second inequality of (7.28) is better than (7.29) whenever the minimal vertex degree is at most the average degree minus one. The inequality (7.29) improves the result of Guo [188]

$$\kappa_n \leq \frac{2m}{n} - \sqrt{\frac{2m}{n(n - 1)}}, \tag{7.30}$$

which again improves the upper bound of (7.24).

We consider the chromatic number with the following theorem.

Theorem 7.17   ([274]) *For a graph G with chromatic number χ,*

$$\kappa_n \leq \left(1 - \frac{1}{\chi - 1}\right) \frac{2m}{n}. \tag{7.31}$$

Equality holds if G is a regular χ-partite graph.

Proof If the colour classes of G are denoted by $V_1, V_2, \ldots, V_\chi$, define a vector $\mathbf{y} \in \mathbb{R}^n$ by $y_i = \chi - 1$ if $i \in V_k$ and $y_i = -1$ otherwise. Since $\|y\|^2 = \chi(\chi - 2)|V_k| + n$, using the Rayleigh principle we get

$$(\chi(\chi - 2)|V_k| + n)\kappa_n \le \chi(\chi - 4)m(V_k, V \setminus V_k) + 4m,$$

for every k $(1 \le k \le \chi)$.

Summing on k, we get

$$\kappa_n \sum_{k=1}^{\chi} (\chi(\chi - 2)|V_k| + n) \le 4m + \sum_{k=1}^{\chi} \chi(\chi - 4)m(V_k, V \setminus V_k),$$

which yields

$$(\chi(\chi - 2)n + \chi n)\kappa_n \le 2\chi(\chi - 4)m + 4\chi m.$$

From the last inequality we easily deduce

$$\chi(\chi - 1)n\kappa_n \le 2\chi m(\chi - 2),$$

which implies the inequality (7.31). $\qquad\square$

The *k-domination number* φ_k is the least cardinality of a set $S \in V(G)$ such that each vertex in $V(G) \setminus S$ is adjacent to at least k vertices in S. The following inequality is proved in a similar way [280]:

$$\kappa_n \le 2\Delta - \frac{nk}{\varphi_k}. \tag{7.32}$$

A simple consequence is

$$\kappa_n \le 2\Delta - \frac{n}{\varphi}. \tag{7.33}$$

Lower bounds

We now provide two lower bounds for connected graphs. The first was obtained by Guo et al. [191] and proved by considering the interaction of $\kappa_n(G)$ and $\kappa_1(\overline{G})$ and use of the third inequality of Table 7.1. For $n \ge 6$,

$$\kappa_n \ge \frac{2m}{n - 2} - n + 1. \tag{7.34}$$

The second lower bound was obtained by Oliveira et al. [352] in the context of determining the bounds for the Q-spread of a graph. It holds for any non-complete graph and includes the number k of vertices with maximal degree $n - 1$:

$$\kappa_n \geq \frac{1}{2} \left(n + 2k - 2 - \sqrt{(n + 2k - 2)^2 - 8k(k - 1)} \right). \qquad (7.35)$$

7.6.2 Small graph perturbations and graphs with extremal least Q-eigenvalue

The next result is analogous to that of Theorem 3.22.

Theorem 7.18 (Wang and Fan [456]) *Let G_1 be a non-trivial connected graph, G_2 a non-trivial connected bipartite graph, and let $v_1, v_2 \in V(G_1)$, $u \in V(G_2)$. Let G (resp. H) be obtained from G_1, G_2 by identifying the vertex v_2 (resp. v_1) with u. If there exists an eigenvector $\mathbf{x}$ of G corresponding to κ_n such that $|x_{v_1}| \geq |x_{v_2}|$ then*

$$\kappa_n(H) \leq \kappa_n(G),$$

where $n = |V(G)| = |V(H)|$. Equality holds only if $x_{v_1} = x_{v_2}$ and, for G_2, $d_u x_u = \sum_{w \sim u} x_w$.

Using the last theorem we easily obtain statement (vi) in Lemma 1.29.

The behaviour of κ_n when a graph is perturbed by deleting a vertex, subdividing an edge or moving edges is considered in [206]. In particular, the following inequalities are obtained:

$$\frac{3 - \sqrt{4n^2 - 20n + 33}}{2} \leq \kappa_n(G) - \kappa_n(G - u) \leq 1, \qquad (7.36)$$

where the left equality holds if and only if the degree of u is 1 and $G - u$ is a complete subgraph of G, while the right equality holds if and only if all non-zero coordinates of the eigenvector of $\kappa_n(G)$ correspond to all neighbours of u. The upper bound holds for any Q-eigenvalue, which will be proved later in Theorem 7.19.

Using the mentioned perturbations, a number of graphs with given order

Signless Laplacian eigenvalues

Figure 7.2 A graph obtained by connecting a triangle and a star by a path.

and some additional properties that attain extremal Q-eigenvalue can be determined. Clearly, only non-bipartite graphs are of interest. Hence, connected non-bipartite graphs of order n with minimal κ_n are determined within the sets of graphs of fixed diameter [528], domination number (whenever it does not exceed $\frac{n+1}{3}$) [151], number of pendant vertices [153], matching number [487] or edge cover number [487]. In all these cases the resulting graph has the same form, as illustrated in Fig. 7.2.

Recalling from the previous section that the absolute minimum for κ_n of non-bipartite graphs of fixed order is attained when the graph of Fig. 7.2 has minimal number of pendant vertices, we deduce that a graph with minimal κ_n (and prescribed additional properties) is expected to be a graph with long diametral path and additional vertices concentrated at its ends. This is confirmed for connected graphs containing a fixed induced non-bipartite subgraph H [456]: the resulting graph is obtained by attaching a path of the appropriate length at a vertex of H.

Concerning connected graphs with maximal κ_n, it is natural to expect that they have a more round shape and smaller diameter. This is confirmed for the graphs with a given number of pendant vertices [153]: the graph of order n with k pendant vertices with maximal κ_n is obtained from K_{n-k} by attaching an almost equal number of pendant vertices at each of its vertices.

7.7 Other Q-eigenvalues

In this section we consider the relations between the Q-eigenvalues of a graph and those of its vertex-deleted subgraph, give some bounds for κ_2 or κ_3, and characterize those graphs with the property $\kappa_2 \leq 3$.

A note on κ_i of vertex-deleted subgraphs

In the following theorem we consider how the deletion of a single vertex affects the Q-eigenvalues of a graph.

Theorem 7.19 (Wang and Belardo [447]) *For a graph G,*

$$\kappa_{i+1}(G) - 1 \leq \kappa_i(G - u) \leq \kappa_i(G) \tag{7.37}$$

for $1 \leq i \leq n - 1$. The right equality holds if and only if u is an isolated vertex.

Proof Let M be the principal submatrix of Q obtained by deleting the row and column related to u and let $N = M - Q(G - u)$. By the Courant–Weyl inequalities, we have

$$\kappa_i(G - u) + \nu_{n-1}(N) \leq \nu_i(M) \leq \kappa_i(G - u) + \nu_1(N).$$

Since N is a $(0,1)$-diagonal matrix, we have $0 \leq \nu_{n-1}(N)$ and $\nu_1(N) \leq 1$, and so

$$\kappa_i(G - u) \leq \nu_i(M) \leq \kappa_i(G - u) + 1.$$

By the Q-interlacing theorem, we have $\nu_i(M) \leq \kappa_i(G)$, which together with the left inequality above gives the right inequality of (7.37).

On the contrary, since M is a principal submatrix of Q, by using the interlacing theorem and Remark 1.7, we get $\nu_i(M) \geq \kappa_{i+1}(G)$, which together with the right inequality above gives the left inequality of (7.37).

The right equality obviously holds whenever u is an isolated vertex. However, if u is not an isolated vertex, since Q is non-negative, we can take G connected and then $\kappa_1(G) > \nu_1(M) \geq \kappa_1(G - u)$. $\square$

Bounds for κ_2

We start with the following lower and upper bounds for κ_2:

$$\delta \leq \kappa_2 \leq n - 2, \tag{7.38}$$

where the lower bound holds for non-complete graphs. It can be obtained by applying the Courant–Weyl inequalities to $Q = A + D$. In this way we get $\kappa_2 \geq \lambda_2 + \delta$. Since $\lambda_2 \geq 0$, the left inequality holds, while equality is possible only if $\lambda_2 = 0$ (i.e., if the graph in question is complete multipartite). Considering such graphs, we get that $\kappa_2 = \delta$ holds exactly for a star, a regular complete multipartite graph, the graph $K_{1,3,3}$, and a complete multipartite graph of the type $K_{1,1,\ldots,1,2,2,\ldots,2}$ (cf. [273]).

Since $\kappa_2(K_n) = n - 2$, the upper bound is a direct consequence of the Q-interlacing theorem. The conditions for $\kappa_2 = n - 2$ are considered in [14, 449], but an elegant characterization of the graphs with $\kappa_{i+1} = n - 2$, for any i $(1 \leq i < n)$, is given by de Lima and Nikiforov [273]: $\kappa_{i+1}(G) = n - 2$ if and only if $\overline{G}$ has either i bipartite components, each of which has equal parts, or $k + 1$ bipartite components. Using this result we get that the equality in the upper bound of (7.38) holds exactly for graphs whose complements either have a bipartite component with equal parts or at least two bipartite components.

Remark 7.20   For a non-complete graph, from the inequalities (6.41) we have $a \leq \delta$, while, as we have just seen, it holds that $\kappa_2 \geq \delta$. Altogether, we have $a \leq \kappa_2$. In other words, any lower bound for the algebraic connectivity is the lower bound for the second largest Q-eigenvalue.   □

We now consider the three lower bounds of Das [125] in terms of vertex degrees.

Theorem 7.21   ([125]) *For a graph* G,

$$\kappa_2 \geq \frac{d_1 + d_2 - \sqrt{(d_1 - d_2)^2 + 4}}{2}, \tag{7.39}$$

where d_1 *and* d_2 *stand for the largest and second largest vertex degree in* G.

Proof   Let $u_1, u_2 \in V(G)$ such that $d_{u_1} = d_1$ and $d_{u_2} = d_2$. The second largest eigenvalue of the matrix Q is at least the second largest eigenvalue of its principal submatrix $\begin{pmatrix} d_1 & 0 \\ 0 & d_2 \end{pmatrix}$ (in case $u_1 \not\sim u_2$) or $\begin{pmatrix} d_1 & 1 \\ 1 & d_2 \end{pmatrix}$ (in case $u_1 \sim u_2$). In the first case we get $\kappa_2 \geq d_2$, while in the second case we get the lower bound (7.39). Since $\frac{d_1 + d_2 - \sqrt{(d_1 - d_2)^2 + 4}}{2} \leq d_2$, the proof is complete.   □

There are two consequences.

Corollary 7.22   ([125]) *For a graph* G,

$$\kappa_2 \geq d_2 - 1, \tag{7.40}$$

where d_2 *is the same as above. If* $\kappa_2 = d_2 - 1$, *then there are at least two vertices of maximal degree.*

Proof   By the previous theorem, we have to show that

$$d_2 - 1 \leq \frac{d_1 + d_2 - \sqrt{(d_1 - d_2)^2 + 4}}{2},$$

that is (after some simple transformations) $d_1 \geq d_2$, which is always obeyed and gives the condition for the equality to hold. □

Corollary 7.23 ([125]) *For a connected graph G,*

$$\kappa_2 \geq \overline{d} - 1, \tag{7.41}$$

where $\overline{d} = \frac{2m}{n}$ is the average vertex degree in G. Equality holds if and only if G is a complete graph.

Proof If $G \cong K_n$, then $\kappa_2(G) = n - 2$ and equality holds. Assume that G is not complete. We have

$$\overline{d} = \frac{\sum_{i=1}^{n} d_i}{n} \leq \frac{d_1 + (n-1)d_2}{n}. \tag{7.42}$$

Using (7.39), we get

$$\kappa_2 - \overline{d} \geq \frac{(n-2)(d_1 - d_2) + 2n - n\sqrt{(d_1 - d_2)^2 + 4}}{2n} - 1.$$

To get (7.41), we have to show that $(n-2)(d_1 - d_2) + 2n - n\sqrt{(d_1 - d_2)^2 + 4} \geq 0$, which reduces to $(d_1 - d_2)(n(n - 2 - d_1 + d_2) + (d_1 - d_2)) \geq 0$, and this is obviously true.

If equality holds in (7.41) then equality must also hold in the last inequality above, which gives $d_1 = d_2$, and also in (7.42), which gives $d_2 = \delta$. In other words, G is a δ-regular graph. If G contains two non-adjacent vertices, by the proof of Theorem 7.21, $\kappa_2 \geq \delta$, and thus $\kappa_2 \geq \overline{d}$, which contradicts the equality condition. Thus, all vertices in G are mutually adjacent (i.e., $G \cong K_n$). □

Concerning the interplay between κ_2 and other invariants of a graph, we mention the following result of Liu et al. [282] giving the condition for edge connectivity c_e. If $2 \leq k \leq \delta$ and

$$\kappa_2 \leq 2\delta - \frac{(k-1)n}{(\delta + 1)(n - \delta - 1)},$$

then $c_e \geq k$.

A lower bound for κ_3

Theorem 7.24 (Wang and Belardo [447]) *Let u, v, and w be the vertices of degree at least d_3 in a graph G.*

(i) *If u, v, and w induce $3K_1$, then $\kappa_3 \geq d_3$.*

(ii) *If u, v, and w induce K_3 or $P_2 \cup K_1$, then $\kappa_3 \geq d_3 - 1$.*

(iii) *If u, v, and w induce P_3, then $\kappa_3 \geq d_3 - \sqrt{2}$.*

$$(7.43)$$

Here, d_3 stands for the third largest degree in G.

Proof Without loss of generality, we may assume that the vertices of G are labelled in such a way that the first three rows and columns of Q are those related to u, v, and w. All three statements are proved by considering the top-left 3×3 principal submatrix. For (i), this matrix is $\begin{pmatrix} d_u & 0 & 0 \\ 0 & d_v & 0 \\ 0 & 0 & d_w \end{pmatrix}$, and thus κ_3 is at least the third largest eigenvalue of this matrix (i.e., $\kappa_3 \geq d_3$). The remaining two situations are considered by the same reasoning. □

Graphs with small κ_2

Graphs with the property $\kappa_2 \leq 3$ are considered in the following two theorems. We need the graph $G(k, 0, q, r)$ illustrated in Fig. 6.9: it consists of k triangles, q pendant paths P_2, and r pendant vertices, all of them sharing a common vertex.

Theorem 7.25 (Aouchiche et al. [14]) *For a connected graph with n ($2 \leq n \leq 6$) vertices:*

- *if $n = 2$, then $\kappa_2 = 0$;*
- *if $n = 3$, then $\kappa_2 = 1$;*
- *if $n = 4$ and G is not a star, then $\kappa_2 = 2$;*
- *if $n = 5$, then $\kappa_2 \leq 3$;*
- *if $n = 6$, then $\kappa_2 \leq 3$ if and only if $G \cong G(k, 0, q, r)$ or G is a subgraph of one of the following graphs – a 3-regular graph, the graph obtained by attaching a pendant edge at each vertex of K_3 or the graph obtained by attaching two pendant edges at the same vertex of C_4.*

Proof Since all graphs considered have at most six vertices, the result is verified by direct computation. □

Theorem 7.26 (Aouchiche et al. [14]) *For a connected graph G with n ≥ 7 vertices:*

- $\kappa_2 = 1$ *if and only if G is a star;*
- $\frac{3+\sqrt{5}}{2} - \frac{1}{n} < \kappa_2 < \frac{3+\sqrt{5}}{2}$ *if and only if $G \cong C(n-2,2)$;*
- $\kappa_2 < \frac{3+\sqrt{5}}{2}$ *if and only if G is a star with k ≥ 2 pendant vertices attached at its pendant vertices;*
- $3 - \frac{2.5}{n} < \kappa_2 \leq 3$ *if and only if G is a graph G(1,0,q,r);*
- $\kappa_2 = 3$ *if and only if G is a graph G(k,0,q,r) with k ≥ 2.*

Proof The proof is obtained by computing the Q-polynomial of the corresponding graphs and using the previous theorem, which provides the forbidden subgraphs for $\kappa_2 \leq 3$ having at most six vertices. □

Exercises

7.1 Use the upper bounds obtained in Subsection 7.1.1 to prove the following results:

$$\kappa_1 \leq n - 4 + 4\frac{m+1}{n} \tag{7.44}$$

for $n \geq 4$ and

$$\kappa_1 \leq n - 1 + 2\frac{m}{n} \tag{7.45}$$

for $n \geq 5$.

7.2 ([92]) Prove that the upper bound

$$\kappa_1 \leq \min_{1 \leq i \leq n} \left\{ \frac{\Delta + 2d_i - 1 + \sqrt{(2d_i - \Delta + 1)^2 + 8\sum_{j=1}^{i-1}(d_j - d_i)}}{2} \right\} \tag{7.46}$$

holds for any connected graph, where $\Delta = d_1 \geq d_2 \geq \cdots \geq d_n = \delta$ is the degree sequence. When does equality hold?

7.3 Prove the following result. For a connected graph with $n \geq 4$ vertices, the independence number α and the domination number φ satisfy

$$\kappa_1 + \alpha \leq \begin{cases} \frac{3n - 2\sqrt{2n^2 - 4n + 4}}{2}, & \text{if } n \text{ is even,} \\ \frac{3n - 2\sqrt{2n^2 - 6n + 3}}{2}, & \text{if } n \text{ is odd,} \end{cases} \qquad (7.47)$$

$$\kappa_1 \alpha \leq n(n-1), \qquad (7.48)$$

and

$$\kappa_1 + \varphi \leq 2n - 1. \qquad (7.49)$$

Hint: For (7.47) and (7.48), prove that, within the set of graphs with given independence number α, the maximal Q-spectral radius is attained for $K_{n-\alpha}\nabla(\alpha K_1)$ and then show that both inequalities hold for this graph. For (7.49), use $\Delta + \varphi \leq n$ and $\kappa_1 \leq 2\Delta$.

7.4 Use the proof of Theorem 2.60 to prove Theorem 7.12.

7.5 Show that a bicyclic graph of fixed order n with maximal Q-spectral radius is obtained by adding two adjacent edges to the star $K_{1,n-1}$.

7.6 Show that a connected graph is degree maximal if and only if it is an NSG.

7.7 Show that a connected bipartite graph of fixed order and size with maximal Q-spectral radius must be a DNG.

7.8 ([147]) Prove that

$$\kappa_n \leq b_v \leq b_e, \qquad (7.50)$$

where b_v (resp. b_e) is the minimal number of vertices (resp. edges) whose deletion yields a bipartite graph.

7.9 If $\mathscr{T}_n^c$ denotes the set of complements of trees with n vertices, then the minimal value of κ_n among all graphs in $\mathscr{T}_n^c$ is attained for $\overline{K_{1,n-1}}$ (since $\overline{K_{1,n-1}}$ contains an isolated vertex, and thus $\kappa_n(\overline{K_{1,n-1}}) = 0$). Prove that the second smallest value of κ_n is attained uniquely for the complement of the comet $C(n-2,2)$.

7.10 If $d_1 \geq d_2 \geq \cdots \geq d_n$ are vertex degrees in a graph G, prove that

$$\kappa_i \leq 2d_i. \tag{7.51}$$

Show that the same inequality holds for μ_i in the role of κ_i.

7.11 ([447]) Let u, v, and w be the vertices of a connected graph G such that $d_u, d_v, d_w \geq d_3$ (d_3 being the third largest degree in G). Prove that if u, v, and w do not share common neighbours, then

$$\kappa_3 \geq d_3 - 1. \tag{7.52}$$

Hint: By Theorem 7.24, the inequality holds whenever u, v, and w do not induce the path P_3. If they do induce P_3, delete some additional vertices so that d_u, d_v, and d_w become equal to d_3, and then consider the corresponding principal submatrix like in the mentioned theorem.

7.12 Prove that for a connected graph G, $\kappa_1 > 2 \geq \kappa_2$ holds if and only if G is one of the following graphs: $K_3, K_4, P_4, C_4, K_{1,n-1}$ $(n \geq 3), K_4 - e$ or the graph obtained by attaching a pendant vertex at two different vertices of K_3.

Hint: Use the method of forbidden subgraphs for $\kappa_2 \leq 2$.

Notes

More inequalities for Q-eigenvalues can be found in the survey papers [104, 106, 114, 115, 116].

The signless Laplacian spectral radius of graphs with given chromatic number is considered in [66]. The upper bounds for κ_n involving the domination number are considered in [208]. The lower bounds for the clique number and independence number of regular graphs in terms of κ_n are given in [285]. Graphs with prescribed degree sequence are considered in [286].

The *Randić index* $R(G)$ is defined as the sum of $\sqrt{d_i d_j}$ over all edges ij of a graph G. Deng et al. [132] considered the interconnection between κ_1 and $R(G)$, and proved in particular that

$$\kappa_1 - R(G) \leq \frac{3}{2}n - 2, \tag{7.53}$$

with equality if and only if G is complete graph.

Concerning the least Q-eigenvalue, note that [171] provides a valuable discussion on the corresponding eigenvector.

Zhou [521] gave conditions on κ_1 for the existence of Hamiltonian paths or cycles. A condition for a graph to be Hamiltonian by means of its arbitrary Q-eigenvalue is given in [448]. The same paper provides all graphs with at most two Q-eigenvalues of at least two, while the graphs with exactly three Q-eigenvalues of at least two (i.e., those with the property $\kappa_4 < 2$) are identified in [451].

8

Inequalities for multiple eigenvalues

In this chapter we consider inequalities that involve at least two graph eigenvalues.

In the first four sections, the inequalities for exactly two eigenvalues of a matrix associated with a graph are considered. Precisely, Section 8.1 deals with the spectral spread, Section 8.2 deals with the spectral gap of a graph, while the inequalities of Nordhaus–Gaddum type are exposed in Section 8.3. Some inequalities that do not fit readily into the previous sections are considered in Section 8.4.

In the last two sections, we consider the inequalities for spectral invariants based on all eigenvalues of a matrix associated with a graph. Section 8.5 deals with the graph energy and Section 8.6 deals with the Estrada index of a graph.

8.1 Spectral spread

The spectral spread of a matrix M is the largest distance in the complex plane between any two eigenvalues of the matrix. If M is a Hermitian matrix, its eigenvalues are real and may be assumed to be in non-increasing order. If so, then the spectral spread is the difference between the extreme eigenvalues.

We have already defined the spectral spread of a graph G to be the spectral spread of its adjacency matrix (page 10), that is, $\sigma(G) = \sigma(A) = \lambda_1 - \lambda_n$.

8.1.1 Upper and lower bounds

A number of bounds for the spectral spread of a graph can be derived by examining the corresponding bounds for λ_1 (Chapter 2) and λ_n (Chapter 3). Here we present some bounds that are not direct consequences of the previous results.

Upper bounds

These bounds were considered by Gregory et al. [174]. We start with the following result.

Theorem 8.1   (cf. [174]) *For a graph G,*

$$\sigma \leq \lambda_1 + \sqrt{2m - \lambda_1^2} \leq 2\sqrt{m}. \tag{8.1}$$

Proof   Since $\sum_{i=1}^{n} \lambda_i^2 = 2m$, we have $\lambda_1^2 + \lambda_n^2 \leq 2m$ and thus $\lambda_1 - \lambda_n \leq \lambda_1 + \sqrt{2m - \lambda_1^2}$. □

The first equality holds if and only if $\lambda_2 = \lambda_3 = \cdots = \lambda_{n-1} = 0$, that is, if and only if G is a totally disconnected or a complete bipartite graph with possible isolated vertices.

Computing $\sum_{i=1}^{n} \lambda_i = \frac{1}{2} \text{tr}(A^4) + m - 4f$, where f denotes the number of quadrangles in G, in a similar way we obtain the following upper bound [279]:

$$\sigma \leq \lambda_1 + \sqrt[4]{2 \sum_{i=1}^{n} \lambda_i - 2m + 8f - \lambda_1^4} \leq 2\sqrt[4]{\sum_{i=1}^{n} \lambda_i - m + 4f}. \tag{8.2}$$

If G is K_4-free, $\sqrt[4]{2\sum_{i=1}^{n} \lambda_i - 2m + 8f - \lambda_1^4} \leq \sqrt{2m - \lambda_1^2}$ and then (8.2) is finer than (8.1).

There follows another similar result based on the number of negative eigenvalues.

Theorem 8.2   ([174]) *For a graph G with n^- $(1 \leq n^- \leq n-1)$ negative eigenvalues,*

$$\sigma \leq \frac{n^- + 1}{n^-} \lambda_1 + \sqrt{2m \frac{n^- - 1}{n^-} - \frac{n^{-2} - 1}{n^{-2}} \lambda_1^2}. \tag{8.3}$$

Proof   If $n^- = 1$, the inequality (8.3) becomes $\lambda_1 \geq -\lambda_n$. For $n^- \geq 2$, using the Cauchy–Schwarz inequality, we compute

$$2m - \lambda_1^2 - \lambda_n^2 \geq \sum_{i=n-n^-+1}^{n} \lambda_i^2 \geq \frac{(\sum_{i=n-n^-+1}^{n} \lambda_i)^2}{n^- - 1} \geq \frac{(\lambda_1 + \lambda_n)^2}{n^- - 1}.$$

Now, (8.3) follows from the above quadratic equation in λ_n.   □

It can be proved that equality holds in (8.3) if and only if G is a union of possible isolated vertices and a complete $(n^- + 1)$-partite graph and, when $n^- \geq 3$, the n^- smallest parts of a non-trivial component all have equal size. We consider the interplay between σ and the clique number ω.

Theorem 8.3 ([174]) *For a graph G with clique number ω,*

$$\sigma \leq \lambda_1 + \sqrt{2m - \lambda_1^2 - \omega + 2}. \tag{8.4}$$

Proof Since K_ω is an induced subgraph of G, by the interlacing theorem, $\omega - 1$ eigenvalues of G are at most -1. Using $\sum_{i=1}^n \lambda_i^2 = 2m$, we get $\lambda_1^2 + \lambda_n^2 + \omega - 2 \leq 2m$, which implies (8.4). $\qquad\square$

Remark 8.4 For Theorems 8.1–8.3, if $m > \left\lfloor \frac{n^2}{4} \right\rfloor$ then $\lambda_1 \geq \frac{2m}{n}$, and thus we can exclude λ_1 from the right-hand side of the inequality. In this way we may obtain three consequences. The first (that follows from Theorem 8.1) is written below, while the remaining two are derived in the same manner.

$$\sigma \leq \frac{2m}{n} + \sqrt{2m - \left(\frac{2m}{n}\right)^2}. \tag{8.5}$$

$\qquad\square$

Concerning the lower bounds for σ, we have the following two results. Both are easily proved by considering the 2×2 quotient matrix induced by specified vertex sets (if necessary, compare the proof of the forthcoming Theorem 8.11). For a graph with two vertex-disjoint induced subgraphs G_1 and G_2,

$$\sigma \geq 2\sqrt{\left(\frac{m_1}{n_1} - \frac{m_2}{n_2}\right)^2 + \frac{(m - m_1 - m_2)^2}{n_1 n_2}}, \tag{8.6}$$

where G_i has n_i vertices and m_i $(1 \leq i \leq 2)$ edges and $n_1 + n_2 = n$ (cf. [279]). If G contains α independent vertices, the average degree of which is $\overline{d}$, then

$$\sigma \geq 2\sqrt{\left(\frac{m - \alpha\overline{d}}{n - \alpha}\right)^2 + \frac{\alpha\overline{d}^2}{n - \alpha}} \geq 2\overline{d}\sqrt{\frac{\alpha}{n - \alpha}}. \tag{8.7}$$

8.1.2 Q-Spread and L-spread

The *signless Laplacian spectral spread* (or Q-*spread*) is defined analogously: $\sigma_Q(G) = \sigma(Q) = \kappa_1 - \kappa_n$. Since $\mu_n = 0$, the *Laplacian spectral spread* (or L-*spread*) is taken to be $\sigma_L(G) = \mu_1 - a$. The latter invariant is mainly considered in the context of extremal graphs (see [25, 150, 154, 290]).

For the Q-spread, note that for regular graphs $\sigma = \sigma_Q$, (see Theorem 1.22). In the general case, these invariants are incomparable. The computational results based on graphs with a small number of vertices show that we often have $\sigma < \sigma_Q$.

Since for any bipartite graph $\kappa_n = 0$, in this case we have $\sigma_Q = \kappa_1$. Moreover, by (7.1), for any connected bipartite graph, $\sigma_Q \geq 2\left(1 + \cos\frac{\pi}{n}\right)$. In addition, Fan and Fallat [149] showed that this lower bound holds only for paths and odd cycles, while for any other connected graph it holds that $\sigma_Q \geq 4$, with equality only for K_4, $K_{1,3}, DK(3,0)$ or even cycles.

The absolute upper bound for σ_Q is again obtained from (7.1): $\sigma_Q \leq 2(n-1)$. Its improvement for connected graphs with at least five vertices is given in [352]: $\sigma_Q < 2(n-2)$.

Concerning the other bounds for σ_Q, it is interesting to mention that

$$\sigma_Q \geq \chi. \tag{8.8}$$

Namely, from $\kappa_1 \leq 2\Delta$ and $\kappa_n \leq \left(1 - \frac{1}{\chi-1}\right)\frac{2m}{n}$ (see (7.31)), we get $\kappa_1 - \kappa_n \geq \Delta + 1 \geq \chi$, giving (8.8).

Two lower bounds that are analogous to those of (8.6) and (8.7) are (cf. [288])

$$\sigma_Q \geq \sqrt{\left(s_1 + s_2 + \frac{2m_1}{n_1} + \frac{2m_2}{n_2}\right)^2 - 16\left(\frac{s_2 m_1}{n_1} + \frac{s_1 m_2}{n_2}\right)} \tag{8.9}$$

(where $s_i = \frac{\sum_{u\in V(G_i)} d_u}{n_i}$, $1 \leq i \leq 2$) and

$$\sigma_Q \geq \frac{1}{n-\alpha}\sqrt{(n\bar{d})^2 + 8(m - \alpha\bar{d})(2m - n\bar{d})}. \tag{8.10}$$

8.1.3 Extremal graphs

The characteristic polynomial of $K_k \nabla (n-k)K_1$ is given by

$$P_{K_k \nabla (n-k) K_1}(x) = x^{n-k-1}(x+1)^{k-1} \left(x^2 - (k-1)x - k(n-k) \right).$$

It follows that $\sigma(K_k \nabla (n-k) K_1) = \sqrt{(k-1)^2 + 4k(n-k)}$. It is straightforward to compute that $\sigma(K_k \nabla (n-k) K_1)$ attains its maximum when $k = \lfloor \frac{2n}{3} \rfloor$. Gregory et al. [174] conjectured that the maximal spectral spread of graphs with n vertices is attained only by $K_{\lfloor \frac{2n}{3} \rfloor} \nabla \left(n - \lfloor \frac{2n}{3} \rfloor \right) K_1$, that is,

$$\max\{\sigma(G) \ : \ |V(G)| = n\} = \sqrt{\left\lfloor \frac{4(n^2 - n + 1)}{3} \right\rfloor}.$$

The conjecture is confirmed for graphs with at most nine vertices, and no counterexamples are known.

On the contrary, since for any bipartite subgraph H of G, $\sigma(G) \geq \sigma(H)$ (see Theorem 3.1), for any connected graph G with n vertices, $\sigma(G) \geq \sigma(P_n)$, with equality if and only if $G \cong P_n$.

So, the connected graph with minimal spectral spread is known and it is the path P_n, while the graph with maximal spectral spread is not. Moreover, since $\lambda_n < 0$ for any non-trivial graph, the graphs with comparatively small spectral spread can be determined by considering the graphs with small spectral radius. For example, from Petrović [361], any connected graph whose spectral spread does not exceed four is an induced subgraph of some Smith graph or an induced subgraph of some of the additional five graphs with at most six vertices.

Concerning the graphs with extremal spectral spread, it is not difficult to show that $K_{1,n-1}$ is the tree with maximal value of σ, while $C(3, n-3)$ is the unicyclic graph with minimal σ. Concerning unicyclic graphs that maximize the spectral spread, we use the result of Subsection 2.3.5 (page 75) on cacti with maximal λ_1 and the similar result on cacti with minimal λ_n given in Theorem 3.32. It appears that the same unicyclic graph maximizes λ_1 and minimizes λ_n, and therefore maximizes σ. This is $P(3, n-3)$.

For the Q-spread, evidently the unique graph with maximal value of σ_Q is $K_1 \cup K_{n-1}$. For connected graphs, that with the maximal value of σ_Q is not known, and there is an actual conjecture, confirmed by a number of computational tests. From Cvetković et al. [104], the maximal value is attained by the pineapple $P(n-1, 1)$. Computing the Q-spread of $P(n-1, 1)$ (see [352]), we get $\sigma_Q(P(n-1, 1)) = \sqrt{4n^2 - 20n + 33}$.

The graphs with minimal or maximal Q-spread within any subset of bipartite graphs coincide with the graphs with minimal or maximal value of κ_1. And so, in the case of trees, these are P_n and $K_{1,n-1}$. As already said, the same graphs have the minimal and maximal spectral spread (within the set of trees). For unicyclic graphs, the graph that maximizes σ_Q is again $P(3, n-3)$, while for a graph with minimal σ_Q we have a different result; contrary to $C(3, n-3)$ (for σ), by the above discussion, here we have C_n.

8.2 Spectral gap

The spectral gap of any Hermitian matrix is the difference between its two largest eigenvalues. According to this, the spectral gap of a graph is defined as $\eta(G) = \lambda_1 - \lambda_2$. Analogously, we have $\eta_L(G) = \mu_1 - \mu_2$ and $\eta_Q(G) = \kappa_1 - \kappa_2$. We have $\eta, \eta_Q > 0$ whenever G is connected.

Consider η. The absolute upper bound is obtained for a graph with maximal value of λ_1 and minimal value of λ_2. Obviously, this is K_n. Concerning lower bounds, we have the following results obtained by Stanić [417].

Theorem 8.5 ([417]) *For a connected non-trivial graph G,*

$$\eta > \frac{2m - \sqrt{mn(n-2)}}{n} \tag{8.11}$$

and

$$\eta \geq \frac{4m - n\sqrt{(n+2)(n-2)}}{2n} + 1. \tag{8.12}$$

Proof We use (2.2) to get $\lambda_1 \geq \frac{2m}{n}$ and (5.2) to get $\lambda_2 < \sqrt{m\frac{n-2}{n}}$. So, we have

$$\eta = \lambda_1 - \lambda_2 > \frac{2m}{n} - \sqrt{m\frac{n-2}{n}} = \frac{2m - \sqrt{mn(n-2)}}{n}.$$

The inequality (8.12) is obtained in a similar way by using the same lower bound for λ_1 and $\lambda_2 \leq \frac{\sqrt{(n+2)(n-2)}}{2} - 1$ (see (4.1)). $\qquad \square$

Comparing the bounds obtained, we get that (8.12) is better than (8.11) whenever $m > \frac{n\left(n^2 - 4\sqrt{(n+2)(n-2)}\right)}{4(n-2)}$.

Theorem 8.6 ([417]) *For a connected graph G,*

$$\eta > \frac{8m - n\sqrt{n(n+4)}}{4n}. \tag{8.13}$$

Table 8.1 *Some graphs with small spectral gap*

n	C_n	P_n	$DK\left(\frac{n-4}{2},4\right)$	$DK\left(\frac{n-6}{2},6\right)$	$DK(4,n-8)$
10	0.3820	0.2365	0.0644	0.2365	0.0640
20	0.0979	0.0665	0.0001	7.2×10^{-6}	2.7×10^{-6}
50	0.0158	0.0114	3.7×10^{-7}	1.1×10^{-9}	1.7×10^{-15}
100	0.0039	0.0029	1.5×10^{-8}	1.1×10^{-11}	*
200	0.0010	0.0007	2.3×10^{-10}	7.1×10^{-14}	*
500	0.0002	0.0001	2.1×10^{-12}	*	*
1000	3.9×10^{-5}	3.0×10^{-5}	4.5×10^{-13}	*	*

Proof By (4.3),

$$\lambda_2 \leq \begin{cases} \left\lfloor \frac{n}{4} \right\rfloor, & \text{if } n = 0 \text{ (mod 4) or } n = 1 \text{ (mod 4)}, \\ \sqrt{\left\lfloor \frac{n}{4} \right\rfloor \left(\left\lfloor \frac{n}{4} \right\rfloor + 1 \right)}, & \text{if } n = 2 \text{ (mod 4) or } n = 3 \text{ (mod 4)}, \end{cases}$$

and therefore we have $\lambda_2 < \sqrt{\frac{n}{4}\left(\frac{n}{4}+1\right)}$, for any $n \geq 1$. Next, we get

$$\eta = \lambda_1 - \lambda_2 > \frac{2m}{n} - \sqrt{\frac{n}{4}\left(\frac{n}{4}+1\right)} = \frac{8m - n\sqrt{n(n+4)}}{4n}.$$

$\square$

For connected bipartite graphs with at least seven vertices, (8.13) gives a better estimate than (8.12).

Considering connected graphs with small spectral gap, Stanić [417] conjectured that the spectral gap is minimized by some double kite graph $D(k,l)$. The conjecture is confirmed for graphs with at most 10 vertices. Moreover, for any n ($1 \leq n \leq 20$), we determine the graph with minimal value of η within the set of double kite graphs with n vertices. For $n \leq 6$ the resulting graph is P_n (any path is a double kite graph). For $7 \leq n \leq 9$, this is $DK(3,l)$; for $10 \leq n \leq 15$, we get $DK(4,l)$; for $16 \leq n \leq 20$, we get $DK(5,l)$.

In Table 8.1 we compare spectral gaps of several graphs. The asterisk stands for a value less than 10^{-16}. According to this table, a putative minimizer, if not $D(k,l)$, must have a very small gap between its two largest eigenvalues. Moreover, it must have another property.

Theorem 8.7 ([417]) *If G is a connected graph with minimal value of η within the set of graphs with n vertices and G is not a tree, then $\lambda_2(G)$ is a simple eigenvalue.*

Proof Assume that $\lambda_2(G) = \lambda_3(G)$. Let uv be an edge which belongs to at least one cycle of G (since G is not a tree, such an edge must exist). Consider the graph $G - u + u'$ obtained from G by removing the vertex u and adding the vertex u', which is not adjacent to v, but is adjacent to all remaining neighbours of u. Clearly, this new graph is connected and has n vertices; in other words, $G - u + u'$ is obtained by deleting the edge uv, and thus we get $\lambda_1(G) > \lambda_1(G - u + u')$.

Next, by the interlacing theorem, we have $\lambda_2(G - u) = \lambda_2(G)$ (since $\lambda_2(G) = \lambda_3(G)$), and then $\lambda_2(G - u + u') \geq \lambda_2(G)$.

Collecting the above inequalities we get $\eta(G) > \eta(G - u + u')$. A contradiction. $\qquad\square$

8.3 Inequalities of Nordhaus–Gaddum type

In 1956, Nordhaus and Gaddum considered the chromatic number of a graph and its complement. They gave a number of lower and upper bounds for the sum and the product of these invariants. Since then, the inequalities that include an invariant of a graph and the same invariant of its complement have been considered by a number of mathematicians. For a survey of this research, we recommend the excellent paper of Aouchiche and Hansen [12].

Concerning spectral invariants, the most attractive ones are the largest eigenvalues of a graph and its complement. One of the first results appeared in 1970, when Nosal [351] gave the following magnitude:

$$n - 1 \leq \lambda_1(G) + \lambda_1(\overline{G}) \leq \frac{1 + \sqrt{3}}{2}(n - 1). \qquad (8.14)$$

The lower bound has been improved by Nikiforov [335] to

$$\lambda_1(G) + \lambda_1(\overline{G}) \geq n - 1 + \sqrt{2}\frac{\left(\sum_{i=1}^{n} \left|d_i - \frac{2m}{n}\right|\right)^2}{n^3}. \qquad (8.15)$$

In contrast, the upper bound of (8.14) has been improved by Csikvári [91] and later by Terpai [445] to

$$\lambda_1(G) + \lambda_1(\overline{G}) \le \frac{4}{3}n - 1. \tag{8.16}$$

The last result is proved by means of mathematical analysis.

Note that any bound for $\lambda_1(G)$ gives an analogous bound for $\lambda_1(\overline{G})$ whenever it is expressed in terms which can be computed for the complementary graph. Here we give just two examples. From the upper bound (2.41) $\lambda_1 \le \sqrt{2m - (n-1)\delta + (\delta-1)\Delta}$ (by replacing m with $\binom{n}{2} - m$, Δ with $n - 1 - \delta$, and δ with $n - 1 - \Delta$), we get the upper bound for $\lambda_1(\overline{G})$. Moreover, using both bounds, we get [400]

$$\lambda_1(G) + \lambda_1(\overline{G}) \le \sqrt{2\big((n-1)^2 - (2n-3)\delta + (2\delta-1)\Delta\big)}. \tag{8.17}$$

Similarly, using the upper bound (2.11), we get

$$\lambda_1(G) + \lambda_1(\overline{G}) \le \sqrt{2m\left(1 - \frac{1}{\omega(G)}\right)} + \sqrt{2\left(\binom{n}{2} - m\right)\left(1 - \frac{1}{\omega(\overline{G})}\right)}. \tag{8.18}$$

Applying the Cauchy–Schwarz inequality, we get [341]

$$\lambda_1(G) + \lambda_1(\overline{G}) \le \sqrt{\left(2 - \frac{1}{\omega(G)} - \frac{1}{\omega(\overline{G})}n(n-1)\right)}. \tag{8.19}$$

This bound improves the result of Hong and Shu [218] with χ instead of ω, and that of Li [270]

$$\lambda_1(G) + \lambda_1(\overline{G}) \le \sqrt{2n(n-1) - 4\delta(n - 1 - \Delta) + 1} - 1. \tag{8.20}$$

The bounds for $\mu_1(G) + \mu_1(\overline{G})$ or $\kappa_1(G) + \kappa_1(\overline{G})$ may be obtained by similar arguments. Some of them are given in Exercises 8.4 and 8.5.

Nikiforov and Yuan [347] considered the upper bounds for an arbitrary eigenvalue (of the adjacency matrix) of a graph and its complement. For example, they proved that

$$|\lambda_i(G)| + |\lambda_i(\overline{G})| \le \frac{n}{\sqrt{2(i-1)}} - 1 \qquad (8.21)$$

holds for any $i \ge 2$ and $n \ge 15(i-1)$, and also that

$$|\lambda_{n-i+1}(G)| + |\lambda_{n-i+1}(\overline{G})| \le \frac{n}{\sqrt{2i}} + 1 \qquad (8.22)$$

holds for $i \ge 1$ and $s \ge 4^i$. If $i = 2^k + 1$ for some integer k, these bounds are asymptotically tight.

8.4 Other inequalities that include two eigenvalues

We start with the following connection between λ_i and μ_{n+1-i}.

Theorem 8.8 (Nikiforov [336]) *For a graph G,*

$$\delta \le \lambda_i + \mu_{n+1-i} \le \Delta \quad (1 \le i \le n). \qquad (8.23)$$

Proof If $\mathbf{x_1}, \mathbf{x_2}, \dots, \mathbf{x_n}$ are orthogonal unit eigenvectors corresponding to the eigenvalues $\mu_1, \mu_2, \dots, \mu_n$ then, for $i \ne n$,

$$\mu_{n+1-i} = \min_{\|\mathbf{y}\|=1, \mathbf{y} \perp \text{Span}\{\mathbf{x_1}, \mathbf{x_2}, \dots, \mathbf{x_{i-1}}\}} \mathbf{y}^T L \mathbf{y}$$

and

$$\lambda_i = \min_{S \subset \mathbb{R}^n, \dim(S) = i-1} \left\{ \max_{\|\mathbf{y}\|=1, \mathbf{y} \perp S} \mathbf{y}^T A \mathbf{y} \right\}$$

(if necessary, consult [223, pp. 178–179]). Let $\mathbf{z} = (z_1, z_2, \dots, z_n)^T$ be such that $\mathbf{z}^T A \mathbf{z}$ is maximal subject to $\|\mathbf{z}\| = 1$ and $\mathbf{z} \perp \text{Span}\{\mathbf{x_1}, \mathbf{x_2}, \dots, \mathbf{x_{i-1}}\}$. We compute

$$\mu_{n+1-i} \leq \mathbf{z}^T L \mathbf{z} = \sum_{u \in V(G)} d_u z_u^2 - \mathbf{z}^T A \mathbf{z}$$

$$\leq \Delta - \max_{\|\mathbf{y}\|=1, \mathbf{y} \perp \mathrm{Span}\{\mathbf{x_1}, \mathbf{x_2}, \ldots, \mathbf{x_{i-1}}\}} \mathbf{y}^T A \mathbf{y}$$

$$\leq \Delta - \min_{S \subset \mathbb{R}^n, \dim(S) = i-1} \left\{ \max_{\|\mathbf{y}\|=1, \mathbf{y} \perp S} \mathbf{y}^T A \mathbf{y} \right\}$$

$$= \Delta - \lambda_i,$$

which gives the right inequality of (8.23). The left one is proved in a similar way, using the dual representations of μ_{n+1-i} and λ_i (cf. [336]). □

Here are two inequalities involving the graph diameter D. The first was obtained by Mohar [323] in a similar way as the inequality (6.51), which bounds D by the algebraic connectivity a. It reads

$$D \leq 2 \left\lceil \sqrt{\frac{\mu_1}{a}} \sqrt{\frac{k^2 - 1}{4k}} + 1 \right\rceil \left\lceil \log_k \frac{n}{2} \right\rceil, \tag{8.24}$$

where k is any real number greater than 1. The question is for which value of k the above inequality gives a good bound for D? Some answers to this question can be obtained by computer search while, by [323], for large values of $\frac{\mu_1}{a}$ a good approximation is obtained for k that minimizes $\sqrt{\frac{k^2-1}{k}} / \ln k$ (this value is close to 6.7870).

In most cases (8.24) is better than (6.51). It also improves the upper bound (6.52).

The second inequality is due to Chung et al. [84], who considered the Chebyshev polynomials $T_n(x) = \cosh\left(\frac{n}{\cosh(x)}\right)$:

$$D \leq \left\lfloor \frac{\cosh^{-1}(n-1)}{\cosh^{-1}\left(\frac{\mu_1 + a}{\mu_1 - a}\right)} \right\rfloor + 1. \tag{8.25}$$

This upper bound holds for connected non-complete graphs. It remains valid for digraphs as well.

To consider the chromatic number, we need the following result.

Theorem 8.9 (Nikiforov [333]) *Let M be a Hermitian matrix of size n partitioned into $k \times k$ blocks so that all diagonal blocks are zero. Then, for every real diagonal matrix N of the same size as M,*

$$\nu_1(N - M) \geq \nu_1 \left(N + \frac{1}{k-1} M \right). \tag{8.26}$$

Proof Set $P = (k-1)N + M$ and let $\mathbf{x} = (x_1, x_2, \ldots, x_n)^T$ denote the unit eigenvector corresponding to $\nu_1(P)$. Defining the vectors $\mathbf{y}_1, \mathbf{y}_2, \ldots, \mathbf{y}_k \in \mathbb{R}^n$, $\mathbf{y_i} = (y_{i1}, y_{i2}, \ldots, y_{in})$, as $y_{ij} = -(k-1)x_j$ if j corresponds to the ith block and $y_{ij} = x_j$ otherwise, we get the following chain of inequalities:

$$k(k-1)\nu_1(N-M) \geq \nu_1(N-M) \sum_{i=1}^{k} \|\mathbf{y_i}\|^2 \geq \sum_{i=1}^{k} \mathbf{y_i}^T (N-M)\mathbf{y_i}$$

$$\geq k(\mathbf{x}^T P \mathbf{x}) = k\nu_1(P),$$

and the proof follows. □

An easy consequence is the well-known result of Hoffman, while another one is left to the reader in Exercise 8.7.

Corollary 8.10 (Hoffman, cf. [201, 333]) *For a graph G with at least one edge and chromatic number χ,*

$$\chi \geq 1 - \frac{\lambda_1}{\lambda_n}. \tag{8.27}$$

The independence number is considered in the next theorem.

Theorem 8.11 (Haemers [201]) *For a graph G with independence number α,*

$$\alpha \leq n \frac{-\lambda_1 \lambda_n}{\delta^2 - \lambda_1 \lambda_n}. \tag{8.28}$$

Proof Let $\overline{d}$ denote the average degree of the vertices in a maximal co-clique. To apply Theorem 1.25, we need the quotient matrix which is induced by the co-clique and which has the form $B = \begin{pmatrix} O & \overline{d} \\ \frac{\overline{d}\alpha}{n-\alpha} & t \end{pmatrix}$, where t is the remaining entry depending on the number of edges outside the co-clique. Interlacing gives $-\lambda_1 \lambda_n \geq -\det(B) = \frac{\overline{d}^2 \alpha}{n-\alpha} \geq \frac{\delta^2 \alpha}{n-\alpha}$. □

At the end, we mention a result of Nikiforov [344] for regular non-bipartite graphs (cf. Theorem 2.42 as well),

$$\lambda_1 + \lambda_n > \frac{1}{nD},\qquad(8.29)$$

where D is the diameter of G.

8.5 Graph energy

The *graph energy* is defined as the sum of the absolute values of the eigenvalues of the adjacency matrix of a graph. In other words,

$$\mathscr{E}(G) = \sum_{i=1}^{n} |\lambda_i|.$$

One of the most significant chemical applications of the spectral graph theory is based on a close connection of molecular orbital energy levels of π-electrons in conjugated hydrocarbons and the energy of the corresponding molecular graphs. According to a recent book [272], chemical research on what we call the energy of a graph started in the 1930s. In the 1970s, Gutman observed that the results obtained in this research are not restricted to molecular graphs but hold for all graphs, which motivated him to define the graph energy (as above). Transferring to the field of spectral graph theory enabled the application of highly developed mathematical tools in this research.

The lower and upper bounds for graph energy and the corresponding extremal graphs are of special interest for us. Several methods for obtaining the bounds for $\mathscr{E}(G)$ are demonstrated in [272]. We single out the following result:

$$\mathscr{E}(H) \leq \mathscr{E}(G) \text{ and } \mathscr{E}(G) - \mathscr{E}(H) \leq \mathscr{E}(G - E(H)) \leq \mathscr{E}(G) + \mathscr{E}(H),\quad(8.30)$$

for any induced subgraph H of a graph G. By the first inequality, if G has at least one edge then $\mathscr{E}(G) \geq \mathscr{E}(K_2) = 2$. Removing the edges from G one by one and applying the second relation from above we get $\mathscr{E}(G) \leq 2m$. Moreover, if we remove all the edges except those that are incident with a vertex with maximal degree, we get [414]

$$\mathscr{E}(G) \leq 2m - 2\left(\Delta - \sqrt{\Delta}\right).\qquad(8.31)$$

So, the absolute numerical lower bound for the energy of an arbitrary graph that contains at least one edge is 2, and it is attained for K_2. In contrast, the graph with n vertices with maximal energy is not known. For $n \leq 7$, this is K_n with $\mathscr{E}(K_n) = 2(n-1)$ but for $n \geq 8$, there are graphs whose energy is greater than $2(n-1)$. Such graphs are called *hyperenergetic*.

One of the most powerful tools in obtaining upper bounds for $\mathscr{E}(G)$ is the application of spectral moments. For example, from the second spectral moment, we have $2m - \lambda_1^2 = \sum_{i=2}^{n} \lambda_i^2$. The application of the Cauchy–Schwarz inequality gives the upper bound obtained by Koolen and Moulton.

Theorem 8.12 ([253]) *For a graph G,*

$$\mathscr{E}(G) \leq \lambda_1 + \sqrt{(n-1)(2m - \lambda_1^2)}. \tag{8.32}$$

Considering the above inequality we get that its right-hand side is a decreasing function in λ_1 whenever $\lambda_1 \in \left(\sqrt{\frac{2m}{n}}, \sqrt{2m} \right)$. Therefore, to exclude λ_1 from (8.32), it is sufficient to replace λ_1 with t, where $\sqrt{\frac{2m}{n}} \leq t \leq \lambda_1$. By taking $t = \frac{w_q + r}{w_q}$ (using the notation of Theorem 2.2), we get another upper bound. Moreover, we can use any of its consequences (2.5)–(2.7) or (2.86) instead.

Using similar arguments but for bipartite graphs, the same authors obtained a slightly modified result.

Theorem 8.13 ([252]) *For a bipartite graph G,*

$$\mathscr{E}(G) \leq 2\lambda_1 + \sqrt{2(n-2)(m - \lambda_1^2)}. \tag{8.33}$$

Again, λ_1 may be replaced with its lower bound.

We have the following lower bound.

Theorem 8.14 (McClelland [309]) *For a graph G,*

$$\mathscr{E}(G) \geq \sqrt{4m + n(n-1)|\det(A)|^{\frac{2}{n}}}. \tag{8.34}$$

The result is proved by considering $\mathscr{E}(G)^2 = \sum_{i=1}^{n} \lambda_i^2 + \sum_{i \neq j} |\lambda_i||\lambda_j|$ and proving $\sum_{i \neq j} |\lambda_i||\lambda_j| \geq n(n-2)|\det(A)|^{\frac{2}{n}}$. A similar argument for bipartite graphs gives the next result.

Theorem 8.15   (Gutman [194]) *For a bipartite graph G,*

$$\mathscr{E}(G) \geq \sqrt{4m + n(n-2)|\det(A)|^{\frac{2}{n}}}. \tag{8.35}$$

Since $\sum_{i=1}^{n} \mu_i = \sum_{i=1}^{n} \kappa_i = 2m$, the *Laplacian graph energy* and the *signless Laplacian graph energy* are usually defined as

$$\mathscr{E}_L(G) = \sum_{i=1}^{n} \left| \mu_i - \frac{2m}{n} \right| \quad \text{and} \quad \mathscr{E}_Q(G) = \sum_{i=1}^{n} \left| \kappa_i - \frac{2m}{n} \right|.$$

Some relevant results on these invariants can be found in the book [272].

8.6 Estrada index

The *Estrada index* was introduced by E. Estrada [142] in 2000 as a measure of the degree of folding of a protein which is represented as a weighted path. In the subsequent period, the Estrada index was usually considered for simple graphs, for which it is defined as

$$EE(G) = \sum_{i=1}^{n} e^{\lambda_i}.$$

There is a wide application of this invariant in computer science, quantum chemistry, thermodynamics, and other branches [144, 145, 143].

The following inequalities follow directly from the definition:

$$EE(nK_1) \leq EE(G) \leq EE(K_n), \tag{8.36}$$

for any graph G with n vertices.

Using the power series expansion of e^x, we get another expression for the Estrada index in terms of spectral moments M_k as

$$EE(G) = \sum_{k=0}^{\infty} \frac{M_k}{k!}. \tag{8.37}$$

Using (8.37), we get the following upper bound.

Theorem 8.16 (Zhou [519]) *For a graph G and an integer $k_0 \geq 2$,*

$$EE(G) \leq n - 1 - \sqrt{2m} + e^{\sqrt{2m}} + \sum_{k=2}^{k_0} \frac{M_k - (\sqrt{2m})^k}{k!}. \qquad (8.38)$$

Equality holds if and only if $G \cong nK_1$.

Setting $k_0 = 2$, we get

$$EE(G) \leq n - 1 - \sqrt{2m} + e^{\sqrt{2m}}, \qquad (8.39)$$

which improves the upper bound obtained by de la Peña et al. [355]:

$$EE(G) \leq n - 1 + e^{\sqrt{2m}}. \qquad (8.40)$$

The following lower bound is also obtained by inspecting the spectral moments.

Theorem 8.17 (Zhou [519]) *For a graph G and an integer $k_0 \geq 2$,*

$$EE(G) \geq \sqrt{n^2 + \sum_{k=2}^{k_0} \frac{2^k M_k}{k!}}. \qquad (8.41)$$

Equality holds if and only if $G \cong nK_1$.

Setting $k_0 = 2$ and $k_0 = 3$, we respectively obtain the lower bound of de la Peña et al. [355]:

$$EE(G) \geq \sqrt{n^2 + 4m} \qquad (8.42)$$

and the lower bound of Deng [131]

$$EE(G) \geq \sqrt{n^2 + 4m + 8t}, \qquad (8.43)$$

where t stands for the number of triangles in G.

We mention another lower bound.

Theorem 8.18 (Das and Lee [129]) *For a graph G,*

$$EE(G) \geq n + \left(\frac{2m}{n}\right)^2 + \frac{1}{12}\left(\frac{2m}{n}\right)^4. \tag{8.44}$$

We conclude with a consequence of (8.36) for particular graphs whose average vertex degree is less than one (observe that any such graph must be disconnected).

Theorem 8.19 (Gutman [195]) *For a graph G with $\frac{2m}{n} < 1$,*

$$EE(G) \geq n - 2m(1 - \cosh(1)). \tag{8.45}$$

Equality holds if and only if $G \cong (mK_2 \cup (n - 2m)K_1)$.

The *Laplacian Estrada index* and the *signless Laplacian Estrada index* are defined analogously by

$$EE_L(G) = \sum_{i=1}^{n} e^{\mu_i} \quad \text{and} \quad EE_Q(G) = \sum_{i=1}^{n} e^{\kappa_i}.$$

Exercises

8.1 Prove that if G has n^+ ($1 \leq n^+ \leq n - 1$) positive eigenvalues and $2m \geq n^+ \lambda_1^2$, then

$$\sigma \geq \lambda_1 + \sqrt{\frac{2m - n^+ \lambda_1^2}{n - n^+}}. \tag{8.46}$$

8.2 Prove that $K(3, n - 3)$ maximizes the spectral spread and Q-spread within the set of unicyclic graphs.

8.3 Prove $\sigma_Q \leq n$.

Hint: Use the upper bound of (7.1) and the lower bound of (7.40).

8.4 Prove the following upper bounds of Nordhaus–Gaddum type:

$$\mu_1(G) + \mu_1(\overline{G}) \leq 2(n - 1) + 2(\Delta - \delta) \tag{8.47}$$

and

$$\mu_1(G) + \mu_1(\overline{G}) \le 2\sqrt{2(n-1)^2 - 3\delta(n-1) + (\Delta + \delta)^2 - \Delta + \delta}. \quad (8.48)$$

Hint: Use an appropriate upper bound for μ_1.

8.5 Prove the following bounds of Nordhaus–Gaddum type:

$$\kappa_1(G) + \kappa_1(\overline{G}) \le 2(n-1) + 2(\Delta - \delta) \quad (8.49)$$

and

$$\kappa_1(G) + \kappa_1(\overline{G}) \ge 2(n-1). \quad (8.50)$$

Hint: For the upper bound, use $\kappa_1 \le 2\Delta$. For the lower one, use $2\lambda_1 \le \kappa_1$ and (8.14).

8.6 Prove that the upper bound for the diameter D in terms of the vertex degree and the second largest eigenvalue (in modulus) of the adjacency matrix of an r-regular graph,

$$D \le \left\lfloor \frac{\cosh^{-1}(n-1)}{\cosh^{-1}\left(\frac{r}{\lambda}\right)} \right\rfloor + 1, \quad (8.51)$$

is a consequence of the upper bound (8.25).

8.7 ([333]) Use Theorem 8.9 to prove

$$\chi \ge 1 + \frac{\lambda_1}{\mu_1 - \lambda_1}. \quad (8.52)$$

8.8 Prove that

$$\chi \ge 1 - \frac{\lambda_{n-\chi+1}}{\lambda_2} \quad (8.53)$$

holds for any graph with positive second largest eigenvalue.

8.9 Prove that for a regular graph with independence number α,

$$\alpha \leq -\frac{-n\lambda_n}{\lambda_1 - \lambda_n}. \tag{8.54}$$

Hint: The quotient matrix induced by a maximal co-clique has the same form as the quotient matrix in the proof of Theorem 8.11.

8.10 ([193]) Prove that for any tree T,

$$\mathscr{E}(K_{1,n-1}) \leq \mathscr{E}(T) \leq \mathscr{E}(P_n). \tag{8.55}$$

8.11 Prove that the diameter of any connected graph satisfies

$$D \leq \bar{n} - 1, \tag{8.56}$$

where $\bar{n}$ denotes the number of distinct eigenvalues. Does the statement holds for L or Q-eigenvalues instead?

8.12 ([98, Theorem 3.14]) Prove that

$$\alpha \leq n^0 + \min\{n^-, n^+\}, \tag{8.57}$$

where α stands for the independence number and n^-, n^0, and n^+ denote the number of eigenvalues smaller than, equal to, and greater than zero, respectively.

8.13 Prove that

$$\sum_{i=1}^{k} v_i \geq \sum_{i=1}^{k} d_i \tag{8.58}$$

holds for any k ($1 \leq k \leq n$), where v_i stands for either μ_i or κ_i and $d_1 \geq d_2 \geq \cdots \geq d_n$ is the degree sequence.

Hint: Use the fact that both matrices L and Q are positive semidefinite and that their main diagonal consists of vertex degrees.

8.14 Prove that for trees or regular graphs,

$$\sum_{i=1}^{k} v_i \leq m + \binom{k+1}{2}$$ (8.59)

holds for any k $(1 \leq k \leq n)$, where v_i stands for either μ_i or κ_i.

Notes

More inequalities for the spectral spread can be found in [174, 254, 279]. Spectral spreads of graphs with few cycles are considered in [364, 365, 404].

More results on the Q-spread can be found in [149, 288, 305] (inequalities) or [290, 438] (extremal graphs).

The Laplacian spectral gap η_L is considered in [260]. For η_Q, see [257].

There is a wide literature for the graph energy or the Estrada index. We give some references, but one should consult the references cited therein as well. The results on graph energy – including its applications, bounds (in particular, those for quadrangle-free graphs, trees or regular graphs) or extremal graphs – can be found in the book [272]. Some notes on $\mathcal{E}_L(G)$ and $\mathcal{E}_Q(G)$ are given in the same book.

The results concerning the Estrada index can be found, say, in the survey paper [197] or in [5, 24, 129, 229, 519].

9

Other spectra of graphs

We consider the spectra of the normalized Laplacian matrix (Section 9.1), the Seidel matrix (Section 9.2), and the distance matrix of a graph (Section 9.3).

9.1 Normalized L-eigenvalues

The *normalized Laplacian matrix* $\widehat{L}$ of a graph G is defined as $\widehat{L} = (\widehat{l}_{i,j})$, where

$$\widehat{l}_{i,j} = \begin{cases} 1, & \text{if } i = j \text{ and } d_i \neq 0, \\ -\dfrac{1}{\sqrt{d_i d_j}}, & \text{if } i \sim j, \\ 0, & \text{otherwise.} \end{cases}$$

If the edges of G are considered as ordered 2-tuples (i,j), let S be the matrix whose rows and columns correspond to vertices and edges of G, respectively, and whose entries are

$$s_{u,e} = \begin{cases} \dfrac{1}{\sqrt{d_i}}, & \text{if } (i,j) \text{ is an edge of } G \text{ and } u = i, \\ -\dfrac{1}{\sqrt{d_j}}, & \text{if } (i,j) \text{ is an edge of } G \text{ and } u = j, \\ 0, & \text{otherwise.} \end{cases}$$

The choice of signs is arbitrary, with the condition that each column has one positive and one negative entry. Then $\widehat{L} = SS^T$, and therefore all eigenvalues of $\widehat{L}$ are non-negative. Moreover, the matrix $\widehat{L}$ is singular, with the eigenvector $(\sqrt{d_1}, \sqrt{d_2}, \ldots, \sqrt{d_n})^T$ corresponding to 0. Hence, we may denote the *normalized Laplacian eigenvalues* (*normalized L-eigenvalues* for short) by

$$\widehat{\mu}_1 \ (= \widehat{\mu}_1(G)) \geq \widehat{\mu}_2 \ (= \widehat{\mu}_2(G)) \geq \cdots \geq \widehat{\mu}_n \ (= \widehat{\mu}_n(G)) = 0.$$

$\widehat{\mu}_1$ is referred to as the normalized L-spectral radius of a graph.

For graphs without isolated vertices we have the following relation between the normalized Laplacian and the Laplacian matrix:

$$\widehat{L} = D^{-\frac{1}{2}} L D^{-\frac{1}{2}},$$

where $D^{-\frac{1}{2}} = \text{diag}\left(\frac{1}{\sqrt{d_1}}, \frac{1}{\sqrt{d_2}}, \ldots, \frac{1}{\sqrt{d_n}}\right)$.

Two simple connections between the eigenvalues of $\widehat{L}$ and those of L or A can be derived from the definition of $\widehat{L}$.

Theorem 9.1   ([72]) *For the eigenvalues of $\widehat{L}$, L, and A,*

$$\frac{\mu_i}{\Delta} \le \widehat{\mu}_i \le \frac{\mu_i}{\delta} \tag{9.1}$$

and

$$\frac{|\lambda_{n+1-i}|}{\Delta} \le |1 - \widehat{\mu}_i| \le \frac{|\lambda_{n+1-i}|}{\delta}, \tag{9.2}$$

where $1 \le i \le n$.

From (9.2), the multiplicity of 0 in the spectrum of A is equal to the multiplicity of 1 in the spectrum of $\widehat{L}$. Moreover, positive eigenvalues of A correspond to the eigenvalues in $[0, 1)$ of $\widehat{L}$ and negative eigenvalues of A correspond to the eigenvalues greater than 1 of $\widehat{L}$.

The Rayleigh quotient (1.1) for the matrix $\widehat{L}$ may be written as $\frac{\sum_{i \sim j}(y_i - y_j)^2}{\sum_{i=1}^n d_i y_i^2}$. From this form we get

$$\widehat{\mu}_1 = \sup_{0 \ne y \in \mathbb{R}^n} \frac{\sum_{i \sim j}(y_i - y_j)^2}{\sum_{i=1}^n d_i y_i^2} \tag{9.3}$$

and

$$\widehat{\mu}_{n-1} = \inf_{0 \ne y \in \mathbb{R}^n, y \perp \mathbf{d}} \frac{\sum_{i \sim j}(y_i - y_j)^2}{\sum_{i=1}^n d_i y_i^2}, \tag{9.4}$$

where $\mathbf{d} = (d_1, d_2, \ldots, d_n)^T$.

Using the last two equalities, we get more properties of normalized L-eigenvalues.

Theorem 9.2 (cf. [72, 83] or [102, Theorem 7.7.2]) *For a graph G with at least two vertices, the following statements hold.*

(i) $\sum_{i=1}^{n} \widehat{\mu}_i \leq n$; *equality holds if and only if G has no isolated vertices.*

(ii) *If G has no isolated vertices then* $\widehat{\mu}_1 \geq \frac{n}{n-1}$; *equality holds if and only if* $G \cong K_n$.

(iii) $\widehat{\mu}_1 \leq 2$; *equality holds if and only if G has non-trivial bipartite component.*

(iv) *If G is connected with diameter D then* $\widehat{\mu}_{n-1} \geq \frac{1}{2mD}$.

(v) *If G is not a complete graph then* $\widehat{\mu}_{n-1} \leq 1$.

(vi) *If G has no isolated vertices then* $\widehat{\mu}_{n-1} \leq \frac{n}{n-1}$; *equality holds if and only if* $G \cong K_n$.

By [72, 83, 102], a motivation for the normalized Laplacian matrix lies in the fact that its eigenvalues can be used to establish the properties of random walks in a graph (the random walk in a graph is a walk which starts at a vertex of a graph and each time randomly takes an edge incident to the current vertex to traverse). Another interesting property arises from the last theorem.

Theorem 9.3 (cf. [72, 83, 102]) *For a graph G the following statements hold.*

(i) *The multiplicity of 0 in the spectrum of* $\widehat{L}$ *is equal to the number of components of G.*

(ii) *If G has no isolated vertices then it is bipartite if and only if the multiplicity of 2 in the spectrum of* $\widehat{L}$ *is equal to the number of components of G.*

9.1.1 Upper and lower bounds for $\widehat{\mu}_1$ and $\widehat{\mu}_{n-1}$

Many results concerning the spectrum of $\widehat{L}$ have analogues in the context of the spectrum of L. In what follows we mention some of them.

For $S \subset V(G)$, let ∂S denote the edge boundary of S as defined on page 111 (recall that we consider this quantity in the context of algebraic connectivity as well). Let also the *volume* of S be the sum of degrees of all vertices in S, that is, $\mathrm{vol}(S) = \sum_{i \in S} d_i$.

Theorem 9.4 (cf. [72, 83]) *For a graph G,*

$$\widehat{\mu}_{n-1} \leq \min_{\emptyset \neq S \subset V(G)} \left(\frac{|\partial S|}{\mathrm{vol}(S)} + \frac{|\partial S|}{\mathrm{vol}(\overline{S})} \right) \leq \max_{\emptyset \neq S \subset V(G)} \left(\frac{|\partial S|}{\mathrm{vol}(S)} + \frac{|\partial S|}{\mathrm{vol}(\overline{S})} \right) \leq \widehat{\mu}_1.$$
$$(9.5)$$

Proof Let $\mathbf{y} = (y_1, y_2, \ldots, y_n)^T$ be defined as follows: $y_i = -\text{vol}(\overline{S})$ if $i \in S$ and $y_i = \text{vol}(\overline{S})$ if $i \in \overline{S}$. Then we compute

$$\frac{\sum_{i \sim j}(y_i - y_j)^2}{\sum_{i=1}^{n} d_i y_i^2} = |\partial S| \left(\frac{1}{\text{vol}(S)} + \frac{1}{\text{vol}(\overline{S})} \right).$$

Using (9.3) and (9.4), we get the assertion. □

For the first equality to hold in (9.5), see Exercise 9.1.

For G connected, let us define $h_G(S) = \frac{|\partial S|}{\min\{\text{vol}(S), \text{vol}(\overline{S})\}}$. Then, by setting

$$h_G = \min_{\emptyset \neq S \subset V(G)} h_G(S),$$

we get

$$\frac{h_G^2}{2} \leq \widehat{\mu}_{n-1} \leq 2h_G. \tag{9.6}$$

The right inequality follows immediately from the previous theorem. For the left inequality, see [83, Theorem 2.2].

Another upper bound for $\widehat{\mu}_1$ was obtained by Rojo and Soto [385] (the proof is based on the Rayleigh principle):

$$\widehat{\mu}_1 \leq 2 - \min_{1 \leq i < j \leq n} \left(\frac{|N(i) \cap N(j)|}{\max\{d_i, d_j\}} \right). \tag{9.7}$$

Finally, we mention an upper bound for the independence number in terms of $\widehat{\mu}_1$.

Theorem 9.5 (Harant and Richter [203]) *For a graph G with no isolated vertices,*

$$\alpha \leq \frac{2(\widehat{\mu}_1 - 1)}{\widehat{\mu}_1 \delta} m. \tag{9.8}$$

Harant and Richter gave a comparison of this result with three similar results that bound α in terms of graph eigenvalues given in (8.28), (6.72), and (8.57). It was proved that (9.8) is generally incomparable with any of these three upper bounds, but the authors constructed an infinite sequence of graphs for which their bound gives a better estimate than either of (8.28) or (6.72). Moreover, it has been shown that the difference between the upper bounds (9.8) and (8.57) can be arbitrarily large.

9.2 Seidel matrix eigenvalues

The *Seidel matrix* of a graph is the matrix $\widehat{S} = J - I - 2A$. This matrix is of particular relevance to *Seidel switching*, described as follows. For any $S \subset V(G)$, the graph H obtained from G by Seidel switching has the same set of vertices; two vertices u and v are adjacent in H if and only if either

(i) they are adjacent in G and both belong to S or both do not belong to S or

(ii) they are non-adjacent in G and exactly one of them belongs to S.

The main property of the Seidel matrix is derived from the described switching. If two regular graphs G and H (where H is obtained from G by Seidel switching) have the same degree, then their Seidel spectra (considered as multisets) coincide [102, Proposition 1.1.8]. According to this property, the Seidel matrix is considered mostly in the context of regular graphs. It can be deduced that for any regular graph, the matrix $\widehat{S}$ has eigenvalues $n - 1 - 2\lambda_1, -1 - 2\lambda_2, \ldots, -1 - 2\lambda_n$, and so, for regular graphs, all the results obtained in Chapters 2–5 can easily be transferred here.

9.3 Distance matrix eigenvalues

The *distance matrix* $\widehat{D}$ of a connected graph G is defined as $\widehat{D} = (\widehat{d}_{i,j})$, where

$$\widehat{d}_{i,j} = d(i,j).$$

That is, the entry (i,j) is equal to the distance between vertices i and j. By Theorem 1.1, the largest eigenvalue of $\widehat{D}$ is simple. Denote the *distance eigenvalues* by

$$\widehat{\delta}_1 \ (= \widehat{\delta}_1(G)) > \widehat{\delta}_2 \ (= \widehat{\delta}_2(G)) \geq \cdots \geq \widehat{\delta}_n \ (= \widehat{\delta}_n(G)).$$

$\widehat{\delta}_1$ is referred to as the distance spectral radius of a graph. Since $\mathrm{tr}(\widehat{D}) = 0$, we have $\sum_{i=1}^{n} \widehat{\delta}_i = 0$.

Some graph invariants can easily be deduced from the distance matrix (the size, all distances in a graph, the graph diameter, vertex degrees, etc.). Further study of the distance matrix and its eigenvalues dates back to the early 1970s when Graham and Pollack [173] showed that $\det(\widehat{D}(T)) = (-1)^{n-1}(n-1)2^{n-2}$ for any tree T. Since this determinant depends only on n, it follows that any tree has exactly one positive eigenvalue and $n-1$ negative ones. In the same period, Graham and Lovás [172] conjectured that the number of positive distance eigenvalues of a graph is never greater than the number of negative ones. In 2014, Azarija [22] confirmed this conjecture for graphs with at most 11 vertices and constructed an infinite family of counterexamples; the smaller of them has 13 vertices.

The distance matrix of a graph is more complex than the adjacency matrix since it is a dense matrix whose entries represent different distances, while the adjacency matrix is just a $(0,1)$-matrix. Therefore, computing the spectra of the distance matrix is a much more intense problem, and in general, there are no simple analytical solutions except for some simple types of graph. The distance eigenvalues of the path, cycle, some regular graphs, and various graph compositions are obtained in [166, 231, 233, 378, 394, 437] (see also Exercises 9.2–9.4). For example, from [437] we know that the distance spectrum of a join of two regular graphs G_1 and G_2 consists of the eigenvalues $-\lambda_{i,j} - 2$ ($1 \le i \le 2$, $2 \le j \le n_i$) and two more eigenvalues of the form

$$n_1 + n_2 - 2 - \frac{r_1 + r_2}{2} \pm \sqrt{\left(n_1 - n_2 - \frac{r_1 - r_2}{2}\right)^2 + n_1 n_2},$$

where $\lambda_{i,1} = r_i \ge \lambda_{i,2} \ge \cdots \ge \lambda_{i,n_i}$ are the eigenvalues (of the adjacency matrix) of G_i ($1 \le i \le 2$).

There is an interlacing inequality involving the distance eigenvalues and the L-eigenvalues of a tree [314]:

$$0 > -\frac{1}{\mu_1} \ge \widehat{\delta}_2 \ge -\frac{1}{\mu_2} \ge \widehat{\delta}_3 \ge \cdots \ge -\frac{1}{\mu_{n-1}} \ge \widehat{\delta}_n. \qquad (9.9)$$

Balaban et al. [23] proposed using $\widehat{\delta}_1$ as a distance descriptor in (molecular) graphs. Another chemical application of $\widehat{\delta}_1$ is given in [199]. For a branch of applications in computer science, and especially in the construction of complex networks, see [307]. We just mention that, from the same reference, the difference between the smallest positive and the largest negative distance eigenvalue can be used as a measure of network complexity.

9.3.1 Upper and lower bounds for $\widehat{\delta}_1$

According to the definition of the distance matrix, an entry of $\widehat{D}$ is smaller if two vertices are closer. In other words, larger connectivity of a graph produces smaller entries in $\widehat{D}$. By virtue of this observation, the following result can be expected. For any connected graph G with n vertices,

$$\widehat{\delta}_1(K_n) \leq \widehat{\delta}_1(G) \leq \widehat{\delta}_1(P_n), \tag{9.10}$$

with equality if and only if $G \cong K_n$ or $G \cong P_n$. See [52] for the first inequality and [394] for the second.

Before we proceed with more complex bounds, we need to mention a few more definitions. For a graph G whose vertices are labelled $1, 2, \ldots, n$, the *distance degree* D_i (of a vertex i) is given by $D_i = \sum_{j=1}^{n} d_{ij}$. For the same graph with a distance degree sequence $D_1, D_2, \ldots, D_n$, the *second distance degree* D_i^* (of a vertex i) is given by $D_i^* = \sum_{j=1}^{n} d_{ij} D_j$. We say that a graph is *k-distance regular* if $D_i = k$ for all i or *pseudo k-distance regular* if $\frac{D_i^*}{D_i} = k$ for all i. The prefix k will usually be suppressed.

Since D_i in fact denotes the ith row sum of the distance matrix $\widehat{D}$, using Theorem 1.2 we get [205]

$$\min\left\{ \sqrt{D_i^*} : 1 \leq i \leq n \right\} \leq \widehat{\delta}_1 \leq \max\left\{ \sqrt{D_i^*} : 1 \leq i \leq n \right\}. \tag{9.11}$$

Following Indulal [232], we give three lower bounds for $\widehat{\delta}_1$.

Theorem 9.6 ([232]) *For a connected graph G,*

$$\widehat{\delta}_1 \geq \sqrt{\frac{\sum_{i=1}^{n} D_i^{*2}}{\sum_{i=1}^{n} D_i^2}}, \tag{9.12}$$

with equality if and only if G is pseudo distance regular.

Proof   Taking the positive vector $\mathbf{y} = \left(\frac{D_1}{\sqrt{\sum_{i=1}^{n} D_i^2}}, \frac{D_2}{\sqrt{\sum_{i=1}^{n} D_i^2}}, \ldots, \frac{D_n}{\sqrt{\sum_{i=1}^{n} D_i^2}} \right)^T$, which is evidently unit, we get

$$\mathbf{y}^T \widehat{D}^2 \mathbf{y} = \frac{\sum_{i=1}^{n} D_i^{*2}}{\sum_{i=1}^{n} D_i^2},$$

and the result follows by the Rayleigh principle.   □

There are two simple corollaries.

Corollary 9.7   ([232]) *For a connected graph G,*

$$\widehat{\delta}_1 \geq \sqrt{\frac{\sum_{i=1}^{n} D_i^2}{n}}. \tag{9.13}$$

The *Wiener index* $W(G)$ of a graph is given by $W(G) = \frac{1}{2}\sum_{i=1}^{n} D_i$. In other words, it denotes the sum of all distances between pairs of vertices.

Corollary 9.8   ([232]) *For a connected graph G,*

$$\widehat{\delta}_1 \geq \frac{2W(G)}{n}. \tag{9.14}$$

Equality holds in both corollaries if and only if G is distance regular.
The next upper bound is obtained by using Theorem 1.2.

Theorem 9.9   (Chen et al. [79]) *For a connected graph G with diameter D and distance degrees $D_1 \geq D_2 \geq \cdots \geq D_n$,*

$$\widehat{\delta}_1 \leq \frac{D_i - D + \sqrt{(D_i + D)^2 + 4D\sum_{j=1}^{i-1}(D_j - D_i)}}{2}, \tag{9.15}$$

where $1 \leq i \leq n$. Equality holds if and only if G is distance regular.

The last bound obviously improves the bound of [79]:

$$\widehat{\delta}_1 \leq \frac{D_i - D + \sqrt{(D_i + D)^2 + 4D(i-1)(D_1 - D_i)}}{2}. \tag{9.16}$$

So far, we have considered the bounds for $\widehat{\delta}_1$ expressed in terms of distance degrees. The following is a lower bound expressed in terms of maximal vertex degrees.

Theorem 9.10   (Zhou and Ilić [523]) *For a connected graph G with maximal vertex degree d_1 and second maximal degree d_2,*

$$\widehat{\delta}_1 \geq \sqrt{(2n - 2 - d_1)(2n - 2 - d_2)}. \tag{9.17}$$

Proof We easily compute $\widehat{\delta}_1 x_i \geq (2n - 2 - d_i)x_j$ and $\widehat{\delta}_1 x_j \geq (2n - 2 - d_j)x_i$, where x_i and x_j denote the minimal and second minimal coordinate of the Perron eigenvector **x**. From this we get (9.17). □

It can be proved that equality holds in (9.17) if and only if G is a regular graph with diameter at most 2.

Using similar reasoning, the same authors obtained the following lower bound for bipartite graphs. Recall that the vertex eccentricity is the maximal distance between a given vertex and any other vertex in a graph.

Theorem 9.11   ([523]) *For a connected bipartite graph with parts X and Y,*

$$\widehat{\delta}_1 \geq n - 2 + \sqrt{n^2 - 4pq + (3q - 2\Delta_X)(3p - 2\Delta_Y)}, \qquad (9.18)$$

where $|X| = p$, $|Y| = q$, and Δ_X is the maximal vertex degree within X. Equality holds if and only if G is the complete bipartite graph $K_{p,q}$ or a semiregular bipartite graph with every vertex eccentricity equal to 3.

The last lower bound improves the lower bound of [126]:

$$\widehat{\delta}_1 \geq n - 2 + \sqrt{n^2 - 3pq}. \qquad (9.19)$$

We mention the lower and upper bound that apply for all connected graphs whose distance matrix has exactly one positive eigenvalue [525]:

$$\sqrt{\frac{\mathrm{tr}(\widehat{D}^2)}{2}} \leq \widehat{\delta}_1 \leq \sqrt{\frac{(n-1)}{n}\mathrm{tr}(\widehat{D}^2)}. \qquad (9.20)$$

The right inequality was initially proved for trees in [520].

Distance Laplacian and signless Laplacian matrix

Recall that D_i is the distance degree of a vertex i. This quantity is sometimes called the *transmission* of i and, according to this terminology, the corresponding diagonal matrix is usually denoted by Tr, so $Tr(G) = \mathrm{diag}(D_1, D_2, \ldots, D_n)$. Aouchiche and Hansen [13] introduced the *distance Laplacian matrix* and the *distance signless Laplacian matrix* of a connected graph by $\widehat{D}_L = Tr - \widehat{D}$ and $\widehat{D}_Q = Tr + \widehat{D}$.

We denote their eigenvalues by

$$\widehat{\delta}_1^L \ (= \widehat{\delta}_1^L(G)) \geq \widehat{\delta}_2^L \ (\widehat{\delta}_2^L(G)) \geq \cdots \geq \widehat{\delta}_n^L \ (\widehat{\delta}_n^L(G))$$

and, since $\widehat{D}_Q$ is positive irreducible, by

$$\widehat{\delta}_1^Q \ (\widehat{\delta}_1^Q(G)) > \widehat{\delta}_2^Q \ (= \widehat{\delta}_2^Q(G)) \geq \cdots \geq \widehat{\delta}_n^Q \ (= \widehat{\delta}_n^Q(G)).$$

There are many results that are similar to those obtained for $\widehat{\delta}_1$. For example,

$$\min\left\{ \sqrt{2(D_i^2 + D_i^*)} \ : \ 1 \leq i \leq n \right\} \leq \widehat{\delta}_1^Q \leq \max\left\{ \sqrt{2(D_i^2 + D_i^*)} \ : \ 1 \leq i \leq n \right\}$$
(9.21)

corresponds to (9.11) and, with the convention that $D_1 \geq D_2 \geq \cdots \geq D_n$,

$$\widehat{\delta}_1^Q \leq \frac{2D_i + D_1 - D + \sqrt{(2D_i - D_1 + D)^2 + 8D(i-1)(D_1 - D_i)}}{2} \qquad (9.22)$$

corresponds to (9.16) [215].

9.3.2 Graphs with small $\widehat{\delta}_2$ or large $\widehat{\delta}_n$

We have seen that $\widehat{\delta}_2 < 0$ for any tree, so it seems natural to consider graphs whose second distance eigenvalue does not exceed a given constant close to zero. Such a problem is considered in the next theorem.

Theorem 9.12 (Xing and Zhou [470]) *A connected graph with at least two vertices and the property $\widehat{\delta}_2 \leq 0.5858$ belongs to the following families of graphs:*

    (i) K_n, where $n \geq 2$;
    (ii) $K_1 \nabla (K_{n_1} \cup K_{n_2})$, where $n_1, n_2 \geq 1$;
    (iii) $K_1 \nabla (K_{n_1} \cup K_{n_2} \cup K_{n_3})$, where $n_1, n_2, n_3 \geq 1$;
    (iv) $K_1 \nabla (K_{n_1} \cup K_{n_2} \cup K_{n_3} \cup K_{n_4})$, where $1 \leq n_1 \leq n_2 \leq n_3 \leq n_4$ and one
    of the following items hold

- $n_1 = n_2 = 1$ and $1 \leq n_3 \leq 873$,
- $n_1 = n_2 = 1, n_3 \geq 874$, and $f(-0.5858) < 0$, where

$$f(x) = x^4 - (n_3 + n_4)x^3 - (3n_3n_4 + 8(n_3 + n_4) + 5)x^2$$
$$- (12n_3n_4 + 11(n_3 + n_4) + 6)x - 6n_3n_4 - 4(n_3 + n_4) - 2,$$

- $n_1 = 1, n_2 = 2$, *and* $n_3 = 2$ *or* $3 = n_3 \leq n_4 \leq 870$ *or* $4 = n_3 \leq n_4 \leq 14$ *or* $5 = n_3 \leq n_4 \leq 8$ *or* $n_3 = n_4 = 6$,
- $n_1 = 1, n_2 = 3$, *and* $3 = n_3 \leq n_4 \leq 7$ *or* $n_3 = n_4 = 4$,
- $n_1 = n_2 = n_3 = 2$ *and* $2 \leq n_4 \leq 5$.

In the next theorem we consider the least distance eigenvalue.

Theorem 9.13 (Lin and Zhou [275], Yu [485]) *The following statements hold.*

(i) *A connected graph G has the property* $\widehat{\delta}_n \geq -2.3830$ *if and only if G is complete or complete multipartite.*

(ii) *A star is the only tree with the property* $\widehat{\delta}_n > -2 - \sqrt{2}$.

(iii) *A tree T has the property* $\widehat{\delta}_n \in [-3 - \sqrt{5}, -2 - \sqrt{2}]$ *if and only if* $T \cong DS(k,2)$ $(2 \leq k \leq 23)$, $T \cong DS(k,3)$ $(3 \leq k \leq 4)$ *or T is obtained from the star* $K_{1,n-1}$ $(n \leq 24)$ *by attaching a pendant vertex at each of* $k \geq 2$ *chosen pendant vertices.*

(iv) *There is no unicyclic graph with the property* $\widehat{\delta}_n > -1$. *A unicyclic graph G has the property* $\widehat{\delta}_n \in (-2 - \sqrt{2}, -1]$ *if and only if* $G \in \{C_3, C_4, C_5, K(3,l)\}$.

Exercises

9.1 ([72]) Let S be a proper subset of $V(G)$. Prove that if

$$\widehat{\mu}_1 = \left(\frac{|\partial S|}{\text{vol}(S)} + \frac{|\partial S|}{\text{vol}(\overline{S})} \right)$$

holds then, for each vertex $i \in S$,

$$\frac{m(\{i\}, \overline{S})}{\text{vol}(\{i\})} = \frac{|\partial S|}{\text{vol}(S)}$$

and also, for each vertex $j \in \overline{S}$,

$$\frac{m(\{j\}, S)}{\text{vol}(\{j\})} = \frac{|\partial S|}{\text{vol}(\overline{S})}.$$

9.2 Prove that the distance spectrum of the complete bipartite graph K_{n_1, n_2} consists of $n_1 + n_2 - 2 \pm \sqrt{n_1^2 - n_1 n_2 + n_2^2}$ and $n_1 + n_2 - 2$ numbers all equal to -2.

9.3 Prove that the distance eigenvalues $\widehat{\delta}_1, \widehat{\delta}_2, \ldots, \widehat{\delta}_n$ of an even cycle are given by

$$\widehat{\delta}_1 = \frac{n^2}{4}$$

and, for $2 \le i \le n$,

$$\widehat{\delta}_i = \begin{cases} 0, & \text{if } i \text{ is even,} \\ -\mathrm{cosec}^2\left(\frac{\pi}{n}i\right), & \text{if } i \text{ is odd,} \end{cases}$$

while for an odd cycle they are given by

$$\widehat{\delta}_1 = \frac{n^2 - 1}{4}$$

and, for $2 \le i \le n$,

$$\widehat{\delta}_i = \begin{cases} -\frac{1}{4}\sec^2\left(\frac{\pi}{2n}i\right), & \text{if } i \text{ is even,} \\ -\frac{1}{4}\mathrm{cosec}^2\left(\frac{\pi}{2n}i\right), & \text{if } i \text{ is odd.} \end{cases}$$

9.4 Determine particular classes of regular graph whose distance spectrum can be derived from the spectrum of the adjacency matrix.

9.5 Prove that $\widehat{\delta}_{n-2} \le -1$ holds for any connected graph with at least seven vertices.

Notes

The normalized Laplacian matrix $\widehat{L}$, its eigenvalues, and other spectral invariants related to $\widehat{L}$ are considered in the book [83]. For similar results one can see the doctoral theses [65, 72].

Another volume-based inequality was derived by Kirkland [239]. Limit points for normalized L-eigenvalues are considered by the same author in [242].

Graphs whose normalized Laplacian has three distinct eigenvalues are characterized in [121].

More inequalities for distance eigenvalues can be found in [126, 520, 523, 524, 525].

Extremal graphs for $\widehat{\delta}_1$ are considered in [293] (graphs with given connectivity, matching number or chromatic number), [52] (graphs with given number of pendant vertices), [230, 350, 436, 458] (various classes of trees), [508] (unicyclic graphs with perfect matchings), [275] (bicyclic graphs), and probably many other papers.

Other distance-based matrices are considered in [524]. Recently, Haemers and Omidi [202] proposed the *universal matrix* of a graph to be any matrix in Span$\{A, D, I, J\}$ with a non-zero coefficient for A.

References

[1] Abreu, N. Algebraic connectivity for subclasses of caterpillars. *Appl. Anal. Discr. Math.*, **4** (2010), 181–196.

[2] Abreu, N., Justel, C. M., Rojo, O., Trevisan, V. Ordering trees and graphs with few cycles by algebraic connectivity. *Linear Algebra Appl.*, **458** (2014), 429–453.

[3] de Abreu, N. M. M. Old and new results on algebraic connectivity of graphs. *Linear Algebra Appl.*, **423** (2007), 53–73.

[4] de Abreu N. M. M., Nikiforov, V. Maxima of the Q-index: Graphs with bounded clique number. *Electron. J. Linear Algebra*, **24** (2013), 121–130.

[5] Aleksić T., Gutman, I., Petrović, M. Estrada index of iterated line graphs. *Bull. Cl. Sci. Math. Nat. Sci. Math.*, **134** (2007), 33–41.

[6] Alon, N. Eigenvalues and expanders. *Combinatorica*, **6** (1986), 83–96.

[7] Alon, N. Eigenvalues, geometric expanders sorting in rounds, and Ramsey theory. *Combinatorica*, **6** (1986), 207–219.

[8] Alon, N., Chung, F. R. K. Explicit constructions of linear sized tolerant networks. *Discrete Math.*, **2** (1988), 15–19.

[9] Alon, N., Spencer, J. H. *The Probabilistic Method* (3rd edition). New York, Wiley, 2008.

[10] Alon, N., Milman, V. D. λ_1, isoperimetric inequalities for graphs, and supercon-centrators. *J. Combin. Theory B*, **38** (1985), 73–88.

[11] Alon, N., Sudakov, B. Bipartite subgraphs and the smallest eigenvalue. *Combin. Probab. Comput.*, **9** (2000), 1–12.

[12] Aouchiche, M., Hansen, P. A survey of Nordhaus-Gadum type relations. *Discrete Appl. Math.*, **161** (2013), 466–546.

[13] Aouchiche, M., Hansen, P. Two Laplacians for the distance matrix of a graph. *Linear Algebra Appl.*, **439** (2013), 21–33.

[14] Aouchiche, M., Hansen, P., Lucas, C. On the extremal values of the second largest Q-eigenvalue. *Linear Algebra Appl.*, **435** (2011), 2591–2606.

[15] Aouchiche, M., Hansen, P., Stevanović, D. A sharp upper bound on algebraic connectivity using domination number. *Linear Algebra Appl.*, **432** (2010), 2879–2893.

[16] Anderson, W. N., Morley, T. D. Eigenvalues of the Laplacian of a graph. *Linear Multilinear Algebra*, **18** (1985), 11–15.

[17] An, C. Bounds on the second largest eigenvalue of a tree with perfect matchings. *Linear Algebra Appl.*, **283** (1998), 247–255.

[18] Anđelić, M., da Fonseca, C. M., Koledin, T., Stanić, Z. Sharp spectral inequalities for connected bipartite graphs with maximal Q-index. *Ars. Math. Contemp.*, **6** (2013), 171–185.

[19] Anđelić, M., da Fonseca, C. M., Simić, S. K., Tošić, D. V. Connected graphs of fixed order and size with maximal Q-index: Some spectral bounds. *Discrete Appl. Math.*, **160** (2012), 448–459.

[20] Anđelić, M., da Fonseca, C. M., Simić, S. K., Tošić, D. V. On bounds for the index of double nested graphs. *Linear Algebra Appl.*, **435** (2011), 2475–2490.

[21] Anđelić, M., Koledin, T., Stanić, Z. Nested graphs with bounded second largest (signless Laplacian) eigenvalue. *Electron. J. Linear Algebra*, **24** (2012), 181–201.

[22] Azarija, J. A short note on a short remark of Graham and Lovás. *Discrete Math.*, **315–316** (2014), 65–68.

[23] Balaban, A. T., Ciubotariu, D., Medeleanu, M. Topological indices and real number vertex invariants based on graph eigenvalues and eigenvectors. *J. Chem. Inf. Comput. Sci.*, **31** (1991), 517–523.

[24] Bamdad, H., Ashraf, F., Gutman, I. Lower bounds for Estrada index and Laplacian Estrada index. *Appl. Math. Lett.*, **23** (2010), 739–742.

[25] Bao, Y.-H., Tan, Y. Y., Fan, Y.-Z. The Laplacian spread of unicyclic graphs. *Appl. Math. Lett.*, **22** (2009), 1011–1015.

[26] Bapat, R. B., Lal, A. K., Pati, S. On algebraic connectivity of graphs with at most two points of articulation in each block. *Linear Multilinear Algebra*, **60** (2012), 415–432.

[27] Behzad, M., Chartrand, G. No graph is perfect. *Amer. Math. Monthly*, **74** (1967), 962-963.

[28] Belardo, F., Li Marzi, E. M., Simić, S. K. Bidegreed trees with small index. *MATCH Commun. Math. Comput. Chem.*, **61** (2009), 503–515.

[29] Belardo, F., Li Marzi, E. M., Simić, S. K. Ordering graphs with index in the interval $(2, \sqrt{2+\sqrt{5}})$. *Discrete Appl. Math.*, **156** (2008), 1670–1682.

[30] Belardo, F., Li Marzi, E. M., Simić, S. K. Path-like graphs ordered by the index. *Internat. J. Algebra*, **1(3)** (2007), 113–128.

[31] Belardo, F., Li Marzi, E. M., Simić, S. K. Some notes on graphs whose index is close to 2. *Linear Algebra Appl.*, **423** (2007), 81–89.

[32] Belardo, F., Li Marzi, E. M., Simić, S. K. Trees with minimal index and diameter at most four. *Discrete Math.*, **310** (2010), 1708–1714.

[33] Belardo, F., Li Marzi, E. M., Simić, S. K., Wang, J. Graphs whose signless Laplacian spectral radius does not exceed the Hoffman limit value. *Linear Algebra Appl.*, **435** (2011), 2913–2920.

[34] Belhaiza, S., de Abreu, N. M. M., Hansen, P., Oliveira, C. S. Variable neighborhood search for extremal graphs. XI. Bounds for algebraic connectivity. In Avis, D., Hertz, A., Macotte, O. (eds.), *Graph Theory and Optimization*. Heidelberg, Springer 2005, pp. 1–16.

[35] Bell, F. K. On the maximal index of connected graphs. *Linear Algebra Appl.*, **144** (1991), 135–151.

[36] Bell, F. A note on the irregularity of graphs. *Linear Algebra Appl.*, **161** (1992), 45–54.

[37] Bell, F., Cvetković, D., Rowlinson, P., Simić, S. K. Graphs for which the least eigenvalue is minimal, I. *Linear Algebra Appl.*, **429** (2008), 234–241.

[38] Bell, F., Cvetković, D., Rowlinson, P., Simić, S. K. Graphs for which the least eigenvalue is minimal, II. *Linear Algebra Appl.*, **429** (2008), 2168–2179.

[39] Bell, F., Rowlinson, P. On the index of tricyclic Hamiltonian graphs. *Proc. Edinburgh Math. Soc.*, **33** (1990), 233–240.

[40] Berman, A., Zhang, X.-D. On the spectral radius of graphs with cut vertices. *J. Combin. Theory B*, **83** (2001), 233–240.

[41] Bermond, J. C., Delorme, C., Farhi, G. Large graphs with given degree and diameter. II. *J. Combin. Theory B*, **36** (1984), 32–48.

[42] Bhattacharya, A., Friedland, S., Peled, U. N. On the first eigenvalue of bipartite graphs. *Electron. J. Combin.*, **15** (2008), R144.

[43] Biggs, N. *Algebraic Graph Theory*. Cambridge, Cambridge University Press, 1974.

[44] Bıyıkoğlu, T., Leydold, J. Algebraic connectivity and degree sequences of trees. *Linear Algebra Appl.*, **430** (2009), 811–817.

[45] Bıyıkoğlu, T., Leydold, J. Graphs of given order and size and minimum algebraic connectivity. *Linear Algebra Appl.* **436** (2012), 2067–2077.

[46] Bıyıkoğlu, T., Leydold, J. Semiregular trees with minimal Laplacian spectral radius. *Linear Algebra Appl.*, **432** (2010), 2335–2341.

[47] Bıyıkoğlu, T., Simić, S. K., Stanić, Z. Some notes on spectra of cographs. *Ars Combin.*, **100** (2011), 421–434.

[48] Bollobás, B., Nikiforov, V. Cliques and the spectral radius. *J. Combin. Theory B*, **97** (2007), 859–865.

[49] Bollobás, B., Nikiforov, V. Graphs and Hermitian matrices: eigenvalue interlacing. *Discrete Math.*, **289** (2004), 119–127.

[50] Borovćanin, B. Line graphs with exactly two positive eigenvalues. *Publ. Inst. Math. (Beograd)*, **72(86)** (2002), 39–47.

[51] Borovćanin, B., Petrović, M. On the index of cactuses with n vertices. *Publ. Inst. Math. (Beograd)*, **79(93)** (2006), 13–18.

[52] Bose, S. S., Nath, M., Paul, S. Distance spectral radius of graphs with r pendant vertices. *Linear Algebra Appl.*, **435** (2011), 2828–2836.

[53] Brand, C., Guidili, B., Imrich, W. Characterization of trivalent graphs with minimal eigenvalue gap. *Croat. Chem. Acta*, **80** (2007), 193–201.

[54] Branković, Lj. *Usability of Secure Statistical Databases* (Doctoral Thesis), University of Newcastle, 2007.

[55] Branković, Lj., Cvetković, D. The eigenspace of the eigenvalue −2 in generalized line graphs and a problem in security of statistical data bases. *Univ. Beograd Publ. Elektrotehn. Fak. Ser. Mat.*, **14** (2003), 37–48.

[56] Brigham, R. C., Dutton, R. D. Bounds on graph spectra. *J. Combin. Theory B*, **37** (1984), 228–234.

[57] Brouwer, A. E., Haemers, W. H. A lower bound for the Laplacian eigenvalues of a graph – Proof of a conjecture by Guo. *Linear Algebra Appl.*, **429** (2008), 2131–2135.

[58] Brouwer, A. E., Haemers, W. H. *Spectra of Graphs*. New York, Springer, 2012.

[59] Brouwer, A. E., Neumaier, A. The graphs with spectral radius between 2 and $\sqrt{2 + \sqrt{5}}$. *Linear Algebra Appl.*, **114/115** (1989), 273–276.

[60] Brualdi, R., Hoffman, A. J. On the spectral radius of $(0, 1)$-matrices. *Linear Algebra Appl.*, **65** (1985), 133–146.

[61] Brualdi, R., Solheid, E. S. On the spectral radius of connected graphs. *Publ. Inst. Math. (Beograd)*, **39(53)** (1986), 45–54.

[62] Bussemaker, F. C., Čobeljić, S., Cvetković, D. M., Seidel, J. J. Cubic graphs on ≤ 14 vertices. *J. Combin. Theory B*, **23** (1977), 234–235.

[63] Bussemaker, F. C., Cvetković, D. There are exactly 13 connected, cubic, integral graphs. *Univ. Beograd, Publ. Elektrothehn. Fak. Ser. Mat. Fiz.*, **544–576** (1976), 43–48.

[64] Buset, D. Maximal cubic graphs with diameter 4. *Discrete Appl. Math.*, **101** (2000), 53–61.

[65] Butler, S. K. *Eigenvalues and Structures of Graphs* (Doctoral Thesis), University of California, 2008.

[66] Cai, G. X., Fan, Y.-Z. The signless Laplacian spectral radius of graphs with given chromatic number. *Math. Appl. (Wuhan)*, **22** (2009), 161–167.

[67] Cameron, P. J., Goethals, J. M., Seidel, J. J., Shult, E. E. Line graphs, root systems, and elliptic geometry. *J. Algebra*, **43** (1976), 305–327.

[68] Caporossi, G., Hansen, P. Variable neighborhood search for extremal graphs. I. The AutoGraphiX system. *Discrete Math.*, **212** (2000), 29–44.

[69] Cao, D. Bounds on eigenvalues and chromatic number. *Linear Algebra Appl.*, **270** (1998), 1–13.

[70] Cao, D., Hong, Y. Graphs characterized by the second eigenvalue. *J. Graph Theory*, **17** (1993), 325–331.

[71] Cardoso, D. M., Cvetović, D., Rowlinson, P., Simić, S. K. A sharp lower bound for the least eigenvalue of the signless Laplacian of a non-bipartite graph. *Linear Algebra Appl.*, **429** (2008), 2770–2780.

[72] Cavers, M. S. *The Normalized Laplacian Matrix and General Randić Index of Graphs* (Doctoral Thesis), University of Regina, 2010.

[73] Chen, Y. F., Fu, H.-L., Kim, I.-J., Stehr, E., Watts, B. On the largest eigenvalues of bipartite graphs which are nearly complete. *Linear Algebra Appl.*, **432** (2010), 606–614.

[74] Chang, A. On the largest eigenvalue of a tree with perfect matchings. *Discrete Math.* **269** (2003), 45–63.

[75] Chang, A., Tian, F. On the spectral radius of unicyclic graphs with perfect matchings. *Linear Algebra Appl.*, **370** (2003), 237–250.

[76] Chang, A., Tian, F., Yu, A. On the index of bicyclic graphs with perfect matchings. *Discrete Math.*, **283** (2004), 51–59.

[77] Chang, T.-J., Tam, B.-S. Graphs with maximal signless Laplacian spectral radius. *Linear Algebra Appl.*, **432** (2010), 1708–1733.

[78] Chen, C. T. *Linear System Theory and Design*, Oxford, Oxford University Press, 1984.

[79] Chen, Y., Lin, H., Shu, J. Sharp upper bounds on the distance spectral radius of a graph. *Linear Algebra Appl.*, **439** (2013), 2659–2666.

[80] Chen, Y., Wang, L. Sharp bounds for the largest eigenvalue of the signless Laplacian of a graph. *Linear Algebra Appl.*, **433** (2010), 908–913.

[81] Chung, F., Lu, L. *Graphs and Networks*. Providence, RI, American Mathematical Society, 2006.

[82] Chung, F. R. K. Diameters and eigenvalues. *J. Amer. Math. Soc.*, **2** (1989), 187–196.

[83] Chung, F. R. K. *Spectral Graph Theory*. Providence, RI, American Mathematical Society, 1997.

[84] Chung, F. R.K̆., Faber, V., Manteuffel, T. A. An upper bound on the diameter of a graph from eigenvalues associated with its Laplacian. *SIAM J. Discrete Math*, **7** (1994), 443–457.

[85] Cioabă, S. The spectral radius and the maximum degree of irregular graphs. *Electron J. Combin*, **14** (2007), R38.

[86] Cioabă, S., van Dam, E. R., Koolen, J. H., Lee, J.-H. A lower bound for the spectral radius of graphs with fixed diameter. *European J. Combin.*, **31** (2010), 1560–1566.

[87] Cioabă, S., van Dam, E. R., Koolen, J. H., Lee, J.-H. Asymptotic results on spectral radius and the diameter of graphs. *Linear Algebra Appl.*, **432** (2010), 722–737.

[88] Cioabă, S., Gregory, D. A., Nikiforov, V. Note: Extreme eigenvalues of nonregular graphs. *J. Combin. Theory B*, **97** (2007), 483–486.

[89] Collatz, L., Sinogowitz, U. Spectren endlicher Grafen. *Abh. Math. Sem. Univ. Hamburg*, **21** (1957), 63–77.

[90] Constantine, G. Lower bounds on the spectra of symmetric matrices with non-negative entries. *Linear Algebra Appl.*, **65** (1965), 171–178.

[91] Csikvári, P. On a conjecture of V. Nikiforov. *Discrete Math.*, **309** (2009), 4522–4526.

[92] Cui, S.-Y., Tian, G.-X., Guo, J.-J. A sharp upper bound on the signless Laplacian spectral radius of graphs. *Linear Algebra Appl.*, **439** (2013), 2442–2447.

[93] Cvetković, D. Chromatic number and the spectrum of a graph. *Publ. Inst. Math. (Beograd)*, **14(28)** (1972), 25–38.

[94] Cvetković, D. On graphs whose second largest eigenvalue does not exceed 1. *Publ. Inst. Math. (Beograd)*, **31(45)** (1982), 15–20.

[95] Cvetković, D. Some possible directions in further investigations of graph spectra. In Lovás, L., Sós, V. T. (eds.) *Algebraic Methods in Graph Theory*. Amsterdam, North-Holland, 1981, pp. 47–67.

[96] Cvetković, D., Čangalović, M., Kovačević-Vujičić, V. Optimization and highly informative graph invariants. In Stanković, B. (ed.), *Two Topics in Mathematics*, Belgrade, Mathematical Institute of SANU, 2004, pp. 5–39.

[97] Cvetković D., Doob, M., Gutman, I. On graphs whose spectral radius does not exceed $\sqrt{2+\sqrt{5}}$. *Ars Combin.*, **14** (1982), 225–239.

[98] Cvetković, D., Doob, M., Sachs, H. *Spectra of Graphs* (3rd edition). Heidelberg, Johann Ambrosius Barth, 1995.

[99] Cvetković, D., Jovanović, I. Network alignment using self-returning walks. *Bull. Cl. Sci. Math. Nat. Sci. Math.*, **38** (2013), 43–61.

[100] Cvetković, D., Rowlinson, P. On connected graphs with maximal index. *Publ. Inst. Math. (Beograd)* **44(58)** (1988), 29–34.

[101] Cvetković, D., Rowlinson, P. The largest eigenvalue of a graph: A survey. *Linear Multilinear Algebra*, **28** (1990), 3–33.

[102] Cvetković, D., Rowlinson, P., Simić, S. K. *An Introduction to the Theory of Graph Spectra*. Cambridge, Cambridge University Press, 2010.

[103] Cvetković, D., Rowlinson, P., Simić, S. K. *Eigenspaces of Graphs*. Cambridge, Cambridge University Press, 1997.

[104] Cvetković, D., Rowlinson, P., Simić, S. K. Eigenvalue bounds for the signless Laplacian. *Publ. Inst. Math. (Beograd)*, **81(95)** (2007), 11–27.

[105] Cvetković, D., Rowlinson, P., Simić, S. K. Graphs with least eigenvalue -2: the star complement technique. *J. Algebr. Comb.*, **14** (2001), 5–16.

[106] Cvetković, D., Rowlinson, P., Simić, S. K. Signless Laplacians of finite graphs. *Linear Algebra Appl.*, **423** (2007), 155–171.

[107] Cvetković, D., Rowlinson, P., Simić, S. K. *Spectral Generalizations of Line Graphs: On Graphs with Least Eigenvalue -2*. Cambridge, Cambridge University Press, 2004.

[108] Cvetković, D., Rowlinson, P., Stanić, Z., Yoon, M.-G. Controllable graphs. *Bull. Cl. Sci. Math. Nat. Sci. Math.* **36** (2011), 81–88.

[109] Cvetković, D., Rowlinson, P., Stanić, Z., Yoon, M.-G. Controllable graphs with least eigenvalue at least -2. *Appl. Anal. Discr. Math.*, **5** (2011), 165–175.

[110] Cvetković, D., Simić, S. K. Graph spectra in computer science. *Linear Algebra Appl.*, **434** (2011), 1545–1562.

[111] Cvetković, D., Simić, S. K. Graph theoretical results obtained by the support of the expert system 'Graph' – An extended survey. In Fajtlowicz, S., Fowler, P. W., Hansen, P., Janovitz, M. F., Roberts F. S. (eds.), *Graphs and Discovery*. Providence, RI, American Mathematical Society, 2005, pp. 39–70.

[112] Cvetković, D., Simić, S. K. On graphs whose second largest eigenvalue does not exceed $(\sqrt{5}-1)/2$. *Discrete Math.* **138** (1995), 213–227.

[113] Cvetković, D., Simić, S. K. The second largest eigenvalue of a graph (a survey). *Filomat*, **9** (1995), 449–472.

[114] Cvetković, D., Simić, S. K. Towards a spectral theory of graphs based on the signless Laplacian, I. *Publ. Inst. Math. (Beograd)*, **85(99)** (2009), 19–33.

[115] Cvetković, D., Simić, S. K. Towards a spectral theory of graphs based on the signless Laplacian, II. *Appl. Anal. Discr. Math.*, **4** (2010), 156–166.

[116] Cvetković, D., Simić, S. K. Towards a spectral theory of graphs based on the signless Laplacian, III. *Linear Algebra Appl.*, **432** (2010), 2257–2272.

[117] Cvetković, D., Simić, S. K., Caporossi, G., Hansen, P. Variable neighborhood search for extremal graphs. 3. On the largest eigenvalue of color-constrained trees. *Linear Multilinear Algebra*, **49** (2001), 143–160.

[118] Cvetković, D. M., Doob, M., Gutman, I., Torgašev, A. *Recent Results in the Theory of Graph Spectra*. New York, Elsevier, 1988.

[119] van Dam, E. R., Kooij, R. E. The minimal spectral radius of graphs with a given diameter. *Linear Algebra Appl.*, **423** (2007), 408–419.

[120] van Dam, E. R. Graphs with given diameter maximizing the spectral radius. *Linear Algebra Appl.*, **426** (2007), 454–457.

[121] van Dam, E. R., Omidi, G. R. Graphs whose normalized Laplacian has three eigenvalues. *Linear Algebra Appl.*, **435** (2011), 2560–2569.

[122] Das, K. C. An improved upper bound for Laplacian graph eigenvalues. *Linear Algebra Appl.*, **368** (2003), 269–278.

[123] Das, K. C. Conjectures on index and algebraic connectivity of graphs. *Linear Algebra Appl.*, **433** (2010), 1666–1673.

[124] Das, K. C. Maximizing the sum of the squares of the degrees of a graph. *Discrete Math.*, **285** (2004), 57–66.

[125] Das, K. C. On conjectures involving second largest signless Laplacian eigenvalue of graphs. *Linear Algebra Appl.*, **432** (2010), 3018–3029.

[126] Das, K. C. On the largest eigenvalue of the distance matrix of a bipartite graph. *MATCH Commun. Math. Comput. Chem.*, **62** (2009), 667–672.

[127] Das, K. C. The largest two Laplacian eigenvalues of a graph. *Linear Multilinear Algebra*, **52** (2004), 441–460.

[128] Das, K. C., Kumar, P. Some new bounds on the spectral radius of graphs. *Discrete Math.*, **281** (2004), 149–161.

[129] Das, K. C., Lee, S. G. On the Estrada index conjecture. *Linear Algebra Appl.*, **431** (2009), 1351–1359.

[130] Delorme, C., Solé, P. Diameter, covering index, covering radius and eigenvalues. *European J. Combin.*, **12** (1991), 95–108.

[131] Deng, H. A proof of a conjecture on the Estrada index. *MATCH Commun. Math. Comput. Chem.*, **62** (2009), 599–606.

[132] Deng, H., Balachandran, S., Ayyaswamy, S. K. On two conjectures of the Randić index and the largest signless Laplacian eigenvalue. *J. Math. Anal. Appl.*, **411** (2014), 196–200.

[133] Desai, M., Rao, V. A characterization of the smallest eigenvalue of a graph. *J. Graph Theory*, **18** (1994), 181–194.

[134] Doob, M., A geometrical interpretation of the least eigenvalue of a line graph. In *Proceedings of the Second Chapel Hill Conference on Combinatorial Mathematics and Its Applications*. Chapel Hill, University of North Carolina, 1970, pp. 126–135.

[135] Doob, M., Cvetković, D. On spectral characterizations and embedding of graphs. *Linear Algebra Appl.*, **27** (1979), 17–26.

[136] Du, X. Shi, L. Graphs with small independence number minimizing the spectral radius. *Discrete Math. Algorithms Appl.*, **5** (2013), N1350017.

[137] Dvořák, Z., Mohar, B. Spectral radius of finite and infinite planar graphs and of graphs with bounded genus. *J. Combin. Theory B*, **100** (2010), 729–739.

[138] Ebrahimi, J., Mohar, B., Nikiforov, V., Ahmady, A. S. On the sum of two largest eigenvalues of a symmetric matrix. *Linear Algebra Appl.*, **429** (2008), 2781–2787.

[139] Edwards, C. S., Elphick, C. H. Lower bounds for the clique and chromatic numbers of graphs. *Discrete Appl. Math.*, **5** (1983), 51–64.

[140] Erdős, P., Fajtlowits, S., Staton, W. Degree sequences in triangle-free graphs. *Discrete Math.*, **92** (1991), 85–88.

[141] Erdős, P., Faudree, R., Pach, J., Spencer, J. How to make a graph bipartite. *J. Combin. Theory B*, **45** (1988), 86–98.

[142] Estrada, E. Characterization of 3D molecular structure. *Chem. Phys. Lett.*, **319** (2000), 713–718.

[143] Estrada, E., Hatano, N. Statistical–mechanical approach to subgraph centrality in complex networks. *Chem. Phys. Lett.*, **439** (2007), 247–251.

[144] Estrada, E., Rodriguez-Velázquez, J. A. Subgraph centrality in complex networks. *Phys. Rev.*, (2005), E71.

[145] Estrada, E., Rodriguez-Velázquez, J. A., Randić, M. Atomic branching in molecules. *Int. J. Quantum Chem.*, **106** (2006), 823–832.

[146] Fajtlowicz, S. On conjectures of Graffiti. *Discrete Math.*, **72** (1988), 113–118.

[147] Fallat, S., Fan, Y.-Z. Bipartiteness and the least eigenvalue of signless Laplacian of graphs. *Linear Algebra Appl.*, **436** (2012), 3254–3267.

[148] Fallat, S., Kirkland, S. Extremizing algebraic connectivity subject to graph theoretic constraints. *Electron. J. Linear Algebra*, **3** (1998), 48–71.

[149] Fan, Y.-Z., Fallat, S. Edge bipartiteness and signless Laplacian spread of graphs. *Appl. Anal. Discr. Math.*, **6** (2012), 31–45.

[150] Fan, Y.-Z., Li, S.-D., Tan, Y.-Y. The Laplacian spread of unicyclic graphs. *J. Math. Res. Exposition*, **30** (2010), 17–28.

[151] Fan, Y.-Z., Tan, Y.-Y. The least eigenvalue of signless Laplacian of non-bipartite graphs with given domination number. *Discrete Math.*, **334** (2014), 20–25.

[152] Fan, Y.-Z., Wang, Y., Gao, Y.-B. Minimizing the least eigenvalues of unicyclic graphs with application to spectral spread. *Linear Algebra Appl.*, **429** (2008), 577–588.

[153] Fan, Y.-Z., Wang, Y., Guo, H. The least eigenvalues of the signless Laplacian of non-bipartite graphs with pendant vertices. *Discrete Math.*, **313** (2013), 903–909.

[154] Fan, Y.-Z., Xu, J., Wang, Y., Liang, D. The Laplacian spread of a tree. *Discrete Math. Theoret. Comput. Sci.*, **10** (2008), 79–86.

[155] Fan, Y.-Z., Yu, G.-D., Wang, Y., The chromatic number and the least eigenvalue of a graph. *Electron. J. Combin.*, **19** (2012), P39.

[156] Fan, Y.-Z., Zhang, F.-F., Wang, Y. The least eigenvalue of the complements of trees. *Linear Algebra Appl.*, **435** (2011), 2150–2155.

[157] Favaron, O., Mahéo, M., Saclé, J.-F. Some eigenvalue properties in graphs (conjectures from Graffiti. II). *Discrete Math.*, **111** (1993), 197–220.

[158] Feng, L., Li, Q., Zhang, X.-D. Some sharp upper bounds on the spectral radius of graphs. *Taiwanese J. Math.*, **11** (2007), 989–997.

[159] Feng, L., Yu, G., Ilić, A. The Laplacian spectral radius for unicyclic graphs with given independence number. *Linear Algebra Appl.*, **433** (2010), 934–944.

[160] Fiedler, M. Algebraic connectivity of graphs. *Czech. Math. J.*, **23** (1973), 298–305.

[161] Fiedler, M. A property of eigenvectors of nonnegative symmetric matrices and its application to graph theory. *Czech. Math. J.*, **25** (1975), 619–633.

[162] Fiedler, M. Some minimax problems for graphs. *Discrete Math.*, **121** (1993), 65–74.

[163] Fiedler, M., Nikiforov, V. Spectral radius and Hamiltonicity of graphs. *Linear Algebra Appl.*, **432** (2010), 2170–2173.

[164] Friedland, S. A proof of Alon's second eigenvalue conjecture. In Larmore, L. L., Goemans, M. X. (eds.), *Proceedings of the 35th Annual ACM Symposium on Theory of Computing*. New York, ACM, 2003, pp. 720–724.

[165] Friedland, S. The maximal eigenvalue of 0-1 matrices with prescribed number of ones. *Linear Algebra Appl.*, **69** (1985), 33–69.

[166] Fowler, P. W., Caporossi, G., Hansen, P. Distance matrices, Wiener indices and related invariants of fullerenes. *J. Phys. Chem. A*, **105** (2000), 6232–6242.

[167] Gantmacher, F. R. *Theory of Matrices* (vol. II). New York, Chelsea, 1960.

[168] Geng, X. Shuchao, L., Simić, S. K. On the spectral radius of quasi-k-cyclic graphs. *Linear Algebra Appl.*, **433** (2010), 1561–1572.

[169] Godsil, C. D., Newman, M. W. Eigenvalue bounds for independent sets. *J. Combin. Theory B*, **98** (2008), 721–734.

[170] Godsil, C. D., Royle, G. *Algebraic Graph Theory*. New York, Springer, 2001.

[171] Goldberg, F., Kirkland, S. On the sign patterns of the smallest signless Laplacian eigenvector. *Linear Algebra Appl.*, **443** (2014), 66–85.

[172] Graham, R. L., Lovás, L. Distance matrix polynomials of trees. *Adv. Math.*, **29** (1978), 60–88.

[173] Graham, R. L., Pollack, H.O. On the addressing problem for loop switching. *Bell System Tech. J.*, **50** (1970), 2495–2519.

[174] Gregory, D. A., Hershkowitz, D., Kirkland, S. J. The spread of the spectrum of a graph. *Linear Algebra Appl.*, **332–334** (2001), 23–35.

[175] Grone, R., Merris, R. Algebraic connectivity of trees. *Czech. Math. J.*, **37** (1987), 660–670.

[176] Grone, R., Merris, R., Sunder, V. S. The Laplacian spectrum of a graph. *SIAM J. Matrix Anal. Appl.*, **11** (1990), 218–238.

[177] Guo, J.-M. A new upper bound for the Laplacian spectral radius of graphs. *Linear Algebra Appl.*, **400** (2005), 61–66.

[178] Guo, J.-M. On limit points of Laplacian spectral radii of graphs. *Linear Algebra Appl.*, **429** (2008), 1705–1718.

[179] Guo, J.-M. On the Laplacian spectral radius of a tree. *Linear Algebra Appl.*, **368** (2003), 379–385.

[180] Guo, J.-M. On the Laplacian spectral radius of trees with fixed diameter. *Linear Algebra Appl.*, **419** (2006), 618–629.

[181] Guo, J.-M. On the second largest Laplacian eigenvalue of trees. *Linear Algebra Appl.*, **404** (2005), 251–261.

[182] Guo, J.-M. On the third largest Laplacian eigenvalue of a graph. *Linear Multilinear Algebra*, **55** (2007), 93–102.

[183] Guo, J.-M. The algebraic connectivity of graphs under perturbation. *Linear Algebra Appl.*, **433** (2010), 1148–1153.

[184] Guo, J.-M. The effect on the Laplacian spectral radius of a graph by adding or grafting edges. *Linear Algebra Appl.*, **413** (2006), 59–71.

[185] Guo, J.-M., Li, J., Shiu, W. C. A note on the upper bounds for the Laplacian spectral radius of graphs. *Linear Algebra Appl.*, **439** (2013), 1657–1661.

[186] Guo, J.-M., Li, J., Shiu, W. C. The smallest Laplacian spectral radius of graphs with a given clique number. *Linear Algebra Appl.*, **437** (2012), 1109–1122.

[187] Guo, J.-M., Shao, J.-Y. On the spectral radius of trees with fixed diameter. *Linear Algebra Appl.*, **413** (2006), 131–147.

[188] Guo, S.-G. *On the Eigenvalues of the Quasi-Laplacian Matrix of a Graph* (Technical Report), Yancheng Teachers College, 2008.

[189] Guo, S.-G. Ordering trees with n vertices and matching number q by their largest Laplacian eigenvalues. *Discrete Math.*, **308** (2008), 4608–4615.

[190] Guo, S.-G. The largest eigenvalues of Laplacian matrix of unicyclic graphs. *Appl. Math. J. Chinese Univ. Ser. A*, **16** (2001), 131–135.

[191] Guo, S.-G., Chen, Y.-G., Yu, G. A lower bound on the least signless Laplacian eigenvalue of a graph. *Linear Algebra Appl.*, **448** (2014), 217–221.

[192] Guo, S.-G., Wang, Y.-F. Ordering cacti with n vertices and k cycles by their Laplacian spectral radii. *Publ. Inst. Mat. (Beograd)*, **92** (2012), 117–125.

[193] Gutman, I. Acyclic systems with extremal Hükel π-electron energy. *Theor. Chem. Acta*, **45** (1977), 79–87.

[194] Gutman, I. Bounds for total π-electron energy. *Chem. Phys. Lett.*, **24** (1974), 283–285.

[195] Gutman, I. Lower bounds for Estrada index. *Publ. Inst. Mat. (Beograd)*, **83** (2008), 1–7.

[196] Gutman, I. The star is the tree with greatest Laplacian eigenvalue. *Kragujevac J. Math.*, **24** (2002), 61–65.

[197] Gutman, I., Deng, H., Radenković, S. The Estrada index: An updated survey. In Cvetković, D., Gutman, I. (eds.). *Selected Topics on Applications of Graph Spectra*. Belgrade, Mathemtical Institute of SANU, 2011, pp. 155–174.

[198] Gutman, I., Gineityte, V., Lepović, M., Petrović, M. The high-energy band in the photoelectron spectrum of alkanes and its dependence on molecular structure. *J. Serb. Chem. Soc.*, **64** (1999), 673–680.

[199] Gutman, I., Medeleanu, M. On the structure-dependence of the largest eigenvalue of the distance matrix of alkane. *Indian J. Chem. A*, **37** (1998), 569–573.

[200] Haemers, W. H. *Eigenvalue Techniques in Design and Graph Theory* (Doctoral Thesis), University of Tilburg, 1979.

[201] Haemers, W. H. Interlacing eigenvalues and graphs. *Linear Algebra Appl.*, **227–228** (1995), 593–616.

[202] Haemers, W.H., Omidi, G. R. Universal adjacency matrices with two eigenvalues. *Linear Algebra Appl.*, **435** (2011), 2520–2529.

[203] Harant, J., Richter, S. A new eigenvalue bound for independent sets. *Discrete Math.*, in press.

[204] He, B., Yin, Y.-L., Zhang, X.-D. Sharp bounds for the signless Laplacian spectral radius in terms of clique number. *Linear Algebra Appl.*, **438** (2013), 3851–3861.

[205] He, C.-H., Liu, Y., Zhao, Z.-H. Some new sharp bounds on the distance spectral radius of graphs. *MATCH Commun. Math. Comput. Chem.*, **63** (2010), 783–788.

[206] He, C.-H., Pan, H. The smallest signless Laplacian eigenvalue of graphs under perturbation. *Electron. J. Linear Algebra*, **23** (2012), 473–482.

[207] He, C.-X., Shao, J. Y., He, J.-L. On the Laplacian spectral radii of bicyclic graphs. *Discrete Math.*, **308** (2008), 5981–5995.

[208] He, C.-H., Zhou, M. A sharp upper bound on the least signless Laplacian eigenvalue using domination number. *Graph Combinator.*, **30** (2014), 1183–1192.

[209] Hoffman, A. J., On limit points of spectral radii of non-negative symmetric integral matrices. In Alavi, Y., Lick, D. R., White, A. T. (eds.), *Graph Theory and Applications*. Berlin, Springer, 1972, pp. 165–172.

[210] Hoffman, A. J. On limit points of the least eigenvalue of a graph. *Ars Combin.*, **3** (1977), 3–14.

[211] Hoffman, A. J., Ray-Chaudhuri, D. K. On the line graph of a symmetric balanced incomplete block design. *Trans. Amer. Math. Soc.*, **116** (1965), 238–252.

[212] Hoffman, A. J., Singleton, R. R. Moore graph with diameter 2 and 3. *IBM J. Research Develop.*, **5** (1960), 497–504.

[213] Hoffman, A. J., Smith, J. H. On the spectral radii of topologically equivalent graphs. In Fiedler, M. (ed.), *Recent Advances in Graph Theory*. Prague, Academia Praha, 1975, pp. 273–281.

[214] Hofmeister, M. Spectral radius and degree sequence. *Math. Nachr.*, **139** (1988), 37–44.

[215] Hong, W., You, L. Some sharp bounds on the distance signless Laplacian spectral radius of graphs, in press.

[216] Hong, Y. Bound of eigenvalues of a graph. *Acta Math. Appl. Sinica*, **432(4)** (1988), 165–168.

[217] Hong, Y. The Kth largest eigenvalue of a tree. *Linear Algebra Appl.*, **73** (1986), 151–155.

[218] Hong, Y., Shu, J.-L. A sharp upper bound for the spectral radius of the Nordhaus–Gaddum type. *Discrete Math.*, **211** (2000), 229–232.

[219] Hong, Y., Shu, J.-L. Sharp lower bounds of the least eigenvalue of planar graphs. *Linear Algebra Appl.*, **296** (1999), 227–232.

[220] Hong, Y., Shu, J.-L., Fang, K. A sharp upper bound of the spectral radius of graphs. *J. Combin. Theory B*, **81** (2001), 177–183.

[221] Hong, Y., Zhang, X.-D. Sharp upper and lower bounds for largest eigenvalue of the Laplacian matrices of trees. *Discrete Math.*, **296** (2005), 187–197.

[222] Hoory, S., Linial, N., Widgerson, A. Expander graphs and their applications. *Bull. Amer. Math. Soc.*, **43** (2006), 439–561.

[223] Horn, R., Johnson, C. *Matrix Analysis*. Cambridge, Cambridge University Press, 1985.

[224] Howes, L. On subdominantly bounded graphs – summary of results. In Capobianco, M., Frechen, J. B., Krolik, M. (eds.), *Recent Trends in Graph Theory, Proceedings of the First New York City Graph Theory Conference*. Berlin, Springer, 1971, pp. 181–183.

[225] Hu, S. A sharp lower bound on the spectral radius of simple graphs. *Appl. Anal. Discr. Math.*, **3** (2009), 379–385.

[226] Hu, S. Sharp bounds on the eigenvalues of trees. *Comp. Math. Appl.*, **56** (2008), 909–917.

[227] Huang, Y., Liu, B., Liu, Y. The signless Laplacian spectral radius of bicyclic graph with prescribed degree sequences. *Discrete Math.*, **311** (2011), 504–511.

[228] Huang, Z., Deng, H., Simić, S. K. On the spectral radius of cactuses with perfect matching. *Appl. Anal. Discr. Math.*, **5** (2011), 14–21.

[229] Ilić, A., Stevanović, D. The Estrada index of chemical trees. *J. Math. Chem.*, **47** (2010), 305–314.

[230] Ilić, A. Distance spectral radius of trees with given matching number. *Discrete Appl. Math.*, **158** (2010), 1799–1806.

[231] Indulal, G. Distance spectrum of graph compositions. *Ars. Math. Contemp.*, **2** (2009), 93–100.

[232] Indulal, G. Sharp bounds on the distance spectral radius and the distance energy of graphs. *Linear Algebra Appl.*, **430** (2009), 106–113.

[233] Indulal, G., Gutman, I. On the distance spectra of some graphs. *Math. Commun.*, **13** (2008), 123–131.

[234] Jovanović, I., Stanić, Z. Spectral distances of graphs. *Linear Algebra Appl.*, **436** (2012), 1425–1435.

[235] Jovović, I., Koledin, T., Stanić, Z. Non-bipartite graphs of fixed order and size that minimize the least eigenvalue. *Linear Algebra Appl.*, **477** (2015), 148–164.

[236] Jin, G., Zuo, L. On further ordering bicyclic graphs with respect to the Laplacian spectral radius. *WSEAS Transactions Math.*, **12** (2013), 979–991.

[237] Kelner, J. A., Lee, J. R., Price, G. N., Teng, S.-H. Higher eigenvalues of graphs. In *Proceedings of the 50th Annual Symposium on Foundations of Computer Science*. Los Alamitos, CA, Computer Society Press, 2009, pp. 735–744.

[238] Kirkland, S. Algebraic connectivity for vertex-deleted subgraphs, and a notion of vertex centrality. *Discrete Math.*, **310** (2010), 911–921.

[239] Kirkland, S. An note on a distance bound using eigenvalues of the normalized Laplacian matrix. *Electron. J. Linear Algebra*, **16** (2007), 204–207.

[240] Kirkland, S. A note on limit points for algebraic connectivity. *Linear Algebra Appl.*, **373** (2003), 5–11.

[241] Kirkland, S. An upper bound on the algebraic connectivity of graphs with many cutpoints. *Electron. J. Linear Algebra*, **8** (2001), 94–109.

[242] Kirkland, S. Limit points for normalized Laplacian eigenvalues. *Electron. J. Linear Algebra*, **15** (2006), 337–344.

[243] Koledin, T. *Reflexive Graphs with a Small Number of Cycles* (Master's Thesis, in Serbian), University of Belgrade, 2007.

[244] Koledin, T., Radosavljević, Z. Unicyclic reflexive graphs with seven loaded vertices of the cycle. *Filomat*, **23** (2009), 257–268.

[245] Koledin, T., Stanić, Z. Reflexive bipartite regular graphs. *Linear Algebra Appl.*, **442** (2014), 145–155.

[246] Koledin, T., Stanić, Z. Regular graphs whose second largest eigenvalue is at most 1. *Novi Sad J. Math.*, **43** (2013), 145–153.

[247] Koledin, T., Stanić, Z. Regular graphs with small second largest eigenvalue. *Appl. Anal. Discr. Math.*, **7** (2013), 235–249.

[248] Koledin, T., Stanić, Z. Some spectral inequalities for triangle-free regular graphs. *Filomat*, **28** (2013), 1561–1567.

[249] Kolotilina, L. Y. Upper bounds for the second largest eigenvalue of symmetric nonnegative matrices. *J. Math. Sci. (N. Y.)*, **191** (2013), 75–88.

[250] König, M. D., Battiston S., Napoletano, M., Schweityer, F. The efficiency and evolution of R&D networks. *CER-ETH Economics Working Paper Series*, (2008), N08/95.

[251] König, M. D., Battiston, S., Napoletano, M., Schweityer, F. On algebraic graph theory and the dynamics of innovation networks. *Networks and Heterogenous Media*, **3** (2008), 201–219.

[252] Koolen, J. H., Moulton, V. Maximal energy bipartite graphs. *Graph Combinator.*, **19** (2003), 131–135.

[253] Koolen, J. H., Moulton, V. Maximal energy graphs. *Adv. Appl. Math.*, **26** (2001), 47–52.

[254] Kumar, R. Bounds for eigenvalues of a graph. *J. Math. Inequal.*, **4** (2010), 399–404.

[255] Lal, A. K., Patra, K. L., Sahoo, B. K. Algebraic connectivity of connected graphs with fixed number of pendant vertices. *Graph Combinator.*, **27** (2011), 215–229.

[256] Lan, J., Lu, L., Shi, L. Graphs with diameter $n - e$ minimizing the spectral radius. *Linear Algebra Appl.*, **437** (2012), 2823–2850.

[257] Li, J., Guo, J.-M., Shiu, W. C. On the second largest Laplacian eigenvalues of graphs. *Linear Algebra Appl.*, **438** (2013), 2438–2446.

[258] Li, J., Guo, J.-M., Shiu, W.C̈. The ordering of bicyclic graphs and connected graphs by algebraic connectivity. *Electron. J. Combin.*, **17** (2010), R162.

[259] Li, J., Guo, J.-M., Shiu, W. C. The smallest values of algebraic connectivity for unicyclic graphs. *Discrete Appl. Math.*, **158** (2010), 1633–1643.

[260] Li, J., Shiu, W. C., Chan, W. H. Some results on the Laplacian eigenvalues of unicyclic graph. *Linear Algebra Appl.*, **430** (2009), 2080–2093.

[261] Li, J., Shiu, W. C., Chan, W. H. The Laplacian spectral radius of some graphs. *Linear Algebra Appl.*, **431** (2009), 99–103.

[262] Li, J., Shiu, W. C., Chan, W. H., Chang, A. On the spectral radius of graphs with connectivity at most k. *J. Math. Chem.*, **46** (2009), 340–346.

[263] Li, J., Shiu, W. C., Chang, A. On the kth Laplacian eigenvalues of trees with perfect matchings. *Linear Algebra Appl.*, **432** (2010), 1036–1041.

[264] Li, J. S., Pan, Y. L. A note on the second largest eigenvalue of the Laplacian matrix of a graph. *Linear Multilinear Algebra*, **48** (2000), 117–121.

[265] Li, J. S., Pan, Y. L. De Caen's inequality and bounds on the largest Laplacian eigenvalue of a graph. *Linear Algebra Appl.*, **328** (2001), 153–160.

[266] Li, J. S., Zhang, X. D. A new upper bound for eigenvalues of the Laplacian matrix of a graph. *Linear Algebra Appl.*, **265** (1997), 93–100.

[267] Li, R., Shi, J. The minimum signless Laplacian spectral radius of graphs with given independence number. *Linear Algebra Appl.*, **433** (2010), 1614–1622.

[268] Li, S., Simić, S. K., Tošić, D., Zhao, Q. On ordering bicyclic graphs with respect to the Laplacian spectral radius. *Appl. Math. Lett.*, **24** (2011), 2186–2192.

[269] Li, S., Zhang, M. On the signless Laplacian index of cacti with a given number of pendant vertices. *Linear Algebra Appl.*, **436** (2012), 4400–4411.

[270] Li, X. The relations between the spectral radius of the graphs and their complements (in Chinese). *J. North China Technol. Inst.*, **17** (1996), 297–299.

[271] Li, X., Li, Y., Shi, Y., Gutman, I. Note on the HOMO–LUMO index of graphs. *MATCH Commun. Math. Comput. Chem.*, **70** (2013), 85–96.

[272] Li, X., Shi, Y., Gutman, I. *Graph Energy*. New York, Springer, 2012.

[273] de Lima, L. S., Nikiforov, V. On the second largest eigenvalue of the signless Laplacian. *Linear Algebra Appl.*, **438** (2013), 1215–1222.

[274] de Lima, L. S., Oliveira, C. S., Abreu, N. M. M., Nikiforov, V. The smallest eigenvalue of the signless Laplacian. *Linear Algebra Appl.*, **435** (2011), 2570–2584.

[275] Lin, H., Zhou, B. On least distance eigenvalues of trees, unicyclic graphs and bicyclic graphs. *Linear Algebra Appl.*, **443** (2014), 153–163.

[276] Liu, B., Bo, Z. On the third largest eigenvalue of a graph. *Linear Algebra Appl.*, **317** (2000), 193–200.

[277] Liu, B., Chen, Z., Liu, M. On graphs with largest Laplacian index. *Czech. Math. J.*, **58** (2008), 949–960.

[278] Liu, B., Li, G. A note on the largest eigenvalue of non-regular graphs. *Electron. J. Linear Algebra*, **17** (2008), 54–61.

[279] Liu, B., Liu, M. On the spread of the spectrum of a graph. *Discrete Math.*, **309** (2009), 2727–2732.

[280] Liu, H., Lu, M. Bounds of signless Laplacian spectrum of graphs based on the k-domination number. *Linear Algebra Appl.*, **440** (2014), 83–89.

[281] Liu, H., Lu, M. On the spectral radius of quasi-tree graphs. *Linear Algebra Appl.*, **428** (2008), 2708–2714.

[282] Liu, H., Lu, M., Tian, F. Edge-connectivity and (signless) Laplacian eigenvalue of graphs. *Linear Algebra Appl.*, **439** (2013), 3777–3784.

[283] Liu, H., Lu, M., Tian, F. On the Laplacian spectral radius of a graph. *Linear Algebra Appl.*, **376** (2004), 135–141.

[284] Liu, H., Lu, M., Tian, F. On the spectral radius of unicyclic graphs with fixed diameter. *Linear Algebra Appl.*, **420** (2007), 449–457.

[285] Liu, J., Liu, B. L. The maximum clique and the signless Laplacian eigenvalues. *Czech. Math. J.*, **58** (2008), 1233–1240.

[286] Liu, M. The (signless Laplacian) spectral radii of connected graphs with prescribed degree seqences. *Electron. J. Combin.*, **19(4)** (2012), P35.

[287] Liu, M., Liu, B. Some results on the spectral radii of trees, unicyclic, and bicyclic graphs. *Electron. J. Linear Algebra*, **23** (2012), 327–339.

[288] Liu, M., Liu, B. The signless Laplacian spread. *Linear Algebra Appl.*, **432** (2010), 505–514.

[289] Liu, R., Zhai, M., Shu, J. The least eigenvalue of unicyclic graphs with n vertices and k pendant vertices. *Linear Algebra Appl.*, **431** (2009), 657–665.

[290] Liu, Y. The Laplacian spread of cactuses. *Discrete Math. Theoret. Comput. Sci.*, **12** (2010), 35–40.

[291] Liu, Y., Liu, Y. Ordering of unicyclic graphs with Laplacian spectral radii (in Chinese). *J. Tongji Univ.*, **36** (2008), 841–843.

[292] Liu, Y., Shao, J.-Y., Yuan, X.-Y. Some results on the ordering of the Laplacian spectral radii of unicyclic graphs. *Discrete Appl. Math.*, **156** (2008), 2679–2697.

[293] Liu, Z. On spectral radius of the distance matrix. *Appl. Anal. Discr. Math.*, **4** (2010), 269–277.

[294] Liu, Z., Zhou, B. On least eigenvalues of bicyclic graphs with fixed number of pendant vertices. *J. Math. Sci. (N. Y.)* **182** (2012), 175–192.

[295] Lovás, L., Pelikán, J. On the eigenvalues of trees. *Periodica Math. Hung.*, **3** (1973), 175–182.

[296] Lu, H., Lin, Y. Maximum spectral radius of graphs with given connectivity, minimum degree and independence number. *J. Discrete Algorithms*, **31** (2015), 113–119.

[297] Lu, M., Liu, H., Tian, F. A new upper bound for the spectral radius of graphs with girth at least 5. *Linear Algebra Appl.*, **414** (2006), 512–516.

[298] Lu, M., Liu, H., Tian, F. An improved upper bound for the Laplacian spectral radius of graphs. *Discrete Math.*, **309** (2009), 6318–6321.

[299] Lu, M., Liu, H., Tian, F. Bounds of Laplacian spectrum of graphs based on the domination number. *Linear Algebra Appl.*, **402** (2005), 390–396.

[300] Lu, M., Liu, H., Tian, F. Laplacian spectral bounds for clique and independence numbers of graphs. *J. Combin. Theory B*, **97** (2007), 726–732.

[301] Lu, M., Zhang, L.-Z., Tian, F. Lower bounds of the Laplacian spectrum of graphs based on diameter. *Linear Algebra Appl.*, **420** (2007), 400–406.

[302] Lubotzky, A., Phillips, R., Sarnak, P. Ramanujan graphs. *Combinatorica*, **8** (1988), 261–277.

[303] Maas, C. Perturbation results for the adjacency spectrum of a graph. *ZAMM*, **67** (1987), 428–430.

[304] Maas, C. Transportation in graphs and the admittance spectrum. *Discrete Appl. Math.*, **16** (1987), 31–49.

[305] Maden, A. D., Çevik, A. S., Habibi, N. New bounds for the spread of the signless Laplacian spectrum. *Math. Inequal. Appl.*, **17** (2014), 283–294.

[306] Maden, A. D., Das, K. C., Çevik, A. S. Sharp upper bounds on the spectral radius of the signless Laplacian matrix of a graph. *Appl. Math. Comput.*, **219** (2013), 5025–5032.

[307] Malarz, K. Spectral properties of adjacency and distance matrices for various networks. In Bubak, M., van Albada, G. D., Dongarra, J., Sloot, P. M. A. (eds.), *Computational Science ICCS*. Heidelberg, Springer, 2008, pp. 559–567.

[308] Maxwell, G. Hyperbolic trees. *J. Algebra*, **54** (1978), 46–49.

[309] McClelland, B. J. Properties of the latent roots of a matrix: The estimation of π-electron energies. *J. Chem. Phys.*, **54** (1971), 640–643.

[310] McKay, B. D. *nauty User's Guide (version 1.5)* (Technical Report TR-CS-90-02), Australian National University, 1990.

[311] Mélot, H. Facet defining inequalities among graph invariants: The system GraPHedron. *Discrete Appl. Math.*, **156** (2008), 1875–1891.

[312] Merikoski, J. K., Kumar, R., Rajput, R. A. Upper bounds for the largest eigenvalue of bipartite graph. *Electron. J. Linear Algebra*, **26** (2013), 168–176.

[313] Merris, R. Laplacian graph eigenvectors. *Linear Algebra Appl.*, **278** (1998), 221–236.

[314] Merris, R. The distance spectrum of a tree. *J. Graph Theory*, **14** (1990), 365–369.

[315] Merris, R. Laplacian matrices of graphs: A Survey. *Linear Algebra Appl.*, **197–198** (1994), 143–176.

[316] Merris, R. Degree maximal graphs are Laplacian integral. *Linear Algebra Appl.*, **199** (1994), 381–389.

[317] Meyer, C. *Matrix Analysis and Applied Linear Algebra*. Philadelphia, PA, SIAM, 2000.

[318] van Mieghem, P. *Graph Spectra for Complex Networks*. Cambridge, Cambridge University Press, 2010.

[319] van Mieghem, P., Wang, H. *Spectra of a New Class of Graphs with Extremal Properties* (Technical Report No. 20091010), Delft University of Technology, 2009.

[320] Mihailović, B., Radosavljević, Z. On a class of tricyclic reflexive cactuses. *Univ. Beograd. Publ. Elektrotehn. Fak. Ser. Mat.* **16** (2005), 55–63.

[321] Milatović, M., Stanić, Z. The nested split graphs whose second largest eigenvalue is equal to 1. *Novi Sad J. Math.*, **42(2)** (2012), 33–42.

[322] Ming, G. J., Wang, T. S. On the spectral radius of trees. *Linear Algebra Appl.*, **329** (2001), 1–8.

[323] Mohar, B. Eigenvalues, diameter, and mean distance in graphs. *Graph Combinator.*, **7** (1991), 53–64.

[324] Mohar, B. Isoperimetric number of graphs. *J. Combin. Theory B*, **47** (1987), 274–291.

[325] Mohar, B., The Laplacian spectrum of graphs. In Alavi, Y., Chatrand, G., Oellermann, O. R., Schwenk, A. J. (eds.), *Graph Theory, Combinatorics, and Applications*. New York, Wiley, 1991, pp. 871–898.

[326] Mohar, B., Poljak, S. Eigenvalues in combinatorial optimization. In Brualdi, F., Friedland, S., Klee, V. (eds.), *Combinatorial and Graph-Theoretical Problems in Linear Algebra*. New York, Springer, 1993, pp. 107–151.

[327] Molitierno, J. J., Neumann, M., Shader, B. L. Tight bounds on the algebraic connectivity of a balanced binary tree. *Electron. J. Linear Algebra* **6** (2000), 62–71.

[328] Morrison, J. L., Breitling, R., Higham, D. J., Gilbert, D. R. A lock-and-key model for protein–protein interactions. *Bioinformatics*, **22** (2006), 2012–2019.

[329] Neumaier, A. The second largest eigenvalue of a tree. *Linear Algebra Appl.*, **46** (1982), 9–25.

[330] Neumaier, A., Seidel, J. J. Discrete hyperbolic geometry. *Combinatorica*, **3** (1983), 219–237.

[331] Nikiforov, V. Bounds on graph eigenvalues I. *Linear Algebra Appl.*, **420** (2007), 667–671.

[332] Nikiforov, V. Bounds on graph eigenvalues II. *Linear Algebra Appl.*, **427** (2007), 183–189.

[333] Nikiforov, V. Chromatic number and spectral radius. *Linear Algebra Appl.*, **426** (2007), 810–814.

[334] Nikiforov, V. Cut-norms and spectra of matrices, *in press.*

[335] Nikiforov, V. Eigenvalue problems of Nordhaus–Gaddum type. *Discrete Math.*, **307** (2007), 774–780.

[336] Nikiforov, V. Eigenvalues and extremal degrees in graphs. *Linear Algebra Appl.*, **419** (2006), 735–738.

[337] Nikiforov, V. Eigenvalues and degree deviation in graphs. *Linear Algebra Appl.*, **414** (2006), 347–360.

[338] Nikiforov, V. Maxima of the Q-index: degenerate graphs. *Electron. J. Linear Algebra*, **27** (2014), 250–257.

[339] Nikiforov, V. More spectral bounds on the clique and independence numbers. *J. Combin. Theory B*, **99** (2009), 819–826.

[340] Nikiforov, V. Revisiting two classical results in graph spectra. *Electron. J. Combin.*, **14** (2007), R14.

[341] Nikiforov, V. Some inequalities for the largest eigenvalue of a graph. *Comb. Probab. Comput.*, **11** (2002), 179–189.

[342] Nikiforov, V. The maximum spectral radius of C_4-free graphs of given order and size. *Linear Algebra Appl.*, **430** (2009), 2898–2905.

[343] Nikiforov, V. The smallest eigenvalue of K_r-free graphs. *Discrete Math.*, **306** (2006), 612–616.

[344] Nikiforov, V. The spectral radius of subgraphs of regular graphs. *Electron. J. Combin.*, **14** (2007), N20.

[345] Nikiforov, V. The spectral radius of graphs without paths and cycles of specified length. *Linear Algebra Appl.*, **432** (2010), 2243–2256.

[346] Nikiforov, V. Walks and the spectral radius of graphs. *Linear Algebra Appl.*, **418** (2006), 257–268.

[347] Nikiforov, V., Yuan, X. More eigenvalue problems of Nordhaus–Gaddum type. *Linear Algebra Appl.*, **451** (2014), 231–245.

[348] Nilli, A. On the second eigenvalue of a graph. *Discrete Math.*, **91** (1991), 207–210.

[349] Ning, W., Li, H., Lu, M. On the signless Laplacian spectral radius of irregular graphs. *Linear Algebra Appl.*, **438** (2013), 2280–2288.

[350] Ning, W., Ouyang, L., Lu, M. Distance spectral radius of trees with fixed number of pendant vertices. *Linear Algebra Appl.*, **439** (2013), 2240–2249.

[351] Nosal, E. *Eigenvalues of Graphs* (Master's Thesis), University of Calgary, 1970.

[352] Oliveira, C. S., de Lima, L. S., Abreu, N. M. M., Kirkland, S. Bounds on the Q-spread of a graph. *Linear Algebra Appl.*, **432** (2010), 2342–2351.

[353] Omidi, G. R. The characterization of graphs with largest Laplacian eigenvalue at most 4. *Australas. J. Combin.*, **44** (2009), 163–170.

[354] Pati, S. The third smallest eigenvalue of the Laplacian matrix. *Electron. J. Linear Algebra*, **8** (2001), 128–139.

[355] de la Peña, J. A., Gutman, I., Rada, J. Estimating the Estrada index. *Linear Algebra Appl.*, **427** (2007), 70–76.

[356] Petrović, M. *A Contribution to The Theory of Graph Spectra* (Doctoral Thesis, in Serbian), University of Belgrade, 1984.

[357] Petrović, M. Graphs with a small number of nonnegative eigenvalues. *Graph Combinator.*, **15** (1999), 221–232.

[358] Petrović, M. On graphs with exactly one eigenvalue less than −1. *J. Combin. Theory B*, **52** (1991), 102–112.

[359] Petrović, M. On graphs whose second largest eigenvalue does not exceed $\sqrt{2}-1$. *Univ. Beograd, Publ. Elektrotehn. Fak., Ser. Mat.*, **4** (1993), 70–75.

[360] Petrović, M. On graphs whose second least eigenvalue is at least −1. *Matematički Vesnik*, **42** (1990), 233–240.

[361] Petrović, M. On graphs whose spectral spread does not exceed 4. *Publ. Inst. Mat. (Beograd)*, **34(48)** (1983), 169–174.

[362] Petrović, M. Some classes of graphs with two, three or four positive eigenvalues. *Collection Sci. Papers Fac. Sci. Kragujevac*, **5** (1984), 31–39.

[363] Petrović, M., Aleksić, T., Simić, S. K. Further results on the least eigenvalue of connected graphs. *Linear Algebra Appl.*, **435** (2011), 2303–2313.

[364] Petrović, M., Aleksić, T., Simić, V. On the least eigenvalue of cacti. *Linear Algebra Appl.*, **435** (2011), 2357–2364.

[365] Petrović, M., Borovćanin, B., Aleksić, T. Bicyclic graphs for which the least eigenvalue is minimum. *Linear Algebra Appl.*, **430** (2009), 1328–1335.

[366] Petrović, M., Gutman, I. The path is the tree with smallest greatest Laplacian eigenvalue. *Kragujevac J. Math.*, **24** (2002), 67–70.

[367] Petrović, M., Gutman, I., Lepović, M. On bipartite graphs with small number of Laplacian eigenvalues greater than two and three. *Linear Multilinear Algebra*, **47** (2000), 205–215.

[368] Petrović, M., Milekić, B. Generalized line graphs with the second largest eigenvalue at most 1. *Publ. Inst. Mat. (Beograd)*, **68(82)** (2000), 37–45.

[369] Petrović, M., Radosavljević, Z. *Spectrally Constrained Graphs.* Kragujevac, Faculty of Science, 2001.

282 *References*

[370] Powers, D. L. Bounds on graph eigenvalues. *Linear Algebra Appl.*, **117** (1989), 1–6.

[371] Powers, D. L. Graph partitioning by eigenvectors. *Linear Algebra Appl.*, **101** (1988), 121–133.

[372] Rad, A. A., Jalili, M., Hasler, M. A lower bound for algebraic connectivity based on the connection-graph-stability method. *Linear Algebra Appl.*, **435** (2011), 186–192.

[373] Radosavljević, Z. On unicyclic reflexive graphs. *Appl. Anal. Discr. Math.*, **1** (2007), 228–240.

[374] Radosavljević, Z., Mihailović, B., Rašajski, M. On bicyclic reflexive graphs. *Discrete Math.*, **308** (2008), 715–725.

[375] Radosavljević, Z., Rašajski, M. A class of reflexive cactuses with four cycles. *Univ. Beograd. Publ. Elektrotehn. Fak. Ser. Mat.*, **14** (2003), 64–85.

[376] Radosavljević, Z., Rašajski, M. Multicyclic treelike reflexive graphs. *Discrete Math.*, **296** (2005), 43–57.

[377] Radosavljević, Z., Simić, S. K. Which bicyclic graphs are reflexive. *Univ. Beograd. Publ. Elektrotehn. Fak. Ser. Mat.*, **7** (1996), 90–104.

[378] Ramane, H. S., Revenkar, D. S., Gutman, I., Walikar, H. B. Distance spectra and distance energies of iterated line graphs of regular graphs. *Publ. Inst. Mat. (Beograd)*, **85(99)** (2009), 39–46.

[379] Reingold, E. M., Nievergelt, J., Deo, N. *Combinatorial Algorithms, Theory and Practice*. Englewood Cliffs, NJ, Prentice Hall, 1977.

[380] Rojo, O. Computing tight upper bounds on the algebraic connectivity of certain graphs. *Linear Algebra Appl.*, **430** (2009), 532–543.

[381] Rojo, O. On the sth Laplacian eigenvalue of trees of order $st + 1$. *Linear Algebra Appl.*, **425** (2007), 143–149.

[382] Rojo, O., Medina, L. Tight bounds on the algebraic connectivity of Bethe trees. *Linear Algebra Appl.*, **418** (2006), 840–853.

[383] Rojo, O., Medina, L., Abreu, N., Justel, C. On the algebraic connectivity of some caterpillars: A sharp upper bound and a total ordering. *Linear Algebra Appl.*, **432** (2010), 586–605.

[384] Rojo, O., Robbiano, M. An explicit formula for eigenvalues of Bethe trees and upper bound on the largest eigenvalue of any tree. *Linear Algebra Appl.*, **427** (2007), 138–150.

[385] Rojo, O., Soto, R. A new upper bound on the largest normalized Laplacian eigenvalues. *Operators and Matrices*, **7** (2013), 323–332.

[386] Rojo, O., Soto, R., Rojo, H. An always nontrivial upper bound for Laplacian graph eigenvalues. *Linear Algebra Appl.*, **312** (2000), 155–159.

[387] Rojo, O., Rojo, H. A decreasing sequence of upper bounds on the largest Laplacian eigenvalue of a graph. *Linear Algebra Appl.*, **381** (2004), 97–116.

[388] Rowlinson, P. Graphs perturbations. In Keedwell, A. D. (ed.), *Surveys in Combinatorics*. Cambridge, Cambridge University Press, 1991, pp. 187–219.

[389] Rowlinson, P. On Hamiltonian graphs with maximal index. *European J. Combin.*, **10** (1989), 489–497.

[390] Rowlinson, P. On the index of certain outerplanar graphs. *Ars Combin.*, **29** (1990), 221–225.

[391] Rowlinson, P. On the maximal index of graphs with a prescribed number of edges. *Linear Algebra Appl.*, **110** (1988), 43–53.

[392] Rowlinson, P. Star sets and star complements in finite graphs: a spectral construction technique. *DIMACS Ser. Discrete Math. and Theoret. Comp. Sci*, **51** (2000), 323–332.

[393] Rowlinson, P., Yuansheng, Y. Tricyclic Hamiltonian graphs with minimal index. *Linear Multilinear Algebra*, **34** (1993), 187–196.

[394] Ruzieh, S. N., Powers, D. L. The distance spectrum of the path P_n and the first distance eigenvector of connected graphs. *Linear Multilinear Algebra*, **28** (1990), 75–81.

[395] Sanchis, L. Maximum number of edges in connected graphs with a given domination number. *Discrete Math.*, **87** (1991), 65–72.

[396] Schwenk, A. J. Exactly thirteen connected cubic graphs have integral spectra. In Alavi, Y., Lick, D. (eds.), *Theory and Applications of Graphs*. Berlin, Springer, 1978, pp. 516–533.

[397] Schwenk, A. J. New derivations of spectral bounds for the chromatic number. *Graph Theory Newsletter*, **5** (1975), 77.

[398] Shao, J.-Y., Guo, J.-M., Shan, H.-Y. The ordering of trees and connected graphs by algebraic connectivity. *Linear Algebra Appl.*, **428** (2008), 1421–1438.

[399] Shearer, J. B. On the distribution of the maximum eigenvalue of graphs. *Linear Algebra Appl.*, **114/115** (1989), 17–20.

[400] Shi, L. Bounds on the (Laplacian) spectral radius of graph. *Linear Algebra Appl.*, **422** (2007), 755–770.

[401] Shao, J. Bounds on the kth eigenvalues of trees and forests. *Linear Algebra Appl.*, **149** (1991), 19–34.

[402] Shu, J.-L., Hong, Y., Wen-Ren, K. A sharp upper bound on the largest eigenvalue of the Laplacian matrix of a graph. *Linear Algebra Appl.*, **347** (2002), 123–129.

[403] Shu, J., Wu, Y. Sharp upper bounds on the spectral radius of graphs. *Linear Algebra Appl.*, **377** (2004), 241–248.

[404] Shu, J., Wu, Y. The spread of the unicyclic graphs. *European J. Combin.*, **31** (2010), 411–418.

[405] Shy, L. Y., Eichinger, B. E. Large computer simulations on elastic networks: Small eigenvalues and eigenvalue spectra of the Kirchhoff matrix. *J. Chem. Phys.*, **90** (1990), 5179–5189.

[406] Simić, S. K. Some notes on graphs whose second largest eigenvalue is less than $(\sqrt{5} - 1)/2$. *Linear Multilinear Algebra*, **39** (1995), 59–71.

[407] Simić, S. K. Some results on the largest eigenvalue of a graph. *Ars Combin.*, **39** (1995), 59–71.

[408] Simić, S. K., Belardo, F., Li Marzi, E. M., Tošić, D. V. Connected graphs of fixed order and size with maximal index: Some spectral bounds. *Linear Algebra Appl.*, **432** (2010), 2361–2372.

[409] Simić, S. K., Li Marzi, E. M., Belardo, F. Connected graphs of fixed order and size with maximal index: Structural considerations. *Le Matematiche*, **LIX** (2004), 349–365.

[410] Simić, S. K., Li Marzi, E. M., Belardo, F. On the index of caterpillars. *Discrete Math.*, **308** (2008), 324–330.

[411] Simić, S. K., Tošić, D. V. The index of trees with specified maximum degree. *MATCH Commun. Math. Comput. Chem.*, **54** (2005), 351–362.

[412] Simić, S. K., Zhou, B. Indices of trees with a prescribed diameter. *Appl. Anal. Discr. Math.*, **1** (2007), 446–454.

[413] Smith, J. H. Some properties of the spectrum of a graph. In Guy, R., Hanahi, H., Sauer, N., Schonheim, J. (eds.), *Combinatorial Structures and Their Applications*. New York, Gordon and Breach, 1970, pp. 403–406.

[414] So, W., Robbiano, M., de Abreu, N. M. M., Gutman, I. Applications of a theorem by Ky Fan in the theory of graph energy. *Linear Algebra Appl.*, **432** (2010), 2163–2169.

[415] Song, H., Wang, Q., Tian, L. New upper bounds on the spectral radius of trees with given number of vertices and maximum degree. *Linear Algebra Appl.*, **439** (2013), 2527–2541.

[416] Stanić, Z. Further results on controllable graphs. *Discrete Appl. Math.*, **166** (2014), 215–221.

[417] Stanić, Z. Graphs with small spectral gap. *Electron J. Linear Algebra*, **26** (2013), 417–432.

[418] Stanić, Z. On determination of caterpillars with four terminal vertices by their Laplacian spectrum. *Linear Algebra Appl.*, **431** (2009), 2035–2048.

[419] Stanić, Z. On graphs whose second largest eigenvalue equals 1 – the star complement technique. *Linear Algebra Appl.*, **420** (2007), 700–710.

[420] Stanić, Z. On nested split graphs whose second largest eigenvalue is less than 1. *Linear Algebra Appl.*, **430** (2009), 2200–2211. (Erratum: *Linear Algebra Appl.*, **434** (2011), xviii.)

[421] Stanić, Z. On regular graphs and coronas whose second largest eigenvalue does not exceed 1. *Linear Multilinear Algebra*, **58** (2010), 545–554.

[422] Stanić, Z. Some graphs whose second largest eigenvalue does not exceed $\sqrt{2}$. *Linear Algebra Appl.*, **437** (2012), 1812–1820.

[423] Stanić, Z. *Some Reconstructions in Spectral Graph Theory and Graphs with Integral Q-Spectrum* (Doctoral Thesis, in Serbian), University of Belgrade, 2007.

[424] Stanić, Z. Some star complements for the second largest eigenvalue of a graph. *Ars Math. Contemp.*, **1** (2008), 126–136.

[425] Stanić, Z. Simić, S. K. On graphs with unicyclic star complement for 1 as the second largest eigenvalue. In Bokan, N., Djorić, M., Rakić, Z., Wegner, B., Wess, J. (eds.), *Proceedings of the Conference Contemporary Geometry and Related Topics*. Belgrade, Faculty of Mathematics, 2006, pp. 475–484.

[426] Stanić, Z., Stefanović, N. *SCL – Star Complement Library* (software). Available at http://www.math.rs/~zstanic/scl.htm .

[427] Stanley, R. P. A bound on the spectral radius of graphs with *e* edges. *Linear Algebra Appl.*, **87** (1987), 267–269.

[428] Stevanović, D. 4-Regular integral graphs avoiding ±3 in the spectrum. *Univ. Beograd. Publ. Elektr. Fak, Ser. Mat.*, **14** (2003), 99–110.

[429] Stevanović, D. Bounding the largest eigenvalue of trees in terms of the largest vertex degree. *Linear Algebra Appl.*, **360** (2003), 35–42.

[430] Stevanović, D. The largest eigenvalue of non-regular graphs. *J. Combin. Theory B*, **91** (2004), 143–146.

[431] Stevanović, D. Resolution of AutoGraphiX conjectures relating the index and matching number of graphs. *Linear Algebra Appl.*, **433** (2010), 1674–1677.

[432] Stevanović, D. *Spectral Radius of Graphs*. Amsterdam, Elsevier, 2015.

[433] Stevanović, D., Aouchiche, M., Hansen, P. On the spectral radius of graphs with a given domination number. *Linear Algebra Appl.*, **428** (2008), 1854–1864.

[434] Stevanović, D., Brankov, V. An invitation to newGRAPH. *Rend. Semin. Mat. Messina Ser. II*, **9** (2004), 211–216.

[435] Stevanović, D., Hansen, P. The minimum spectral radius of graphs with a given clique number. *Electron. J. Linear Algebra*, **17** (2008), 110–117.

[436] Stevanović, D., Ilić, A. Distance spectral radius of trees with fixed maximum degree. *Electron. J. Linear Algebra*, **20** (2010), 168–179.

[437] Stevanović, D., Indulal, G. The distance spectrum and energy of the compositions of regular graphs. *Appl. Math. Lett.*, **22** (2009), 1136–1140.

[438] Sun, F., Wang, L. The signless Laplacian spectral radii and spread of bicyclic graphs. *J. Math. Res. Appl.*, **34** (2014), 127–136.

[439] Tam, B.-S., Fan, Y.-Z., Zhou, J. Unoriented Laplacian maximizing graphs are degree maximal. *Linear Algebra Appl.*, **429** (2008), 735–758.

[440] Tan, Y.-Y., Fan, Y.-Z. The vertex (edge) independence number, vertex (edge) cover number and the least eigenvalue of a graph. *Linear Algebra Appl.*, **433** (2010), 790–795.

[441] Tannner, R. M. Explicit concentrators from generalized n-gons. *SIAM J. Algebra Discr.*, **5** (1984), 287–293.

[442] Taniguchi, T. On graphs with the smallest eigenvalue at least $-1 - \sqrt{2}$, part I. *Ars. Math. Contemp.*, **1** (2008), 81–98.

[443] Teranishi, Y. Main eigenvalues of a graph. *Linear Multilinear Algebra*, **49** (2002), 289-303.

[444] Teranishi, Y., Yasuno, F. The second largest eigenvalues of regular bipartite graphs. *Kyushu J. Math.*, **54** (2000), 39–54.

[445] Terpai, T. Proof of a conjecture of V. Nikiforov. *Combinatorica*, **31** (2011), 739–754.

[446] Wang, H., Kooij, R. E., van Mieghem, P. Graphs with given diameter maximizing the algebraic connectivity. *Linear Algebra Appl.*, **433** (2010), 1889–1908.

[447] Wang, J., Belardo, F. A note on the signless Laplacian eigenvalues of graphs. *Linear Algebra Appl.*, **435** (2011), 2585–2590.

[448] Wang, J., Belardo, F. Signless Laplacian eigenvalues and circumference of graphs. *Discrete Appl. Math.*, **161** (2013), 1610–1617.

[449] Wang, J., Belardo, F., Huang, Q., Borovćanin, B. On the two largest Q-eigenvalues of graphs. *Discrete Math.*, **310** (2010), 2858–2866.

[450] Wang, J., Belardo, F., Huang, Q., Li Marzi, E. M. On graphs whose Laplacian index does not exceed 4.5. *Linear Algebra Appl.*, **438** (2013), 1541–1550.

[451] Wang, J., Belardo, F., Wang, W., Huang, Q. On graphs with exactly three Q-eigenvalues at least two. *Linear Algebra Appl.*, **438** (2013), 2861–2879.

[452] Wang, J., Huang, Q. Maximizing the signless Laplacian spectral radius of graphs with given diameter of cut vertices. *Linear Multilinear Algebra*, **59** (2011), 733–744.

[453] Wang, X.-K., Tan, S.-W. Ordering trees by algebraic connectivity. *Linear Algebra Appl.*, **436** (2012), 3684–3691.

[454] Wang, Y., Fan, Y.-Z. The least eigenvalue of a graph with cut vertices. *Linear Algebra Appl.*, **433** (2010), 19–27.

[455] Wang, Y., Fan, Y.-Z. The least eigenvalue of graphs with cut edges. *Graphs and Combin.*, **28** (2012), 555–561.

[456] Wang, Y., Fan, Y.-Z. The least eigenvalue of signless Laplacian of graphs under perturbation. *Linear Algebra Appl.*, **436** (2012), 2084–2092.

[457] Wang, Y., Qiao, Y., Fan, Y.-Z. On the least eigenvalue of graphs with cut vertices. *J. Math. Research Exposition*, **30** (2010), 951–956.

[458] Wang, Y., Zhou, B. On distance spectral radius of graphs. *Linear Algebra Appl.*, **438** (2013), 3490–3503.

[459] Wei, F.-Y., Liu, M. Ordering of the signless Laplacian spectral radii of unicyclic graphs. *Australas. J. Combin.*, **49** (2011), 255–264.

[460] Wei, J., Liu, B. The index of tricyclic Hamiltonian graphs with $\Delta(G) = 3$. *Ars Combin.*, **73** (2004), 187–192.

[461] Wilf, H. S. Graphs and their spectra – old and new results. *Congressus Numerantium*, **50** (1985), 37–42.

[462] Wilf, H. S. Spectral bounds for the clique and independence numbers of graphs. *J. Combin. Theory B*, **40** (1986), 113–117.

[463] Wilf, H. S. The eigenvalues of a graph and its chromatic number. *Journal London Math. Soc.*, **42** (1967), 330–332.

[464] Wolk, E. S. A note on the comparability graph of a tree. *Proc. Amer. Math. Soc.*, **16** (1965), 17–20.

[465] Wolkovicz, H., Styan, P. H. Bounds for eigenvalues using traces. *Linear Algebra Appl.*, **29** (1980), 471–506.

[466] Woo, R., Neumaier, A. On graphs whose smallest eigenvalue is at least $-1 - \sqrt{2}$. *Linear Algebra Appl.*, **226–228** (1995), 577–591.

[467] Woo, R., Neumaier, A. On graphs whose spectral radius is bounded by $\frac{3}{2}\sqrt{2}$. *Graphs Combin.*, **23** (2007), 713–726.

[468] Wu, Y., Yu, G., Shu, J. Graphs with small second largest Laplacian eigenvalue. *European J. Combin.*, **36** (2014), 190–197.

[469] Wu, B., Xiao, E., Hong, Y. The spectral radius of trees on k pendant vertices. *Linear Algebra Appl.*, **395** (2005), 343–349.

[470] Xing, R., Zhou, B. Graphs characterized by the second distance eigenvalue, *in press*.

[471] Xing, R., Zhou, B. On least eigenvalues and least eigenvectors of real symmetric matrices and graphs. *Linear Algebra Appl.*, **438** (2013), 2378–2384.

[472] Xing, R., Zhou, B. On the least eigenvalue of cacti with pendant vertices. *Linear Algebra Appl.*, **438** (2013), 2256–2273.

[473] Xu, G. H. On the spectral radius of trees with perfect matchings. In *Combinatorics and Graph Theory*. Singapore, World Scientific, 1997.

[474] Xu, G. H. Ordering unicyclic graphs in terms of their smaller least eigenvalues. *J. Inequalities Appl.*, (2010), ID 591758.

[475] Yan, C. Properties of spectra of graphs and line graphs. *Appl. Math. J. Chinese Univ. Ser. B*, **3** (2002), 371–376.

[476] Ye, M.-L., Fan, Y.-Z., Liang, D. The least eigenvalue of graphs with given connectivity. *Linear Algebra Appl.*, **430** (2009), 1375–1379.

[477] Ye, M.-L., Fan, Y.-Z., Wang, H.-F. Maximizing signless Laplacian or adjacency spectral radius of graphs subject to fixed connectivity. *Linear Algebra Appl.*, **433** (2010), 1180–1186.

[478] Yong, X. On the distribution of eigenvalues of a simple undirected graph. *Linear Algebra Appl.*, **295** (1999), 70–83.

[479] Yoon, M.-G., Cvetković, D., Rowlinson, P., Stanić, Z. Controllability of multi-agent dynamical systems with a broadcasting control signal. *Asian J. Control*, **16** (2014), 1066–1072.

[480] Yu, A., Lu, M., Tian, F. Characterization on graphs which achieve a Das' upper bound for Laplacian spectral radius. *Linear Algebra Appl.*, **400** (2005), 271–277.

[481] Yu, A., Lu, M., Tian, F. On the spectral radius of graphs. *MATCH Commun. Math. Comput. Chem.*, **51** (2004), 97–109.

[482] Yu, A., Tian, F. On the spectral radius of unicyclic graphs. *MATCH Commun. Math. Comput. Chem.*, **51** (2004), 97–109.

[483] Yu, A., Lu, M. Laplacian spectral radius of trees with given maximum degree. *Linear Algebra Appl.*, **429** (2008), 1962–1969.

[484] Yu, A., Lu, M., Tian, F. Ordering trees by their Laplacian spectral radii. *Linear Algebra Appl.*, **405** (2005), 45–59.

[485] Yu, G. On the least distance eigenvalue of a graph. *Linear Algebra Appl.*, **439** (2013), 2428–2433.

[486] Yu, G. On the maximal signless Laplacian spectral radius of graphs with given matching number. *Proc. Japan Acad.*, **84** (2008), 163–166.

[487] Yu, G., Guo, S., Xu, M. On the least signless Laplacian eigenvalue of some graphs. *Electron. J. Linear Algebra*, **26** (2013), 560–573.

[488] Yu, G., Fan, Y. The least eigenvalue of graphs whose complements are 2-vertex or 2-edge connected. *Operation Research Transactions*, **17** (2013), 81–88.

[489] Yu, G., Wu, Y., Shu, J. Signless Laplacian spectral radii of graphs with given chromatic number. *Linear Algebra Appl.*, **435** (2011), 1813–1822.

[490] Yuan, X.-J., Liu, Y., Han, M. The Laplacian spectral radius of trees and maximum vertex degree. *Discrete Math.*, **311** (2011), 761–768.

[491] Yuan, X.-J., Shan, H.-Y., Liu, Y. On the Laplacian spectral radii of trees. *Discrete Math.*, **309** (2009), 4241–4246.

[492] Yuan, X.-J., Shao, J.-Y., Liu, Y. The minimal spectral radius of graphs of order n with diameter $n-4$. *Linear Algebra Appl.*, **428** (2008), 2840–2851.

[493] Yuan, X.-J., Shao, J.-Y., Zhang, L. The six classes of trees with the largest algebraic connectivity. *Discrete Appl. Math.*, **156** (2008), 757–769.

[494] Zhai, M., Lin, H., Wang, B. Sharp upper bounds on the second largest eigenvalues of connected graphs. *Linear Algebra Appl.*, **437** (2012), 236–241.

[495] Zhai, M., Liu, R., Shu, J. Minimizing the least eigenvalue of unicyclic graphs with fixed diameter. *Discrete Math.*, **310** (2010), 947–955.

[496] Zhai, M., Liu, R., Shu, J. On the spectral radius of bipartite graphs with given diameter. *Linear Algebra Appl.*, **430** (2009), 1165–1170.

[497] Zhai, M., Shu, J., Lu, Z. Maximizing the Laplacian spectral radii of graphs with given diameter. *Linear Algebra Appl.*, **430** (2009), 1897–1905.

[498] Zhai, M., Yu, G., Shu, J. The Laplacian spectral radius of bicyclic graphs with a given girth. *Comput. Math. Appl.*, **59** (2010), 376–381.

[499] Zhang, F., Chen, Z. Ordering graphs with small index and its applications. *Discrete Appl. Math.*, **121** (2002), 295–306.

[500] Zhang, F.-F. *The Least Eigenvalues of the Complements of Graphs* (Master's Thesis), Anhui University, 2011.

[501] Zhang, F.J., Chang, A. Acyclic molecules with greatest HOMO–LUMO separation. *Discrete Appl. Math.*, **98** (1999), 165–171.

[502] Zhang, J.-M., Huang, T.-Z., Guo, J.-M. The smallest signless Laplacian spectral radius of graphs with a given clique number. *Linear Algebra Appl.*, **439** (2013), 2562–2576.

[503] Zhang, X., Zhang, H. Some results on Laplacian spectral radius of graphs with cut vertices. *Discrete Math.*, **310** (2010), 3494–3505.

[504] Zhang, X., Zhang, H. The Laplacian spectral radius of some bipartite graphs. *Linear Algebra Appl.*, **428** (2008), 1610–1619.

[505] Zhang, X.-D. Bipartite graphs with small third Laplacian eigenvalue. *Discrete Math.*, **278** (2004), 241–253.

[506] Zhang, X.-D. Eigenvectors and eigenvalues of non-regular graphs. *Linear Algebra Appl.*, **409** (2005), 79–86.

[507] Zhang, X.-D. Graphs with fourth Laplacian eigenvalue less than two. *European J. Combin.*, **24** (2003), 617–630.

[508] Zhang, X.-D. On the distance spectral radius of unicyclic graphs with perfect matchings. *Electron. J. Linear Algebra*, **27** (2014), 569–587.

[509] Zhang, X.-D. On the two conjectures of Graffiti. *Linear Algebra Appl.*, **385** (2004), 369–379.

[510] Zhang, X.-D. Ordering trees with algebraic connectivity and diameter. *Linear Algebra Appl.*, **427** (2007), 301–312.

[511] Zhang, X.-D. The Laplacian spectral radii of trees with degree sequences. *Discrete Math.*, **308** (2008), 3143–3150.

[512] Zhang, X.-D. Two sharp upper bounds for the Laplacian eigenvalues. *Linear Algebra Appl.*, **376** (2004), 207–213.

[513] Zhang, X.-D., Li, J. S. On the k-th largest eigenvalue of the Laplacian matrix of a graph. *Acta Appl. Math. Sinica (English Ser.)*, **17** (2001), 183–190.

[514] Zhang, X.-D., Li, J. S. The two largest eigenvalues of Laplacian matrices of trees (in Chinese). *J. China Univ. Sci. Technol.*, **28** (1998), 513–518.

[515] Zhang, X.-D., Luo, R. Non-bipartite graphs with third largest Laplacian eigenvalue less than three. *Acta Math. Sinica (English Ser.)*, **22** (2006), 917–934.

[516] Zhang, X.-D., Luo, R. The Laplacian eigenvalues of mixed graphs. *Linear Algebra Appl.*, **362** (2003), 109–119.

[517] Zhang, X.-D., Luo, R. The spectral radius of triangle-free graphs. *Australas. J. Combin.*, **26** (2002), 33–39.

[518] Zhou, B. Bounds for index of a modified graph. *Discuss. Math. Graph Theory*, **24** (2004), 213–221.

[519] Zhou, B. On Estrada index. *MATCH Commun. Math. Comput. Chem.*, **60** (2008), 485–492.

[520] Zhou, B. On the largest eigenvalue of the distance matrix of a tree. *MATCH Commun. Math. Comput. Chem.*, **58** (2007), 657–662.

[521] Zhou, B. Signless Laplacian spectral radius and Hamiltonicity. *Linear Algebra Appl.*, **432** (2010), 566–570.

[522] Zhou, B., Cho, H. H. Remarks on spectral radius and Laplacian eigenvalues of a graph. *Czech. Math. J.*, **55** (2005), 781–790.

[523] Zhou, B., Ilić, A. On distance spectral radius and distance energy of graphs. *MATCH Commun. Math. Comput. Chem.*, **64** (2010), 261–280.

[524] Zhou, B., Trinajstić, N. Further results on the largest eigenvalues of the distance matrix and some distance based matrices of connected (molecular) graphs. *Internet Electron. J. Mol. Des.*, **6** (2007), 375–384.

[525] Zhou, B., Trinajstić, N. On the largest eigenvalue of the distance matrix of a connected graph. *Chem. Phys. Lett.*, **447** (2007), 384–387.

[526] Zhu, B.-X. On the signless Laplacian spectral radius of graphs with cut vertices. *Linear Algebra Appl.*, **433** (2010), 928–933.

[527] Zhu, D. On upper bounds for Laplacian graph eigenvalues. *Linear Algebra Appl.*, **432** (2010), 2764–2772.

[528] Zhu, M., Guo, M., Tian, F., Lu, L. The least eigenvalues of the signless Laplacian of non-bipartite graphs with fixed diameter. *Int. J. Math. Eng. Sci.*, **2** (2013), 1–12.

Inequalities

290

Index

Printed in the United States
By Bookmasters